AF356554

Power Electronic Converter Configuration and Control for DC Microgrid Systems

Power Electronic Converter Configuration and Control for DC Microgrid Systems

Special Issue Editors

Jens Bo Holm-Nielsen
P. Sanjeevikumar

MDPI • Basel • Beijing • Wuhan • Barcelona • Belgrade • Manchester • Tokyo • Cluj • Tianjin

Special Issue Editors
Jens Bo Holm-Nielsen
Aalborg University
Denmark

P. Sanjeevikumar
Aalborg University
Denmark

Editorial Office
MDPI
St. Alban-Anlage 66
4052 Basel, Switzerland

This is a reprint of articles from the Special Issue published online in the open access journal *Energies* (ISSN 1996-1073) (available at: https://www.mdpi.com/journal/energies/special_issues/PECC_CDCMS).

For citation purposes, cite each article independently as indicated on the article page online and as indicated below:

LastName, A.A.; LastName, B.B.; LastName, C.C. Article Title. *Journal Name* **Year**, *Article Number*, Page Range.

ISBN 978-3-03936-431-2 (Hbk)
ISBN 978-3-03936-432-9 (PDF)

Contents

About the Special Issue Editors

Jens Bo Holm-Nielsen received his M.Sc. degree in Agricultural Systems, Crops & Soil Science, from KVL, Royal Veterinary & Agricultural University, Copenhagen, Denmark, in 1980 and Ph.D. degree in Process Analytical Technologies for Biogas Systems, Aalborg University, Esbjerg, Denmark, in 2008. He is currently with the Department of Energy Technology, Aalborg University, Esbjerg, Denmark, and Head of the Esbjerg Energy Section. He is Head of the research group at the Center for Bioenergy and Green Engineering, established in 2009. He has vast experience in the field of biorefinery concepts and biogas production–anaerobic digestion. He has implemented bioenergy system projects in provinces of Denmark and in other European states. He has been the Technical Advisor for many industries in this field. He has executed many large-scale European Union and United Nation projects in research aspects of bioenergy, biorefinery processes, and the full chain of biogas and green engineering. He has authored more than 300 scientific papers. His current research interests include renewable energy, sustainability, and green jobs for all. Dr. Holm-Nielsen was a member on invitation with various capacities in the committees of over 500 various international conferences and organizer of international conferences, workshops, and training programs in Europe, Central Asia, and China.

P. Sanjeevikumar (Senior Member, IEEE) received the Bachelor's degree in Electrical Engineering from the University of Madras, Chennai, India, in 2002; the Master's Degree (Hons.) in Electrical Engineering from Pondicherry University, Puducherry, India, in 2006; and the Ph.D. degree in Electrical Engineering from the University of Bologna, Bologna, Italy, in 2012. He was an Associate Professor with VIT University from 2012 to 2013. In 2013, he joined the National Institute of Technology, India, as a Faculty Member. In 2014, he was invited as a Visiting Researcher at the Department of Electrical Engineering, Qatar University, Doha, Qatar, funded by the Qatar National Research Foundation (Government of Qatar). He continued his research activities with Dublin Institute of Technology, Dublin, Ireland, in 2014. He was an Associate Professor with the Department of Electrical and Electronics Engineering, University of Johannesburg, Johannesburg, South Africa, from 2016 to 2018. Since 2018, he has been a Faculty Member with the Department of Energy Technology, Aalborg University, Esbjerg, Denmark. He has authored more than 300 scientific papers. Dr. Padmanaban received awards for Best Paper or Most Excellent Research Paper from IET-SEISCON 2013, IET-CEAT 2016, IEEE-EECSI 2019, and IEEE-CENCON 2019, along with five best paper awards from ETAEER 2016. He also sponsored Lecture Notes in Electrical Engineering, a Springer book. He is a Fellow of the Institution of Engineers, India; the Institution of Electronics and Telecommunication Engineers, India; and the Institution of Engineering and Technology, U.K. He is an Editor/Associate Editor/Editorial Board Member for refereed journals, in particular for IEEE Systems Journal, IEEE Transactions on Industry Applications, IEEE Access, IET Power Electronics, and Wiley's International Transactions on Electrical Energy Systems. He is the Subject Editor for IET Renewable Power Generation; IET Generation, Transmission & Distribution; and FACTS journal (Canada).

Preface to "Power Electronic Converter Configuration and Control for DC Microgrid Systems"

Writing a preface is always a challenging task, but I always enjoy recommending works in my field of power electronics and renewable energy technologies—an application which benefits society and solves power demand crises. It was therefore my pleasure to read and recommend this book, *Power Electronic Converter Configuration and Control for DC Microgrid*, authored/edited by my colleagues Jens Bo Holm-Nielsen and Sanjeevikumar Padmanaban from Aalborg University, Esbjerg, Denmark, as it has shown exciting and cutting-edge research findings in power electronics for microgrids. Readers can find four different sections in the book: conceptual review detailing the state of the art, research investigation outcomes, new configuration technologies, and real-time hardware-in-loop testing for validation. This approach makes this book unique and user-friendly in reading and understanding the complexity of the topics discussed.

Microgrids and renewables are continuously gaining attention as crucial empowering technologies with smart control approaches to enhance system reliability and efficiency. These challenges are solvable through modern power electronics converters, with higher flexibility, adjustable electronic loads, and energy storage systems (batteries). From my reading, the topics which I anticipate will draw strong reader attention are the new configuration of the DC-to-DC converter, multiport converters, multilevel inverters, load sharing concept, wireless charging network, battery management schemes, large-scale renewable integration challenges and issues, selection of permanent magnet machines for wind energy technology, and hardware-in-loop test findings towards microgrid system and electric vehicle charging through renewable energy systems.

Finally, I congratulate and thank the editors, authors, *Energies* journal, MDPI publishers, reviewers, and press production team. This book is a result of their support and effort on all levels.

I hope the readers will enjoy reading this book and will be able to apply the research findings herein for further future enhancement in technology and skills!

Prof. Frede Blaabjerg

Fellow IEEE Villum Investigator Professor
Aalborg University, Denmark.

Review

Conceptual Framework of Antecedents to Trends on Permanent Magnet Synchronous Generators for Wind Energy Conversion Systems

K. Padmanathan [1,*], N. Kamalakannan [1], P. Sanjeevikumar [2,*], F. Blaabjerg [3], J. B. Holm-Nielsen [2], G. Uma [4], R. Arul [5], R. Rajesh [6], A. Srinivasan [7] and J. Baskaran [8]

[1] Department of Electrical and Electronics Engineering, Agni College of Technology, Thalambur, Chennai, Tamil Nadu 600130, India

[2] Center for Bioenergy and Green Engineering, Department of Energy Technology, Aalborg University, Esbjerg 6700, Denmark

[3] Department of Energy Technology, Aalborg University, Aalborg, Denmark

[4] Department of Electrical and Electronics Engineering, College of Engineering, Guindy, Anna University, Chennai, Tamilnadu 600025, India

[5] School of Electrical Engineering, Vellore Institute of Technology, Chennai Campus, Chennai, Tamil Nadu 600127, India

[6] Department of Automobile Engineering, Madras Institute of Technology, Anna University, Chennai, India

[7] Department of Electrical and Electronics Engineering, Sri Krishna College of Technology, Coimbatore, Tamil Nadu 641042, India

[8] Department of Electrical and Electronics Engineering, Adhiparasakthi Engineering College, Melmaruvathur, Tamil Nadu 603319, India

* Correspondence: padmanathanindia@gmail.com (K.P.); san@et.aau.dk (P.S.); Tel.: +45-7168-2084 (P.S.)

Received: 1 April 2019; Accepted: 3 July 2019; Published: 8 July 2019

Abstract: Wind Energy Conversion System (WECS) plays an inevitable role across the world. WECS consist of many components and equipment's such as turbines, hub assembly, yaw mechanism, electrical machines; power electronics based power conditioning units, protection devices, rotor, blades, main shaft, gear-box, mainframe, transmission systems and etc. These machinery and devices technologies have been developed on gradually and steadily. The electrical machine used to convert mechanical rotational energy into electrical energy is the core of any WECS. Many electrical machines (generator) has been used in WECS, among the generators the Permanent Magnet Synchronous Generators (PMSGs) have gained special focus, been connected with wind farms to become the most desirable due to its enhanced efficiency in power conversion from wind energy turbine. This article provides a review of literatures and highlights the updates, progresses, and revolutionary trends observed in WECS-based PMSGs. The study also compares the geared and direct-driven conversion systems. Further, the classifications of electrical machines that are utilized in WECS are also discussed. The literature review covers the analysis of design aspects by taking various topologies of PMSGs into consideration. In the final sections, the PMSGs are reviewed and compared for further investigations. This review article predominantly emphasizes the conceptual framework that shed insights on the research challenges present in conducting the proposed works such as analysis, suitability, design, and control of PMSGs for WECS.

Keywords: permanent magnet synchronous generators; wind energy conversion system; finite element analysis; soft computing techniques.

---

1. Introduction

Energy is predominantly the driving factor of human life and the economy of global countries. Henceforth, the research investigation in this area is highly critical and the need lot of time to invest for in-depth study [1,2]. Due to the fast depletion of the natural conventional resources, sustainable alternative energy sources, for instance tidal wave, solar, wind, biogas/biomass and hydro energy, must be tap together for developmental activities. Therefore, there is currently a tremendous increase in the lookout for sustainable and alternative energy sources to generate electricity. Wind energy seems to be a promising and potential alternative renewable energy source with its enhanced sustainability and eco-friendly nature. According to 'Global Energy Outlook and the Increasing Role of India', in the year 2040, the electricity generation capacity of India will be equivalent to what is produced by today's European Union [3]. Figure 1 shows a summary of electricity generation by selected region and its electricity generation by 2040. The Global Wind Report (GWP, 2018) mentioned that the wind energy is one of the cheapest forms of electricity in a number of markets. Has it is a cost-effective option for countries which have ever-growing power demands and distribution challenges with centralized grid system [3].

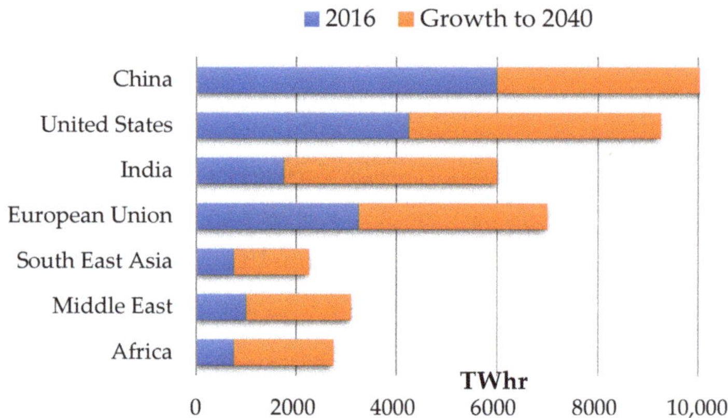

Figure 1. Electricity generation by selected region up to 2040. Source: International Energy Agency [3].

The Global Wind Energy Council (GWEC) suggested that wind energy sector (both the on-shore and off-shore) supplies 300 GW of wind power capacity to come online by 2024 for global consumption. The global wind energy capacity increased with 51.3 GW in 2018. In spite of the fact, it is less than 2017 in about 4.0%; it is still a good achievement in wind energy capacity addition. From the year 2014, there is a 50 GW capacity addition occurring for every year though some markets behave differently. Thus, wind energy may contribute to electricity generation in India about 34,046 MW, which was 49.3) compared with all other renewable energy mix in the end of year 2018. By the year 2030, the wind power capacity is expected to generate 2300 GW power, fulfilling 22% of the global electricity demands. The report published by Global Wind Energy Outlook 2018 [4] predicted the future of the wind energy industry until 2050. In 2018, 50,100 MW was added, which was lesser than that of the 2017's capacity addition 52,552 M). It is viewed in 2018 as the consecutive year with increased new installations accounting to 9.1%, but this is lesser than the previous year's data i.e., 10.8% growth in 2017. The global electricity demand met by 6% of the wind turbines installed in 2018. In Figure 2, the cumulative production based on wind sources for the year 2018 shown along with the newly added capacity for the year 2018 [4].

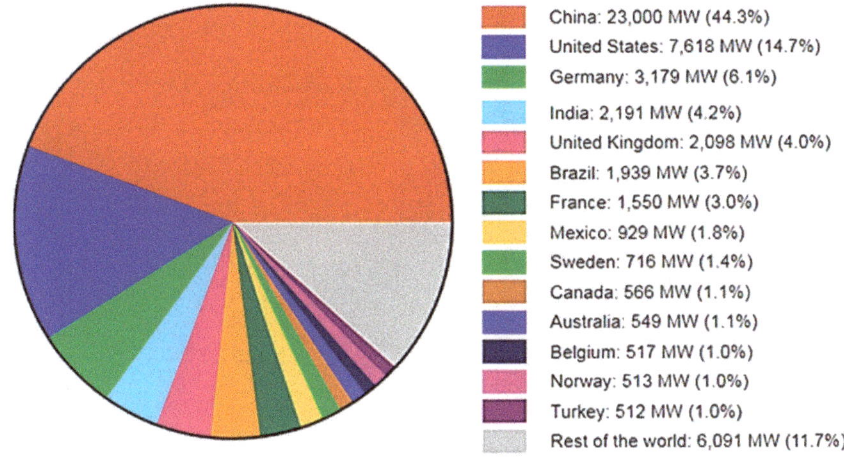

a) New installed capacity (MW) by different country in 2018

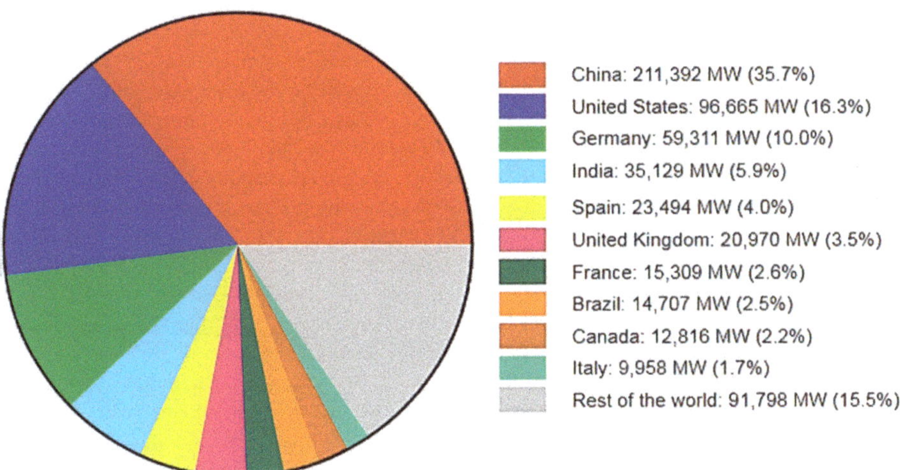

b) Installed capacity (MW) by different country in 2018

Figure 2. Cumulative installed capacity of wind energy in the world end-of-year by 2018 and newly added capacity by different country in 2018 [4].

Figure 3 presents the overall baseline information of various settings, such as the new polices, moderate, and advanced scenarios. A global status report, published at the end of 2018, reported that global installed wind capacity was approximately 590 GW, which meant that Asia topped the regional market scale for the 9th consecutive year. It accounts for a whopping 48% of the added capacity (a total that exceeds 235 GW by the end of the year 2019) followed by Europe (over 30%), North America (14%), and Latin America and the Caribbean (almost 6%). In case of new installations, China retained the top position, though there was a contraction for two years. This was followed by US, Germany, UK, and India in respective positions.

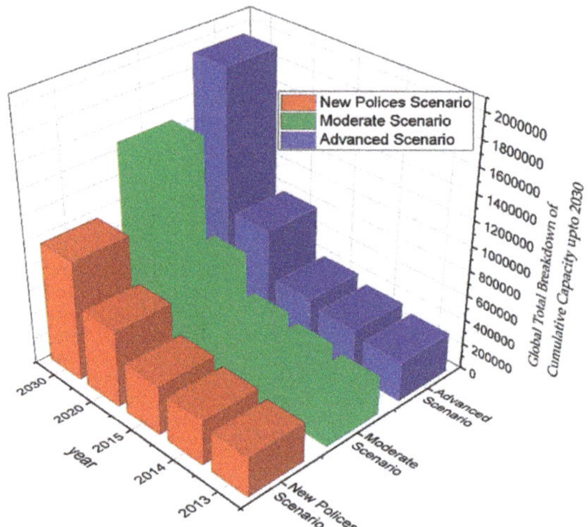

Figure 3. Global total breakdown of cumulative capacity up to 2030. Source: Global Wind Energy Outlook [4].

Globally, the energy demands were 282.5 GW and 318.105 GW in the years 2012 and 2013, respectively. This denotes that there was a strong market growth of more than 19% and 12.5% in the years 2012 and 2013, respectively. However, this seems to be the lowest growth rate i.e., 22% and 21% of global electricity, when compared with annual average growth rate in the past decade. This is predicted to increase in the range of 8%–12% by the year 2020. The wind penetration level increased up to 10% in the year 2016, in alignment with the guidelines for international agreements on environmental commitment. By the years 2030 to 2035, the predicted saturation level is about 1.9×10^9 kW. The work by International Renewable Energy Agency (IRENA) titled 'Global energy transformation: A roadmap to 2050 (2019 edition)' inferred that by the year 2050, electricity would be the central energy carrier with growth up to 50% share from its current 20% share on final consumption. This would make the consumption of gross electricity double. The power demand across the globe (accounting to 86%) will be met by renewable resources-based power. Overall, the final energy will have two-thirds of contribution from renewable energy [5]. According to the literature [6], the current study focuses on the hypothesis subjects such as Wind Energy Conversion System (WECS) history, transformation of Permanent Magnet Synchronous Generators (PMSG), Finite Element Method (FEM) leveraging, Soft Computing (SC) applications, and the upgradation of Computer Aided Design (CAD) which looks to be a novel perspective as the first step. Generally, the wind turbine is moved by the wind pressure as in step-like method, though its design is different. In wind energy production, low (cut-in) and abundant (cut-out) wind speeds are labelled as risk potentials. On the basis of size and design parameters, the risk potential of every turbine is decided. Generally, the electricity yield of a wind turbine ranges from 3 to 25 m/s whereas high generation is examined once it crosses 10–15 m/s values. Each turbine has cut-in as well as cut-out values that are contingent on size as well as design parameters [7]. Therefore, the wind turbine design plays an important role in energy production. Dai et al. (2019) stressed that, in recent years, the incorporation of wind turbine generators, such as Permanent Magnet Synchronous Generator (PMSG), and Doubly Fed Induction Generator (DFIG), in which the former is predominantly utilized in wind energy conversion system's has been commonly seen, since it is cost-effective, highly reliable, and has flexibility in control [7]. This paper aims to address the technical issues and fitness of WECS components and integration with electrical grid. Furthermore, it will explore the study of PMSG comprehensive comparisons with other topologies of generator. In addition, this paper will also shed insights on the gaps in research and areas to further enhance research, in the context of WECS.

2. A Brief Review of WECS

In 2004 article discussed wind engineering in general and wind power meteorology with special reference to turbine and generator technology. Further, they discussed the economics, which are involved in this regard [1]. In a study conducted in 2007, the researchers stressed that the conversion of wind electricity is currently a green technology factor due to (1) structural design improvements, (2) design and manufacturing of blades, and (3) efficient power processing techniques, on the bases of power-electronics followed by new generator design, to achieve variable-speed operation [8]. In 2013, [9] discussed a list of possible changes in the methodology towards the implementation of utility-scale wind energy into the power grid and follow up in accordance to the updated research with their obtainable alleviation techniques. Figure 4 disseminates the growth in size of wind turbines since 1980 and for predicted future prospects. The scaling up of turbines to lower cost has been effective so far, but it is not clear that the trend can continue forever [10].

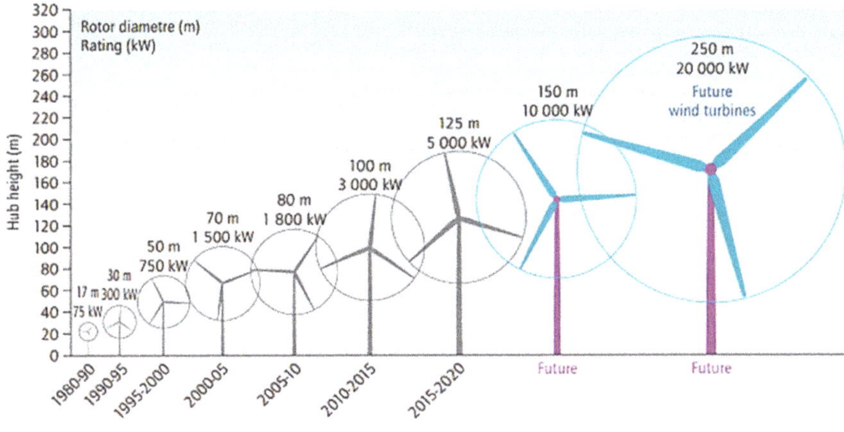

Figure 4. Growth in size of wind turbines since 1980 and future prospects [10].

In 2012, [11] developed a 5 MW baseline design in deep wind concept with more than 150 deep Darrieus-type floating wind turbine systems. In this research article, the technology used in previous works employing various generator types and manufacturers of large power direct drive wind turbines were detailed. In Figure 4, the developments that occurred in the tower, blades, rotor diameter, power rating, and wind turbine hubs heights are illustrated. Amongst the available turbines, the 7.5 MW turbine seems to be the most powerful one with a 126 m rotor diameter. The global wind report published in 2012 cited the new Alston Haliade 6 MW turbine to be the world's large turbine with a 150.8 m rotor diameter [12]. In the future, the next-generation wind turbines are predicted to hold 20,000 kW capacity with a 250 m rotor diameter.

In 2010, [13] investigated the power output density functions of different WECS for a variety of operating wind regimes with the help of a probabilistic approach. In 2007, [14] conducted a review of information regarding global wind energy scenarios, performance, and stability of wind turbines, sizes of wind turbine, wake effects, evaluation of wind resourced, site selection, wind turbine aerodynamics, and challenges faced in wind turbines followed by wind turbine technology. Which is inclusive of control system, design, loads, blade behavior, generators, transformers, and grid connection. In 2014, a review of notable technical as well as environmental impacts of wind farms, wind power resource assessment techniques, control strategies, and grid integration techniques, were conducted [15]. A comparative investigation was conducted using a Maximum Power Point Tracking (MPPT) control device in 2009 [16] between the optimized configurations of passive wind turbine generators with that of the active ones that operate at optimal wind power.

3. Wind Turbine, Types, and Generator Technologies

In the past decade, there has been a tremendous growth observed in wind turbine technologies and that have resulted in the development of new-age wind turbine concepts. With developments in wind generator systems, cost-effectiveness of the systems has become the new mandate. In a wind power generator system, there is a tower which supports rotating as well as the stationary parts. The nacelle that has the generator in it, power converter, grid side step-up transformer, monitoring and control equipment are present in the stationary part. In 2014, [17] developed a summary about compact and lightweight wind turbines along with the technical hindrances with special reference to Horizontal Axis Wind Turbines (HAWT). There are two broad categories of wind turbine technology at present; such has the HAWT and the Vertical Axis Wind Turbines (VAWT). The HAWT main rotor shaft rotates in alignment with the wind direction, whereas it is perpendicular to the ground, generator, transformer, converters, and other equipment in the case of the VAWT rotor shaft.

In HAWT, the nacelle is placed at the top position in the tower. The HAWT showcase better aerodynamic performance when compared to VAWT, due to which the former is largely deployed in large-sized offshore wind farms [17]. According to [18], there are approximately 8000 different components present in a typical wind turbine. This information is based on a RE power MM92 turbine with the blades' lengths being 45.3 m and the tower height being 100 m.

Figure 5 shows the major components in a wind turbine and the share of the overall wind energy system parts cost. A direct-drive radial flux permanent magnet generator was checked for its suitability [19] to act as a drive-train runner. FEM software was used to test the generator fitness, based on structural design (or in other terms the stability of the air-gap present between the rotor and the stator) as per PMSG. So as to deduce the differences in flux density and force along the periphery of the rotor. In this study, the researchers used a simple analytical model. Further, 2D magneto-static simulations where also used to check the validity of the analytical model by making use of FEM software carried out [19].

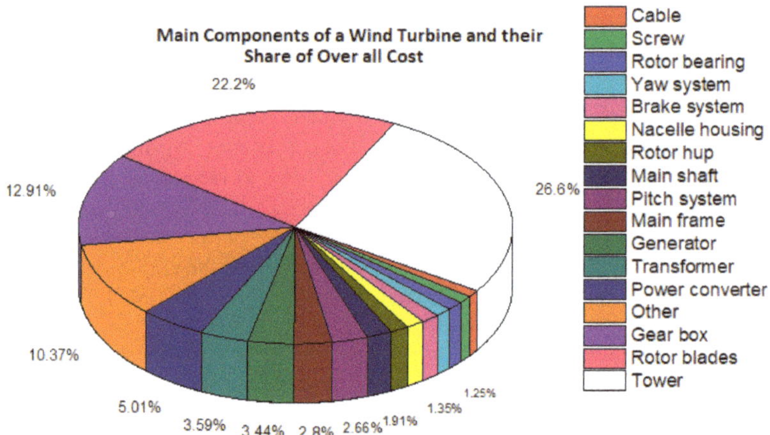

Figure 5. Main components of a wind turbine and their share of the overall cost [9].

According to the literature [20], the induction and synchronous generator models are general candidates used to convert wind energy to electrical energy. In 2009, [20] listed Danish wind power status and various topologies of other wind farm configurations. A classification was done by [21] to differentiate the wind turbine technology schemes. To be specific, the different categories are Full Rate Converter Wind Turbine (FRCWT), PMSG, Fixed Speed Wind Turbine-Squirrel Cage Induction Generator (FSWT-SCIG), Variable Speed Wind Turbine-Direct Drive Synchronous Generator (VSWT-DDSG), Squirrel Cage Induction Generator-Wind Turbine (SCIG-WT), Full Rate Converter Induction Generator (FRCIG), Direct Drive Synchronous Generator (DDSG), Variable Speed Wind Turbine-Doubly Fed Induction Generator (VSWT-DFIG), Squirrel Cage Induction Generator (SCIG), Fixed Speed Wind Turbine-Permanent Magnet Synchronous Generator (FSWT-PMSG), Fixed Speed Wind Turbine (FSWT), Doubly Fed Induction Generator (DFIG) and Variable Speed Wind Turbine-Full Rate Converter Induction Generator (VSWT-FRCIG) [21].

This segregation is done on the basis of power level, working principle, application type, and the usage in a number of commercial applications. The research and development in this area is still happening, and various novel configurations and advanced applications are in the testing stage. In 2006 compared different classification types and explained them in detail [22]. In general, based on the working principle, three electric generators are considered as main types: induction, synchronous machines. Parametric which are associated with magnetic anisotropy and permanent magnets. The study further mentioned that the parametric generators in most cases be called as doubly salient electric generators [22]. Since they are mostly equipped with doubly salient magnetic circuit structures. When classified according to the magnetic flux penetration, there are three types of permanent magnet generators present: transversal-flux, axial flux, and radial-flux machines [22].

Since the efficiency provided is better, most of the high-power direct-driven wind power applications prefer low-speed and high-torque PMSGs [23]. These are generally applied in a wide range of applications due to cost-effective Permanent Magnets (PM). According to the literature [23], Permanent Magnets can provide high-power densities, higher efficiency, and chances of compactness which eventually results in the reduction of turbine size. The advantages of Permanent Magnet generators are when it excludes the exciter field winding, slip rings, and brushes in association with the capability to self-excite making option, so as to achieve good efficiency as well as the high power factor. In a standalone system, the PMSG has overloading and full torque capability, a highly competitive feature, due to which it is unique when compared to other traditional electrical machines. The PMSG is capable of self-excitation, another exciting feature which makes it the best option for operating at higher power factors and efficiencies. Further, PM machines possess the ability of overloading and full torque at zero speed, as well as at lower speeds [24]. To be specific, the standalone power systems are utilized in the isolated areas. When compared with the traditional electrical machines, this is inevitably effective.

In 2009, [25] studied the prospective site matching of direct-drive wind turbine models on the basis of electromagnetic design optimization of PM generator systems. In this study, a three-phase radial-flux PM generator was developed with a back-to-back power convertor. The study had a total of 45 PM generator systems which were designed, optimized, and grouped as a collage of five-rated rotor speeds in the 10–30 rpm range and nine-power ratings in the range of 100 kW to 10 MW, respectively. Following this, the study also determined the rotor diameter and the rated wind speed of a direct-drive wind turbine under optimum PM generator on the basis of the maximum wind energy capture design principle. This study also calculated the Annual Energy Output (AEO) with the help of the Weibull density function. At last, at eight potential sites, the maximum AEO Per Cost (AEOPC) of the optimized wind generator systems was calculated along with yearly mean wind speeds ranging between 3 and 10 m/s [25].

In 2008, [26] developed a concept of Permanent Magnet Generators Design. In this study, the researcher discussed the geared as well as direct-driven PM generators. Further, they also classified the direct-driven PM generators and the researchers dealt with various topologies of design aspects and unique nature in PM generators [26]. In 2012, [27] conducted a techno-economic evaluation of the basic assembly and magnetic topographies of the Salient Pole Synchronous Machine and Permanent Magnet Synchronous Machine. The study also provided the economic analyses of the machines that accompanied wind turbines.

4. Various Aspects of Comparison for PMSG's

The design of electrical machines is important for any kind of applications. The basic design of an electrical machine involves certain procedures and analytical strategies. For calculation of magnetic circuit, electrical circuit, efficiency, insulation type, number of slots/poles combinations, winding dimensions, cogging torque analysis, control strategies, usage of materials, cost of products, thermal and structural design of electric machines, and manufacturing techniques etc. Finite Element Analysis (FEA) software can provide support for design and optimization tools to determine the best performance parameters. In 2008, [28] elaborately briefed and further used a deterministic global mathematical optimization which became a vital tool in the processes of design. Several mathematical models and optimization techniques could handle such problems associated with multi-faceted design. Figure 6 describes a complex range of ideas and significances of parameters for electrical machine design, analysis and characteristics studies, it has been simplified with partial adoption [28]. The studies conducted so far in this research areas, and various viewpoints have been established [28].

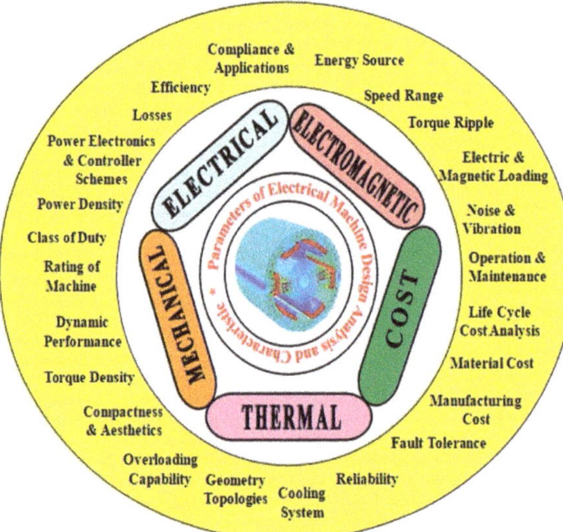

Figure 6. Electrical machine design parameters for analysis and characteristics studies.

In 2012, [29] conducted a general, as well as magnetic, analysis of various parameters, such as size, topology, voltage, magnetic field air-gap flux, weight, torque, losses, and efficiency between Permanent Magnet Synchronous Machines (PMSMs) and Conventional Salient Pole Synchronous Machine (CSPSMs) with the help of FEM. Figure 7, the weights of active material and costs are compared, and analyzed. Based on the comparison, it is observed that the total weight of the active material in the PMSM is reduced by 6.55% more than the conventional salient pole machine. In Figure 8, the losses at full load are presented [27].

With the same output power generated by the Permanent Magnet used in the machine, there will be reduction in machine weight which eventually becomes lighter to produce and so it increases the efficiency. Once the investigation was complete, it was observed that the CSPSM expressed less efficiency when compared to PMSM's. Further, when it comes to enhancement of magnet and semi-conductor expertise, the PMSMs reaped a cost-based benefit. Therefore, at the time of designing electrical machines, it is advised to follow their strategy in terms of machine efficiency and efficient use of energy [29].

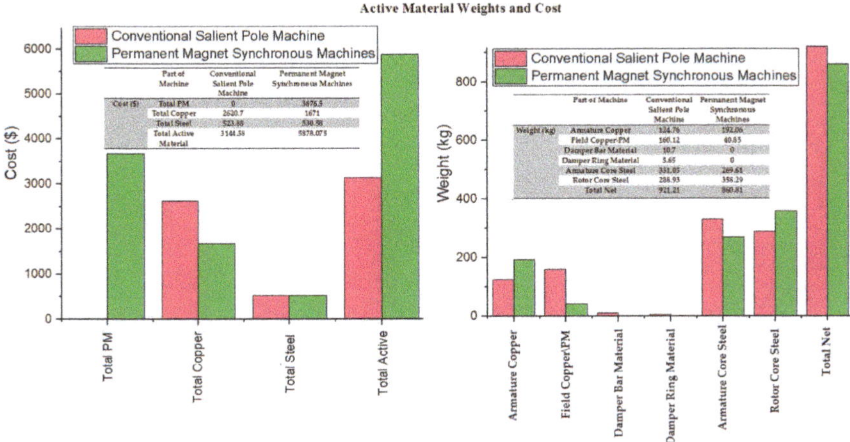

Part of Machine	Conventional Salient Pole Machine	Permanent Magnet Synchronous Machines
Cost ($) Total PM	0	3676.5
Total Copper	2620.7	1671
Total Steel	523.88	530.58
Total Active Material	3144.58	5878.075

Part of Machine	Conventional Salient Pole Machine	Permanent Magnet Synchronous Machines
Weight (kg) Armature Copper	124.76	192.06
Field Copper-PM	160.12	40.85
Damper Bar Material	10.7	0
Damper Ring Material	5.65	0
Armature Core Steel	331.05	289.61
Rotor Core Steel	286.93	358.29
Total Net	921.21	860.81

Figure 7. Active material weights and Cost comparison of PMSM and conventional machines [29].

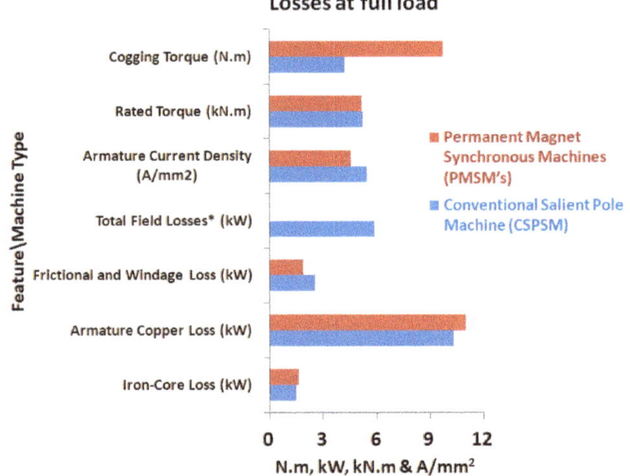

Figure 8. Losses comparison for PMSM and conventional machines at full load conditions [27].

A seven type of systems such as variable-speed constant frequency (VSCF) wind generator system, PMSGDD, PMSG1G, PMSG3G, DFIG3G, DFIG1G, EESG_DD (Electricity-Excited Synchronous Generator with direct-driven), and SCIG_3G (Squirrel Cage Induction Generator with three-stage gearbox) has been compared. In this comparative study, the researcher made optimization designs for different wind generator systems in the range of 0.75, 1.5, 3.0, 5.0, and 10 MW [30,31]. The results inferred that the PMSG_DD was cost-effective when compared to EESG_DD systems due to the cost incurred in lower generator system and enhanced Annual Energy Production (AEP) per cost. When there is an increase in wind turbine, the cost spent on direct-drive wind generator seems to be reduced. However, when there is an increase in the rated power, there is an enhanced performance exhibited by the PMSG_DD system when compared to the EESG_DD system.

Following is the description for a single-stage gearbox drive train concept. Due to the low-cost generator system and high AEP per cost, the focus shifted to the DFIG_1G system which seems to be the best alternative. Further, when viewed from AEP per cost perspective, the DFIG_1G system seems to be the most cost-effective and is close to 1.5 MW. Following is the concept behind three-stage behavior drive-train. Due to the least cost generator system and high AEP per cost, the DFIG_3G system was considered as the best solution among other three wind generator systems. Additionally, in terms of AEP per cost aspect, more emphasis is given to the PMSG_3G system compared to the SCIG_3G system [31]. Figure 9 compares all five various wind generator systems of respective manufacturers in a wide range of aspects.

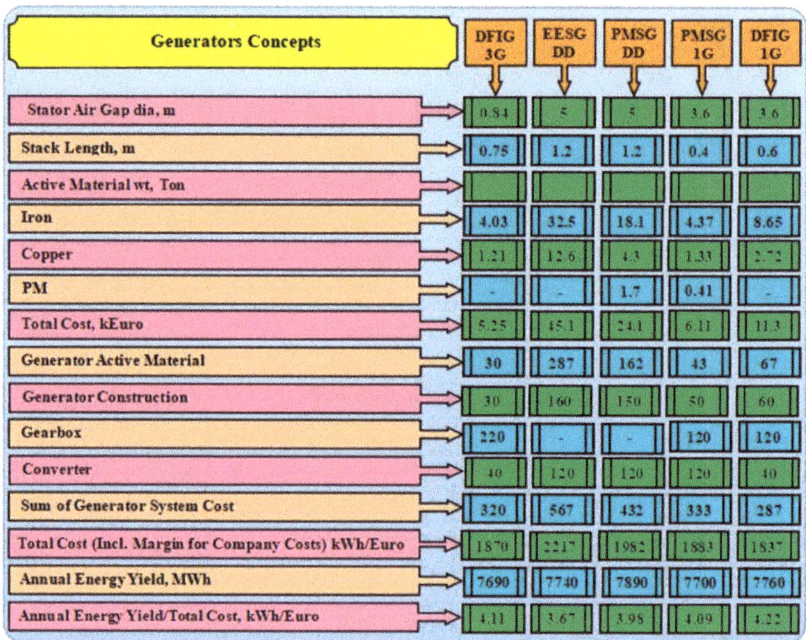

Generators Concepts	DFIG 3G	EESG DD	PMSG DD	PMSG 1G	DFIG 1G
Stator Air Gap dia, m	0.84	5	5	3.6	3.6
Stack Length, m	0.75	1.2	1.2	0.4	0.6
Active Material wt, Ton					
Iron	4.03	32.5	18.1	4.37	8.65
Copper	1.21	12.6	4.3	1.33	2.72
PM	-	-	1.7	0.41	-
Total Cost, kEuro	5.25	45.1	24.1	6.11	11.3
Generator Active Material	30	287	162	43	67
Generator Construction	30	160	150	50	60
Gearbox	220	-	-	120	120
Converter	40	120	120	120	40
Sum of Generator System Cost	320	567	432	333	287
Total Cost (Incl. Margin for Company Costs) kWh/Euro	1870	2217	1982	1883	1837
Annual Energy Yield, MWh	7690	7740	7890	7700	7760
Annual Energy Yield/Total Cost, kWh/Euro	4.11	3.67	3.98	4.09	4.22

Figure 9. Comparison of five different wind generator systems [31].

When compared in terms of cost between a multi-hybrid PM wind generator system loaded in single-stage behavior and the direct-drive concept, the former seems to be cost-effective. When there is an increase in the size of the wind turbine, then the adoption of gear ratios may also widely vary. Based on the rated power levels, the optimum gear ratio may vary from 4:1 to 10:1. In the case of larger power ratings, the literature [17] suggests making use of higher gear ratios would be better performance. In 2014 mentioned that PMSGs are predominantly employed by giants such as the manufactures as follows GE energy, Vestas, Siemens, Gamesa and Goldwind. The stator of the PMSG is wound where the rotor is present with the PM pole system and may possess salient cylindrical poles. At most of the time, the low-speed synchronous machines project the salient-poly type with predominantly numerous poles. One can develop a direct drive system based on a synchronous generator with an ideal number of poles (a multi-pole PMSG). Some common types are transversal flux machine, axial flux machine, and the radial flux machine. The PMSG machine expressed highest the efficiency in an induction machine since the excitation was supplied excluding any energy flow. However, it is difficult to manufacture the PMs, whereas its inventory is cost-consuming too [17].

The long-term-unaddressed issue comes with the mandate to maintain the rotor temperature less than the magnet's threshold temperature. This may further be influenced by the magnetic material's Curie point and the binding material's thermal criterion in the case of power metallurgy composites. In turn, the synchronous process generates the issue according to the start-up, synchronization, and voltage regulation [32]. In 2011, Sandra Eriksson et al. performed an excellent comparison of direct-driven PMSGs. A total of six different-range generators were compared among each other [33].

Figure 10 clearly depicts the considerations of various factors with respect to the fixed and variable parameters for different ranges of generators.

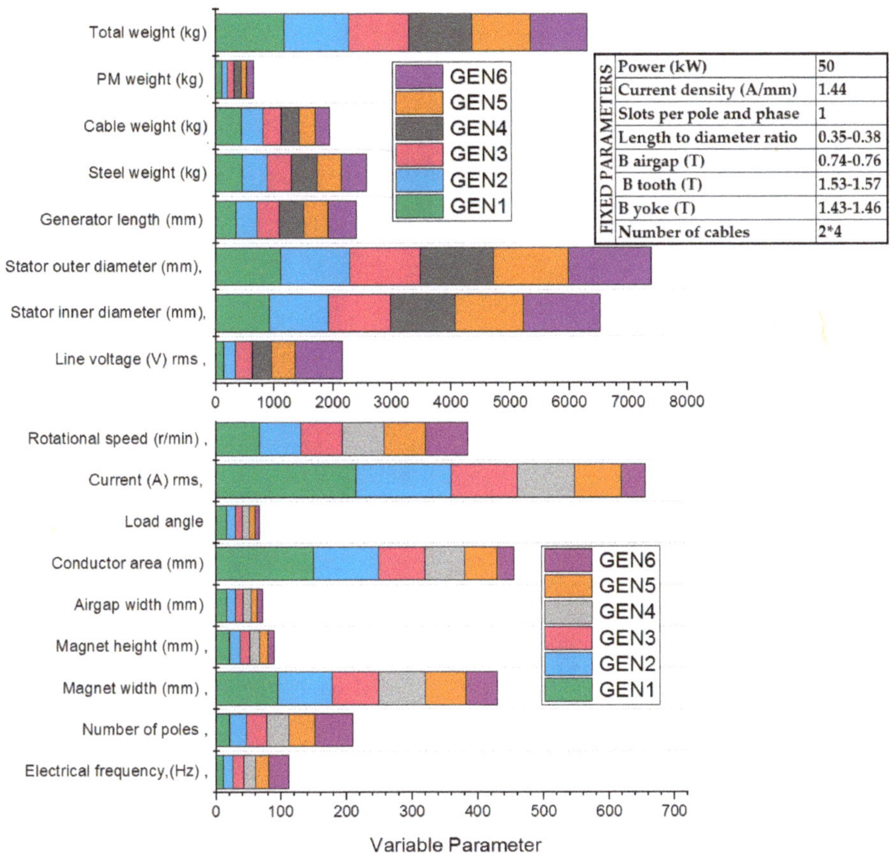

Figure 10. Characteristics of rated speed and power for stationary simulations [33].

In 2013, [34] compared three configurations such as gearless-drive Permanent magnet induction generator PMIG-WECS, gearless-drive PMSG, geared-drive squirrel-cage induction generator (SCIG), and in the index every system was allocated with a number such as 1, 2, and 3 to position itself in the rank in accordance to other two systems. According to Table 1, the geared-SCIG system seems to be prominent in 61.5% of the indexes, while at the same time 38.5% of indices where dominated by the gearless-PMSG system. There was a 60% similarity in advantages between gearless-PMIG and gearless-PMSG. Therefore, the geared-SCIG system exists in alignment with the number of indexes. However, there is a domination of gearless-PMSG in the three top priority indexes such as generation efficiency, Operation & Maintenance (O&M) cost, and the duration of failure behavior. Further, there was a domination of geared-SCIG in the four top priority indexes such as kWh production at low speed, frequency of failure, generator O&M cost, and capital cost. In order to achieve the results with best accuracy, the weight of an index should be considered as per the order. Among the different configurations considered for the study, the results concluded that the gearless-drive PMSG-based and geared-drive SCIG-based systems seem to be the most desirable solutions. From Table 1, it is identified that the gear less PMSG is the only machine, which has the best option in efficiency, as there is no gearbox and copper loss [34].

Table 1. Comparison of Geared-Drive Squirrel Cage Induction Generator (SCIG), Gearless-Drive Permanent Magnet Synchronous Generator (PMSG), and Gearless-Drive PMIG-Wind Energy Conversion System (WECS) configurations [34].

Order of Index	Index Name	Details of Index	Geared SCIG	Gearless PMSG	Gearless PMIG	Best Options	Comments and Justifications
1	Reliability	Duration of failure	2	1	1	PMSG and PMIG	No
2		Frequency of failure	1	2	2	SCIG	There may be high failure rate present in direct-drive due to the fact that there is a direct transfer of wind rotor fluctuation to generator and power electronics
3	O&M cost	Gearbox	2	1	1	PMSG and PMIG	Absence of gearbox in direct-drive turbines
4		Generator	1	2	2	SCIG	Generator failures are costly in direct-drive turbines. In direct-drive turbines, the generator failures prove to be cost-consuming processes
5	kWh production	Affected by cut-in, rated and cut-out wind speed	1	2	2	SCIG	There is a negative association present between the cogging torque in PM generators and the cut-in speed of the turbine. This eventually is reflected in the generation of kWh
6	Capital cost	Cost of generator and gearbox (if any)	1	3	2	SCIG	Due to the magnets, the PM machines are high cost-consuming
7	Efficiency	Accounts for gearbox and generator loss	3	1	2	PMSG	There is neither gearbox nor rotor copper losses occur in gearless PMSG and it also works at high PF
8	Excitation Requirements	Reactive power source	3	1	2	PMSG	Full external excitation is observed in SCIG whereas it is partial in PMIG. In case of PMSG, it is completely internally-excited
9	Magnet problems	Demagnetization and future availability	1	2	2	SCIG	There are no magnets in SCIG
10	Control simplicity		1	2	3	SCIG	In spite of SCIG being simple in control, the fixed magnet excitation in PM machines complicates the controls
11	Construction simplicity	Number of poles, diameter size, and rotor design	1	2	3	SCIG	Due to the multi-pole development, the direct-drive PMSG is large and heavy. Due to double-rotor design, the PMIG seems to be complicated
12	Noise level	Drive train	2	1	1	PMSG and PMIG	There is no gearbox noise in PMSG and PMIG
13	Generator		1	2	2	SCIG	There is no notable cogging torque is SCIG

Order of index (1–13) denotes degree of significance/priority (1: highest priority). The index (1, 2 or 3) denotes superiority (1: the best option).

In 2010, [35] with the field-circuit method for rapid calculation of load characteristics for stand-alone PM synchronous generators (PMSGs) that were developed with various rotor structures. The study results were compared with load characteristic calculations and results. The field-circuit method was defined, and utilized to determine the load characteristics of PMSGs with surface-mounted, inset or interior mounted permanent magnets and with inner or outer rotors [35]. In a comparative study conducted 2013, two PM generator types such as radial flux PM (RFPMG) and axial flux PM (AFPMG) generators were compared. To compare the generator performance during mechanical energy storage, the study measured the output powers of both RFPMG and AFPMG [36]. Results shown in Figures 11 and 12, concludes that there was a better performance exhibited by RFPMG when the machine's electrical parameters were in very similar condition and in relatively small power applications. It was inferring that the RFPMG has fewer copper, core, and rotor losses with respect to the varying generator and wind speed when compared to AFPMG [36].

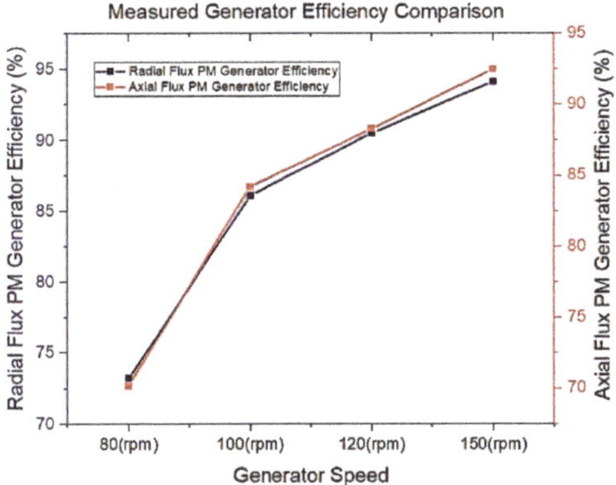

Figure 11. Measured generator efficiency comparison [34].

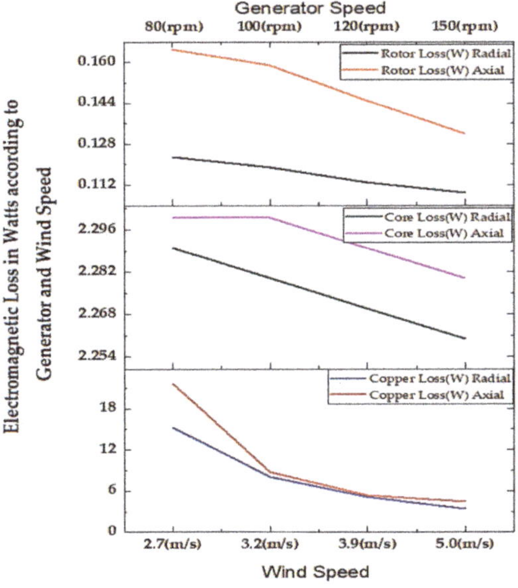

Figure 12. Electromagnetic loss according to generator speed [33]. R: Radial Flux PM Generator, A: Axial Flux PM Generator.

5. Different Design Perspectives

Designing PMSG has several challenges, which make it complicated when compared to conventional machine design procedure. The combination called 'Slot and Pole' poses various other challenges, which include reducing eddy current losses and cogging on permanent magnets. In 2013, [37] a technique was proposed to improve the air-power gap apparently transferred under the constraint of tangential stress using analytical optimization algorithms. The processes of optimization have been optimized for expressions that are relevant for the design of main variables, external derivations, and operational restrictions for the formulation of mathematical derivations.

In general, terms, during PMSG design, the optimum design includes various mandatory requisites, which are to improve profitability, and mitigate utilization of material to reduce cost and weight [38]. In addition, the design considerations should also take into account availability, high reliability, and low serviceability and maintainability for TC Ia that is wind class [38]. Furthermore, the utilization of gearless or semi-geared drive machines improves efficiency and reliability of wind power generators. Additionally, such requisites are associated characterization of compactness in terms of weight and dimensions. In addition, during the design of PMSG, the mechanical forces and voltage waveforms are quite imperative in several applications [38].

The design of machines is generally concerned with the electric and magnetic circuits; however, there are several losses which are measured using empirical equations [39]. In 2011, [39] explored the various design aspects concerned with the radial and axial field of synchronous machines with permanent magnets. In addition, the analysis of three fractional-slot and concentrated winding permanent magnet synchronous machine topologies are suited especially for specific applications [39]. According to a study [40], which explored the performance of wind power generators fitted with external permanent magnet rotors. The authors analyzed the FEM and electromagnetic results that examined the turbine characteristics and variations of the nominal wind speeds; various systematic methods were employed in previous research. For the calculation of the electrical characteristics, such as synchronous inductance, Electromotive force (EMF) constant, and phase resistance, an electromagnetic analytical and magnetic field distribution method was applied. In this study, a d-q model coordinate transformation theorem was employed for the analysis of performances. In addition, FEMs and curve fitting are used for the analysis of core losses [40]. Furthermore, a dissertation [41] presented a transformation theorem that developed a technique for the optimization and design of machines mounted with Surface Mount Permanent Magnet (SMPM), as impacted by mechanical loads, energy source, thermal effects, and state-of-the-art developments in manufacturing and material capabilities. A method was proposed for the design and development of cage rotor induction machines that can be optimized for better performance. Both genetic algorithm GA and particle swarm optimization (PSO) were used for optimization of the machines. Different integrated methods were applied and the Electromagnetic-Thermo-Mechanical method was used for the fabrication of Surface Mount Permanent Magnet (SMPM) machines [41]. An iron-less brushless permanent magnet machine was proposed and designed in 2013 [42] for the design and optimization of generator applications. The proposed approach constituted a dimensioning technique that involves comprehensive geometric techniques; both electrical and magnetic methods were used followed by the use of a detailed 3-D finite element (FE). In addition, the machine configurations used were both circular and rectangular designs, and were compared against each other. Furthermore, the performance of ironless stator designs configurations and the effectiveness of materials used were compared [42]. Tangential magnetic flux and stator concentric windings were incorporated in wind power generators in 2009 [43] with the rotation frequencies of 75–300 rpm. The parameters associated with the developed generators were depicted in the research. The intention of the previous research was to analyze the working of synchronous generators fitted with permanent magnets, which is in line with the concept of mitigating the problem of magnetic field distribution that was studied separately using FEM. During the development of such models, as given below in Tables 2 and 3 the following parameters to acquire synchronous machines should be considered and varied:

In addition, the following parameters should be considered for mathematical simulation.

Table 2. Parameters need to be considered for acquiring synchronous machines [43].

S.No	Parameters
1	Rated project parameters such as voltage, power, rotation frequency, and power coefficient
2	Basic sizes such as active length and stator boring diameter
3	Machine's winding parameters such as phase number, number of poles, number of slots and turns, and parallel branches of stator winding
4	Magnet's geometrical sizes and the Air gap between stator and rotor

Table 3. Parameters need to be considered for mathematical simulation.

S.No	Parameters
1	Efficiency and distribution of losses in a generator
2	Diagram of magnets and reserve coefficient over magnets
3	Static overloading in generators
4	Inductive scattering resistances along longitudinal and cross axes
5	Generator's parameters under loading and short circuit
6	Generator's performance (external and no load)

In 2012, [44] examined and designed PMSG using FEM simulation software that involves low speed three phase generators associated with external rotors. The aim of the research paper was to obtain sinusoidal voltages that are induced in stator windings which are espoused magnetization and arrangement path of permanent magnets within the rotor structure [44]. Again in 2012 [45], used the multi-physics approach for the design and development of a 10-MW doubly fed induction generator (DFIG). The optimal design and analyses were considered for the operation of direct drive of wind turbines with a conversion that has reduced size. In 2005, [45] performed a study that comprised of PMSGs that were used in wind power generation systems that are small. The output voltage was examined using FEM wherein both no-load and load conditions were considered. The influence of shapes and magnetic dimensions was examined. The previous research is a novel study wherein the outcomes of FEM were analyzed that revealed the PMSG's cogging torque frequency was influenced by number of poles and stator slots. However, the performance was influenced by factors such as magnet dimension, air-gap length, and cogging torque magnitude [46]. Research conducted by [47] (2008) depicted the design, prototyping, and analysis of relatively small and cheap axial-flux three-phase coreless permanent magnet generator. In the previous research, the FEM approach was used for the measurement of equivalent circuit inductances. In addition, the end winding inductance calculation and equivalent resistance of eddy-current loss where calculated using traditional methods. In 2002, [48] proposed a method for performance improvement using soft magnetic composite inter poles in drive permanent magnet machines. Several factors such as suitable pole arc shapes, magnet dimensions' influence, material usage efficiency, and labor costs where considered. In 2011, [49] examined the design considerations of double rotor radial flux permanent-magnet wind generators in terms of the mechanical and electromagnetic non-overlap air-cored (ironless) stator windings. The developed model was examined using finite-element analysis. The results of the analysis revealed that the electromagnetic design determines the mass, cost, rotor yoke dimensions, and leakage flux paths. In 2012, [50] examined the axial flux PM generator performances using wind turbine characteristics and electromagnetic field. The analytical approach could mitigate the analysis time required when compared with the FEM that is three-dimensional, which could use for the calculation of performances in the preliminary design phase. In 2010, [51] proposed and developed an optimal design high-speed DC generation system that uses a slot-less PM machine. In the previous research, the researcher used soft magnetic composite (SMC) stator yoke and a controlled rectified fitted to the stator winding [51]. In [52] (1997) further examined the multi pole PMSG with the radial field. PMSG machines have been used as direct-coupled grid-connected generators with ratings between 100 KW and 1 MW. However, the previous research revealed that the poles that are between 100 and 300 are found to render better performance in terms of efficiency and reactance. The stator and rotor section design present the suitable pole and power number. Standard ferrite magnet wedges are used in the rotor sections. The stator sections however are made up of E cores with a single rectangular coil in each core. The researcher also developed a lumped-parameter magnetic model that permits the calculation of machine parameters in a rapid manner [52]. In [53] (2007) examined the direct-coupled an Axial Flux PMSG (AFPMSG) that is appropriate for a wind turbine system. Furthermore, the researcher used horizontal-axis and vertical-axis wind turbine generator systems. FEM analysis was undertaken for the analysis of the AFPMSG magnetic flux density distribution. The results analyzed were compared with the proposed machine configuration wherein the voltage from the output line was found to be of sinusoidal pattern. AFPMSG design feasibility was confirmed using a prototype generator [53]. In 2010, [54] further displayed an Axial-Flux Permanent-Magnet Generator for Induction Heating Gensets whereas ([55], 1997) and ([56], 1994) proposed a straightforward approach for the design of brushless permanent-magnet machines; the results are supported by several analytical results. The main difference between sine wave and square wave motors are detailed and described in terms of EMF, self-inductance, flux density and so on. A stage by stage method is involved with the design of computer-aided systems which are elaborated in detail. The previous research detailed the information such as torque, shape, magnet poles and phases, slots, poles, teeth, energy and co-energy, magnetic circuit concepts, yokes, basic relationships, magnetic materials, flux linkage and inductance, influence of stator slots, tooth flux, back-EMF, need for the field analysis based design FEM, cogging torque, series and parallel connections, and loss modeling [56]. Though machines achieve infinitely, the core of the machine that operates under unsaturated conditions and deep rectangular slots are not appropriate and not suitable for the design of today electrical machines with non-linear materials. The machine's performance should be predicted with great accuracy to solve non-linear equations which is expressed in terms of the Magnetic vector potential. The irregular machine geometry confirmation makes the analytic method configuration challenging. Hence, there is a need to use appropriate field computation, and modeling techniques utilizing electromagnetic fields such as the energy minimization. Includes, differential/integral functions, variational method, discretization, shape functions, stiffness matrix, 1D and 2D planar and axial symmetry problem and computation of electric and magnetic field intensities, capacitance and inductance, force, torque, and energy for basic configurations of electrical machines [57].

In Figure 13, various electromagnetic analytical methods are illustrated. Every method contains a set of advantages as well as disadvantages. In this scenario, the finite elements were found to be robust in nature to conduct general electromagnetic analyses [57].

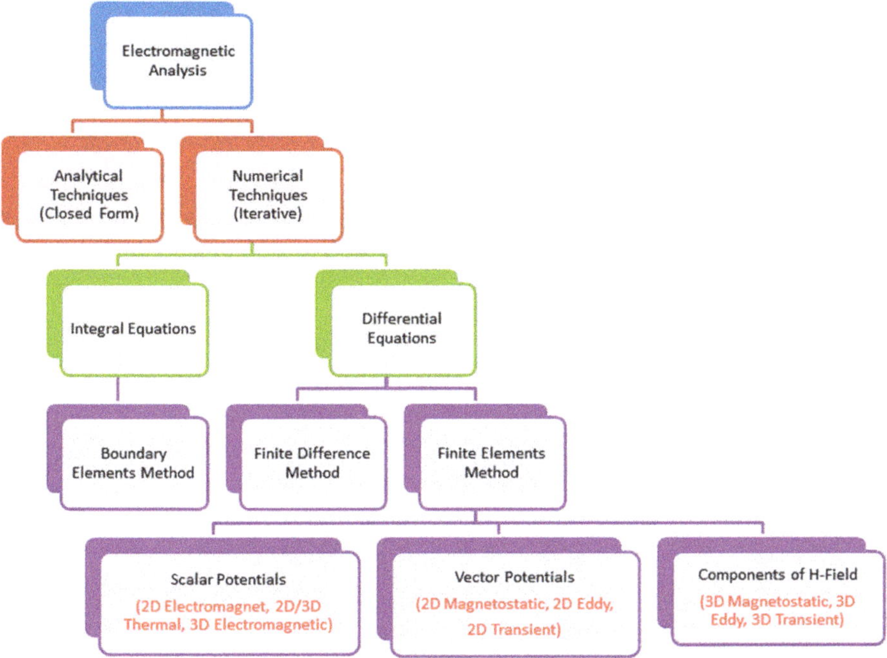

Figure 13. Different methods of electromagnetic analysis [57].

6. Consideration of Losses Calculation for PMSGs

One of the important design factors discussed in this study is the determination of losses in PMSGs. In 2010, published a model with an elaborate loss computation and calculation method with updated analytical loss calculation. In this model, conventional losses, for instance, stator core iron losses, ventilation losses, I^2R losses, and other detailed losses like stator end region losses, were discussed. However, being a separate engine, the cooling caused by the bearing friction and the losses incurred via excitation system were not considered. The components which were lost are discussed here in detail [58].

(a). Iron Losses

Excluding the stator and rotor windings, there seemed to be losses in eddy current as well as some more losses in entire metallic parts, which can be segregated as follows.

1. Iron losses in (teeth and yoke) stator core which included the impact of rotating fields and harmonics.
2. Eddy current losses on pole shoe surface because of the tooth ripple pulsation and stator winding armature reaction magneto motive force.
3. Eddy current losses in the stator clamping plates.
4. Eddy current losses in the stator clamping fingers.
5. Eddy current losses in the stator core end laminations.
6. Eddy current losses in external metallic air guides.

(b). Winding Losses

Various types of losses in stator, rotor, and damper windings are inclusive in winding losses

1. Stator winding copper I^2R losses.

2. Rotor winding copper I^2R losses.
3. Due to the tangential slot leakage field, the occurrence of eddy current losses in the stator winding.
4. Due to the radial slot leakage field, the occurrence of eddy current losses in the stator winding.
5. Due to end leakage field, the overhanging of the circulating current and eddy current losses.
6. Damper winding losses due to tooth-ripple pulsation and the stator winding armature reaction magneto-motive force.
7. On the basis of statistical measurements, the rest of the losses were calculated with basic equations.

(c). Ventilation Losses

Ventilation losses segmented further into following parts

1. Friction losses of rotating parts.
2. Air friction losses of the forced cooling airflow.
3. Pressure generation losses (e.g., fan, inter polar gap, rotor rim) [58].

In 2011, a study [59] of extreme excellence was conducted by on electromagnetic losses which were incurred in direct-driven PMSGs. Using electromagnetic model, the solutions were obtained from FEM. By utilizing a MATLAB-driven model, the researchers performed the simulations. The results obtained inferred that the iron and copper losses were completely based on the rated voltage and rated current. In terms of a fixed output power, the experiment achieved larger machine volume with an increase in rated voltage. Further, higher frequency and increased iron loss were observed in parallel to decrease rated current and reduce copper losses. At the time of simulations, the generator losses were determined for various wind speeds, using which the loss distribution was calculated. Furthermore, they tested an analytical model to predict the eddy current losses in PMSG rotor magnets by feeding a rectifier load. The eddy current loss achieved during time stepping resulted in the coherence of 2-D FEM and coupled-circuit when performing the investigation. In 1997, conducted an experiment and designed losses for the model of a 1 MW machine design prepared in alignment with the parasitic losses. These were stator back-iron reluctance, rotor and stator slotting, rotor reluctance, stator back-iron reluctance, stator module weld loss, rotor eddy-current loss, stator beam loss, the polygon effect, and stator structure cage loss [60]. In 2014, [61] experimented on eddy current losses in PMs of surface-mounted magnet synchronous machines. This study introduced a true analytical method on the basis of magneto-dynamic problem of a conductive ring. The results were obtained and compared with the information retrieved from 3-D FEM analysis. In the analytical model, the effect of the width on magnet loss was considered. The axial effect was considered via a correction coefficient. In the comparison executed, the researchers included impact of the circumferential segmentation, instantaneous losses, effect of the frequency on magnet losses, and induced current density. Through stressing the criticality of the skin effect and magnetic reaction due to magnet currents, this analytical model yielded an accurate measurement of magnet eddy current losses [61].

7. Faults and Protection

At the time of designing PMSGs, researchers must be considered for the chance of fault occurrences and protection schemes methods. In 2013, [62] listed the influence of asymmetrical magnet faults upon PMSG rotors. Mechanical looseness, eccentricity, and damage in any one magnet are the most commonly found attributes that result in rotor faults. Further, the rotor eccentricity is caused by unequal distribution of static, dynamic, or mixed air-gap. In the presence of static eccentricity, the air-gap seems to be the least and positioned as per the stator. On the contrary, in the case of dynamic eccentricity, there seems to be no coincidence between the rotor's centers and the center of rotation. Therefore, the minimum air-gap position rotates in line with the rotor. There are notable reasons behind the cause of eccentricities such as looseness, incorrect assembly, load unbalances, misalignment, and sometimes the bending of the rotor. At the time of analysis, the study conducted series as well as parallel-connected windings. In order to quantify the demagnetization in a single magnet, the study defined the faulty severity factor. As per the study investigations, one can conclude that, for a generator where all windings are series-connected, the induced EMF value gets decreased due to the demagnetization of a single magnet. Likewise, if the load is a resistive type, the current also may decrease. Therefore, one may not be able to identify the frequency components which are in association with the fault whereas one can observe only the decreased total flux linked to the windings [62]. In 2012, Rodrigues et al expressed his ideas on direct or indirect lightning strokes after thoroughly reviewing the over-voltages and electromagnetic transients [63]. The transient behavior can easily be explained via the lightning protection of the wind turbines accurately, for which the modified version of EMTP (Electro-Magnetic Transient Programme) was utilized. In this study, the researcher adopted a case study model in which two interconnected wind turbines were used so as to study the direct lightning stroke to the blade or the lightning strikes which happens in the soil near a building. Further, this study also conducted a holistic computer simulation in addition to EMTP-RV [63]. Investigation in 2011 [64] which evaluated the fault conditions and identified efficient fault ride-through and protections schemes in electrical systems of both small-scale (land) and large-scale (offshore) wind farms. In their study, the researcher considered two variable-speed generation systems such as PMSGs and DFIGs. After discussing the protection issues associated with DFIGs, the research proposed a new protection scheme as well. Following this proposal, the protection scheme options for fully rated converter and direct-driven PMSGs were analyzed and simulation results were compared.

The development in magnetic materials and its impact on the electric machine design investigated (2007) [65]. In addition to that, few potential faults were also selected using a fault-tolerant system design. Two fault types may occur in the system, of which the electromagnetic faults are as follows:

1. Winding open circuit;
2. Winding short circuit (phase/phase);
3. Winding short circuit at terminals;
4. Turn-to-turn fault in a phase.

   The power converter faults are listed herewith

1. Power device open circuit;
2. Power device short circuit;
3. DC link capacitor failure.

One should focus on development of a fault-tolerant system, if the operation needs to be continued even in the presence of faults, if any. In this design, every phase should have a stand-alone single-phase PWM inverter that has a modular system in which the modules are isolated by every phase fault. When a module has less thermal interaction or electrical/magnetic interactions, then the system is likely to proceed with the operations excluding the faulty phase [65].

By 2012, inducted a rotor core design and FEA simulation, to diminish the mechanical stress put upon the core bridge. After considering rotor speed variations, the researcher performed the mechanical transient analysis. The experimental result was presented for the S-N curve (S-N curve is deduced from material test data) of rotor care material so as to assure the validity of the model against fatigue failure [66].

8. Damping and Oscillation

In order to handle damping and oscillation, the PMSG-based stability issues in WECS should be taken into consideration. In 2011, a torque compensation strategy was devised [67] based on DC-link current determination of the converters, after the stability challenges faced in PMSG-WECS were studied. In general, the instability issue is caused at the time when generators are in direct connection with the wind turbine during which the speed oscillations occur because of the lack of damper in design, and torsional vibration. With the purpose of reducing the oscillation amplitude and enhancement of the system stability, one can make use of generator torque controller. However, due to limited ability, it may impact the WECS' power response. The torque compensation strategy, when deployed with the sole purpose of enabling positive damping of the oscillations, may lead to enhancements in small signal and transient stabilities of the WECS [67]. In 2018, [68] opined the influence on system oscillations because of the grid-connected wind farms. The study focused on the contributions made by the damping of power system oscillations and the assistance rendered by the inner wind turbine oscillations upon the changes in several aspects of power system behavior, which is inclusive of stability. The stability of the power systems gets connected with electro-mechanical interactions and the generator's behavior which is already in connection with the grid. Therefore, the influence of wind power penetration over the power system becomes a key challenge to tackle. In the literature [68], an elaborate investigation was conducted about the oscillations in power systems and its influence and control schemes in the wind farms for various wind turbine technologies [68]. The growing technologies that focus on magnetic gears were the primary theme of the study conducted [69] (2011). The concept of magnetic gears has an advantage of dealing with inherent overload capability surpassing the mechanical gears. However, there is a less amount of torsional stiffness found in magnetic gears than their counterparts i.e., mechanical gears. This leads to oscillations at the time of transient changes in load and speed alike, and the damper windings utilized in synchronous generators to alleviate the oscillations that occur due to transients [69]. In a study conducted in 1996, the researchers display the damping of PMSG power-angle oscillations in terms of wind turbine applications [70]. The small pole pitch present in the generator allows it to work in every low speed and this is conjoined with the wind turbine, thus a direct electrical grid connection is maintained. In this research paper, an alternative damping system was proposed in which the stator is allowed to confine the rotational movement through a connection with the wind turbine that is located near a spring and mechanical damper. This proposed method enables high damping of power-angle oscillations when compared to conventional damper windings. The design's efficiency can be illustrated via the generator's response to initiate the changes in driving torque. In order to showcase the new design's vibrant nature and viability, the generator's behavior on synchronization as well as on operation front, where there is a difference in wind occurs, is described [70].

Short Circuit

In 2011, [68] stressed the occurrence of sudden short circuit when applied in large PMG machines thus denoting the differences in short-circuit behavior amongst the would-field generator and the PMG. With the help of FEM analytics, the researcher calculated the sub-transient reactance and time constants of the PMG and utilized it at the typical circuit theory simulation in short-circuit fault. The FEM was then used in the risk evaluation of magnetization loss in magnets. Being complex, the transient magnetic field looks for transient non-linear circuit-coupled FEA in 3-D in association with voltage-source excitation. Various calculation methods where summarized in this research paper with further discussions on implications of futuristic design and PMG application after considering the attributes that are relevant to application of standard tests and specifications [71].

9. Several Aspects of Cost Factor

In the study conducted by Salem Alshibani et al. (2014) [72], the high CAPEX (Capital Expenditure) issue was taken into account since, at the beginning of a project, it is always a hindrance for such techniques, especially in case of PMSGs. The study proposed a method, which utilized to assess typical PMSGs designed and reported in this article. The results of the proposed method were compared with the results of the traditional methods. The results inferred that the lifetime cycle assessment (LCA) seemed to favor the gearless PMSGs that incur high CAPEX. Further, in the case of inclusion of lifetime cost in the design optimization, the scenario develops machines which can yield significantly higher lifetime revenues than the extra CAPEX required [72]. Figures 14 and 15 compare the CAPEX values of geared as well as gearless PMSGs in a range of power ratings with percentage.

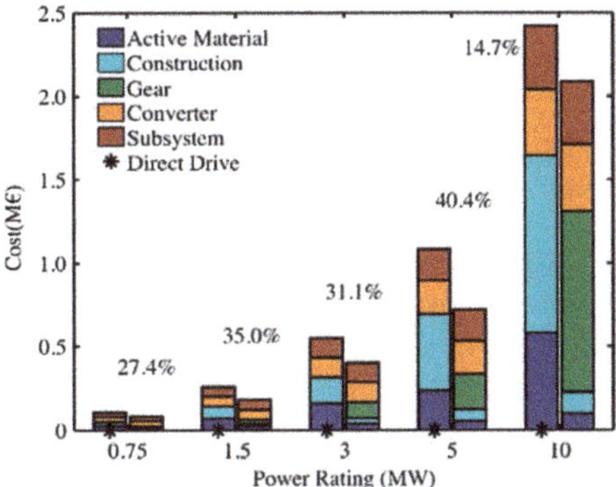

Figure 14. Depicted CAPEX (Capital Expenditure) comparison of geared and gearless PMSGs at a range of power ratings with percentage difference in cost shown at each power level [72].

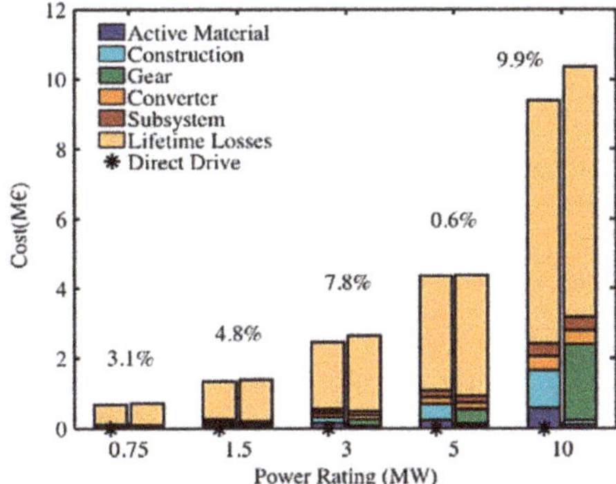

Figure 15. Cost comparisons of the machines with lifetime losses cost added and gear cost calculated twice. The percentage difference in cost is shown at each power level [72].

To conclude, it can be inferred that higher power rating-based wind turbines are the most preferred ones in reducing the development and maintenance time and eventually increasing the energy yield [72].

10. Soft Computing Technique Based Optimization Used for PMSGs

There are two critical issues that influence an electrical machine's optimal design considering the usage of FEM, the computation time from FEM simulation, and the different parameters concerned with the electrical machine. In the present day scenario, the use of soft computing techniques-based optimization has gained momentum owing to the use of the statistical analysis with multiple correlation coefficients and moving least squares (MLS) approximation as proposed (2007) which are compatible with the electrical machines [73]. In general parlance, the process of optimization includes several computations which are all dependent on parameters; the effort of computation is very minimal when compared to the time that is saved. Such a method is assessed by the same application to synchronous machine's optimal design. The results of such analysis reveal the increase in the torque per weight ratio by 13% when compared with the results that are acquired from traditional optimization techniques [73]. In 2010, [74] used the Fuzzy and FEM method for the analysis of the comparison that includes leakage field analysis witnessed in the electrical generator. The process of leakage field analysis is performed by developing a fuzzy model of the generator with the technology called adaptive neuro-fuzzy inference system (ANFIS). In this regard, the researcher performed a comparative evaluation on fuzzy model and FEM model wherein a good correlation was found to be present between them [74]. Furthermore, a study (2008) [75] revealed the new and novel approaches towards automating optimization processes that are manual, and examined the implementation obstacles that are witnessed by the engineering community. Based on the effort for design evaluation and the degrees of freedom viewpoints, engineering design optimization was subjected to classification. In the previous research, the researchers presented a holistic view on the various design optimization approaches. Furthermore, the major challenge witnessed was scalability for the techniques of design optimization considered in the study. Large-scale optimization requires effective algorithms such as swarm intelligence and a considerable computing power [75]. However, 2001, [76] proposed the use of a neural network in comparison with the Finite Element Technique (FET) based sensitivity analysis for the optimization of permanent magnet generators. In 2012, [77] further identified the challenges that were witnessed during design optimization for minimizing or maximizing the fitness function which positively influence the design purpose. Genetic algorithm which is incorporated in the optimization technique that is population based does not consider certain inferences, such as the magnet, copper, and magnetic laminations, and raw active materials. The main intention towards the reduction of the fitness function is based on the cost of energy that is generated by the system which further accounts for the variables that are uncertain in nature [77]. In 2009, [25] further explored the use of direct-drive PM wind generation system optimum design models wherein the PM was designed and developed using enhanced genetic algorithm with a PM generator fitted with 500 kW direct drive wherein the minimization of the active material cost tends to improve the design optimization effectiveness.

In 2009 used the concept of direct-drive PM wind generation system optimum design models in which the PM is developed using an improvised genetic algorithm along with a 500 kW direct drive PM generator; this actually reduces the cost of generator active material which further illustrates design optimization effectiveness [25]. Furthermore, [78] (2007) proposed a novel approach for the design of electrical rotating machines wherein a rational solution of predesign was done by integrating exact global optimization algorithms and analytical model. However, prior to developing an extensive prototype, validation of previous solutions should be performed using FEM. The purpose of the previous research was to extend the accurate global optimization algorithm through the introduction of an automatic numeric tool. Such a novel technique is used in resolving rationally the design problems. Furthermore, several examples were evaluated to examine the effectiveness of the novel technique [78]. In 2008, [79] further established a new hybrid machine with 36/24 pole outer rotor permanent magnet (PM) that is directly coupled with a wind power generator. For effective control of the flux control, two excitation (PMs and DC field windings) hybridization in the double-layer stator is utilized. This result in constant output with wide range of speeds and a load varying where examined. In 2001, [80] further used genetic algorithms wherein a new algorithm called orthogonal genetic algorithm along with quantization/quantification for global numerical optimization was used with continuous variables. Furthermore, a quantization technique and orthogonal design were used for the development of a new crossover operator; this crossover operator generates representative sample points which are small, however are a potential offspring. Such a proposed algorithm solves 15 benchmark problems with 30–100 dimensions belonging to the local minima [80]. It was 2005 arrived at new dimensions in this research area of evolutionary computation and structural design [81]. Furthermore, [82] (2008) examined soft computing (SC) techniques associated with the design of engineering concepts. Through the inspection of soft computing methods, techniques, and their competence, to further address the high complexity issues and design tasks, the researcher reviewed Fuzzy logic (FL), artificial neural networks (ANN), and Genetic Algorithms (GA) [82]. In 2012, [83] further made an overview to compare research that was conducted to optimize the parametrization of machining process of modern and conventional machining. Following are the most important techniques used: genetic algorithm (GA), particle swarm optimization (PSO), simulated annealing (SA), artificial bee colony (ABC) algorithm, and ant colony optimization (ACO). Amongst the aforementioned algorithms, GA is widely applied in the literature [83]. In 2004, [84] proposed a new solution called the multi-agent genetic algorithm (MAGA) which is an integration of the genetic algorithms and multi-agent systems to solve the problem of global numerical optimization. In 2008, [85] further proposed a disagreement versus randomness in the various SC techniques. In 2011, [86] further reviewed the state-of-the-art research developments associated with the use of soft computing techniques used for the optimization of problems associated with design, planning, and control in the field of sustainable and renewable energy. Furthermore, several soft computing methods were reviewed and presented regarding the current state of the art in computational optimization methods applied to renewable and sustainable energy, wherein a vibrant visualization of the state-of-the-art research progresses was proposed [86]. It is important to generate random numbers using soft computing methods, as random numbers are used during the beginning of the estimation or during the processes of learning and searching. When compared between simultaneous randomness consideration and opposition and pure randomness, it was revealed that the former is better than the recent results acquired from evolutionary algorithms, neural networks, and reinforcement learning. To further increase the performance of soft computing algorithms, it was revealed that opposition-based learning provides an inclining effect. This was experimentally and mathematically proven that SC has better merits when applied to improve the differential evolution (DE) [86]. In 2010, [87] also presented the Genetic algorithm (GA) with memetic algorithm and MADS (Mesh Adaptive Direct Search) for the optimal design of an electric machine. To acquire an effective optimal design of an electric machine with longer computation time and many local optima, the previous research proposed a hybrid algorithm to acquire global optimum. To maximize further Annual Energy Production (AEP), the prospective algorithm was referred. By 2006, [88] classified the modelling and optimization techniques for process problems shows in Figure 16, which displays the conventional and non-conventional optimization techniques and tools used in this regard.

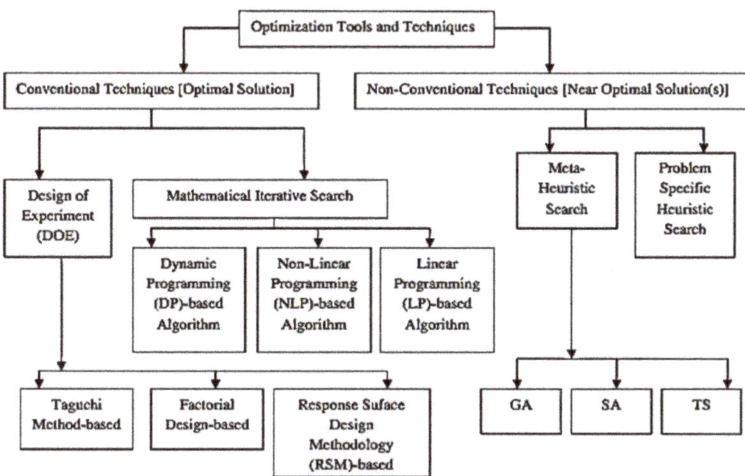

Figure 16. Conventional and non-conventional optimization tools and techniques [88].

To further conclude, it was deemed that MADS is combined with GA as an effective computation time reduction method for optimal PM wind generator design and is considered over other parallel computing methods [89]. Further offered a type of multidisciplinary design and optimization (MDO) of a diffuser for an incompressible and steady magneto-hydrodynamic (MHD) method. The design problem can be resolved using GA-based programme that is optimized with the FEM based MHD simulation technique for which least-square FEM was used and developed in later research [89]. In 2017, [90] presented about Multiple Criteria Decision Making (MCDM) concepts and has been used for economics analysis. Similarly, this concept can be used for lifecycle cost analysis of machine design [90]. In (2002) [91] presented the non-dominated sorting GA to mitigate the performance related problems wherein the performance was analyzed through the comparison of the results from the other four algorithms. Further discussion was performed on the multi-objective optimization process solution using evolutionary algorithms, wherein the findings of the research revealed effectiveness against analytical and electro-magnetic problems [91]. Furthermore, [92] (2001) displayed an approach that was used to design PM for wind power applications wherein the approach was made up of two phases: preliminary design stage and optimization stage. In 2008, [93] further examined the use of Differential Evolution (DE) and Particle Swarm Optimization (POS) Algorithms with technical analysis. It was ascertained that the Artificial Bee Colony (ABC) algorithm could be used as an innovative swarm optimization algorithm with fine results of numerical optimization.

Furthermore, [94] (2011) proposed an enhanced algorithm called the fast mutation artificial bee colony algorithm (FMABC). In 2012 proposed an improved ABC algorithm, which was used to solve numerical optimization issues, which further improved the capability of the ABC algorithm's exploitation feature. An alternate search mechanism and a varying probability function were proposed by the previous researcher. Seven numerical optimization problems were tested on the enhanced ABC algorithm [95].

In 2012, further utilized genetic algorithm (GA) for the achievement of an optimal design for an axial-flux PMSG (AFPMSG) [96]. In 2009, [97] proposed an approach based on a numerical optimization algorithm wherein a generalized receding horizon control of fuzzy systems was proposed. To further resolve generic fuzzy dynamic systems' optimal control problem, a numerical method was developed. Fine optimization was developed in the previous research.

The researcher made a thoughtful understanding of soft computation techniques in the electrical engineering field applications, with integrated pseudo-code operational summaries [98]. In 2010, [99] considered population-based algorithm and its application to solve numerical optimization problems. In certain cases, there are complexities in computing search problems which is associated with high dimensionality of search spaces. Until there is an employment of appropriate approaches, a search process could reduce effectiveness and increase cost. The use of nature inspired algorithms could tackle such difficulties. For example, fish schools tend to increase the mutual survivability since a large number of constituent individuals are deployed.

In 2008, first to introduce a method that searches high dimensional spaces that consider account behaviors that are obtained from fish schools. The derived algorithm—Fish-School Search (FSS) was made up of three operators: feeding, breeding, and swimming. In a cumulative scale, these operators tend to afford the evoked computation: (i) wide-ranging search abilities, (ii) automatic capability to switch between exploitation and exploration, and (iii) self-adaptable search process for global guidance [100].

S.L. Ho et al. (2006) examined the use of particle swarm optimization (PSO) methods wherein the previous research considered several variables such as age; new strategies were figured out to examine the optimum particle solutions, the original formula for velocity updating, and intensified search phase integration with enhanced PSO method. The findings of the previous research revealed that the proposed method contains a refined ability to perform a pinpointing search and the overall global ability improved when compared to traditional PSOs [101].

It was [102] offered the use of support vector machine (SVM) classifier for the detection of broken electrical induction machines. Furthermore, the previous researchers also considered the analysis of Gaussian, linear and quadratic kernel function as opposed to the error rate and the support vector numbers. The findings of the previous researchers revealed the successful detection of broken bars in different situations wherein there also evidences fast, precise, and robust load changes which tend to qualify for the right use of such techniques in real-time online applications in industrial drives.

Furthermore, in 2002 proposed a tabu-search algorithm to identify multi objective optimal design problems' pareto solutions from which there is a utilization of the contact algorithm to assess the previous aspects. During the initiation of the iteration cycle, identification of the new current points, fitness sharing function, and ranking selection approaches are introduced. A more detailed explanation of the numerical results is displayed in the previous research to highlight the power of the proposed algorithm to ensure that there is uniform sampling performed which yields Pareto optimal front of the multi-objective design problems. Furthermore, effective execution strategies for the proposed algorithms were also displayed [103].

In 2011, further displayed the Improved Discrete Particle Swarm Optimization (IDPSO) searching technique, which is applied on the head of an electromagnet and for the optimization of the magnetic field gradient. For the previous research, COMSOL software was used for the measurement of the magnetic forces and field. The aim of the optimization algorithm is the search of optimal pole shape geometry in a refined manner, which results in the distribution of the homogeneous magnetic field with the desired holding force in the specific area of interest [104].

Furthermore, [105] (2007) displayed an innovative recursive fuzzy logic categorizing (R-FL-C) strategy for the PM generators design approach that is utilized to mitigate search space and for expelling the local minima in due course of the process of optimization. In the previous research, finite element state space models are used to examine the space database with the knowledge that is acquired from off-line.

In 2012, [106] assessed the numerical functional optimization wherein the use of artificial bee colony optimization led the researcher to derive the use of the same bee swarm foraging behavior in their approaches. Furthermore, the ABS's efficacy was found to be high when compared with the genetic algorithm (GA), ant colony optimization (ACO), and the Particle swarm optimization (PSO). Though the ABC technique is found to be pretty important and efficient during exploration, the capacities associated with exploitation are found to be poor with issues regarding convergence speed in several instances. To mitigate this, the researcher further introduced the improved ABC algorithm or the I-ABC, which during the process of search with refining used the acceleration and inertia weight as the fitness functions.

In addition, [107] (2012) provided a heuristic structural optimization for the Surface Mounted PMSG. The use of structural optimization is the process of identifying the material distribution in an optimal way in every machine part; this technique is very prevalent in the field of mechanical engineering. Similarly, the use of structural optimization can also be witnessed in the field of electrical engineering. When compared to the other methods reported to deploy the continuous models for the elaboration of the material properties with Heuristic Search Algorithm [107], it gives a solution to the structural optimization issues.

By 2019, [108] proposed an identification method on K-means-singular value decomposition and least squares support vector machine which the simulations were proposed for voltage sags based upon an annealing algorithm for multi-objective optimization. To gain the pareto solutions in a significant manner. This is completely dependent on Pareto and can successfully be introduced in addition to parameter and objective space strings. The novel method proposed in this study questions the stop criterion, new rank formula, fitness sharing functions, and other such enhancements. For the purpose of validating, the proposed method's robustness, the study validated two numerical examples [108].

In 2001, [109] proposed an enhanced tabu-search algorithm to practically applied, it when finding optimal designs for electromagnetic devices. In parallel, the study also conducted team workshops and mathematical test functions. Based on the numerical results, it was inferred that there is a less significant iteration number achieved for the proposed method when compared with simulated annealing and other such algorithms.

In 2008, proposed a novel methodology with reference to PSO in order to find out the parametrically non-linear model structure. In this study, an existing method used in PMSM's dq-model to identify the parameters. Both the disturbed load torque as well as the motor stator resistance was established for PMSM variable-frequency drive system application. In order to question the efficiency of the identification method, the study conducted a simulation and the experimental results were provided. The results inferred excellent precision in terms of time-varying parameters when the PSO algorithm was used [110].

In the study conducted 2000), an auto-learning simulated annealing algorithm was proposed. This algorithm was developed by collaborating simulation annealing as well as the characteristics of the domain elimination method. This study utilized the standard mathematical function to assess the algorithm in addition to optimization of the power transformer practical end region [111].

In 2005, [112] demonstrated single as well as multi-objective optimizations by experimenting with a PMSM with rotor feedback with the help of a Genetic Algorithm. This artefact's extensions are nothing but the implementation of core losses cited with the help of the Steinmetz approach. A few other up-front changes are the modifications in tooth shape (especially the base), addition of voltage drawbacks, and changes in the volume expression for addition of end turns.

In the study conducted by [113] (2005), an improved Ant Colony Optimization Algorithm was proposed to be used in Electro-Magnetic Device Designs. The experiment deployed the algorithm in an inverse problem along with a mathematical function where its performance was contrasted with other better-designed methods.

A comparative study was conducted (2007) between the performance of ABCs upon the optimization of numerical function with swarm intelligence and population-based algorithms such as PSO, GA, and Particle-Swarm Inspired Evolutionary Algorithm (PS-EA) [114]. In order to explore the performance of the ABC, a total of five high dimensional benchmark functions that consisted of multi-modality were deployed. From the simulation results, the authors made a strong recommendation that the proposed algorithm is capable of expelling local minimum and can be used well in multi-variable multi-modal function optimization. The scope for future researchers in this study was the investigation of influence exerted by the control parameters in the convergence speed and performance of ABC [114].

In 2009, a comparative study conducted to assess the performance of ABC algorithm with Evolution Strategy (ES), DE, GA, and PSO using a large set of unconstrained test functions. From the results, it was concluded that there was an excellent performance exhibited by the ABC algorithm when compared to other algorithms, though the study made use of only less-control parameters thereby efficiently solving multi-dimensional as well as multi-model optimization problems [115]. The results further inferred that the performance of the ABC algorithm is superior compared to other such algorithms.

A beneficial design procedure was proposed (2012) for the controller utilized in the frequency converter of a variable speed wind turbine (VSWT)-driven PMSG with GA and RSM [116]. A mess-less technique was recommend by the study conducted in 2004, which focused on connecting the radial basis functions (RBFs) as well as wavelets. This new method proposed in this study leveraged the advantages of RBFs as well as the wavelets. In order to maintain the linear independence as well as consistency, the bridging scales were utilized so as to safeguard the mathematical properties. With the purpose of validating the proposed method, a numerical example was utilized [117].

A hybrid Genetic Algorithm (GA) was proposed in 2003 [118], in order to optimize the electromagnetic topology. After taking a 2-D encoding technique into account, the geometrical topology was at first applied to electromagnetic topology. In the later stages for the crossover operator, the study utilized a 2-D geographic crossover. In order to enhance the convergence features, the study used a novel local optimization algorithm, otherwise called an on/off sensitivity method, which is hybridized with 2-D encoded GA. Once the algorithm was verified with different case studies, the results were published [118].

11. Novel Topology Development in PMSGs

In 2012 stated the assessment of low maintenance slip-synchronous, PM wind generator, which was developed using the concept of PM induction generator [119]. In 1926 introduced the PMIG (Permanent Magnet Induction Generator) concept upon which the slip-synchronous permanent magnet generator (SS-PMG) was constructed. In generator design, there exists an induction machine cage-rotor, traditional stator winding along with an add-on of second free-rotating PM-rotor. The second PM-rotor runs synchronous speed while the cage-rotor operates at a relative slip speed in accordance to the PM rotor and rotating synchronous stator field. This is a gearless wind turbine generator that is connected with the grid directly i.e., no power electronic convertor or such behavior is required in the drive train. In the summary developed by [17] (2014), a comparison was performed between large-sized wind turbines which can produce more electricity at less cost with small-sized turbines. This comparison was executed since the costs involved in experimental set-up and maintenance do not impact the size of the machine. Therefore, more than 7 MW output power is being achieved from today's wind generators. For instance, from 2011, Enercon manufacturing an E-126/7500 wind turbine with 7.5 MW power capacity. At present, Sway Turbine and Windtec Solutions are in the process of developing 10 MW wind turbine generators which might hit the commercial markets in 2015 [17]. Figure 17 shows the voltage ratings of seven various models of common wind turbine generators with respect to the turbine power which clearly depicts the model performance.

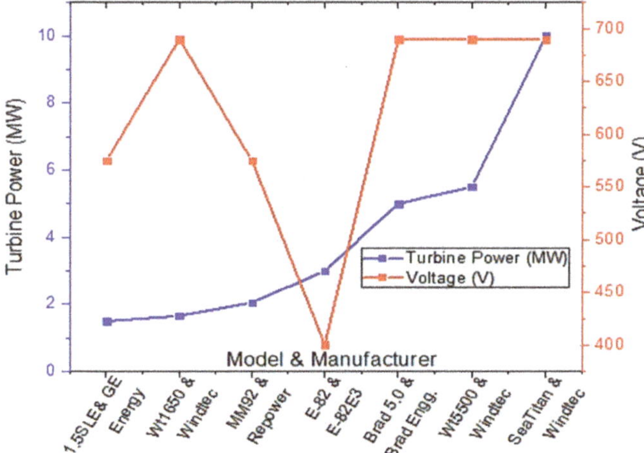

Figure 17. Summary of the voltage ratings of a few common wind turbine generators [15].

An innovative model with a Surface-Inserted Permanent Magnets Synchronous Generator was proposed in 2011, with air slots in the rotor that can be adjusted. This model removes the disadvantage present in PMSGs i.e. fluctuation of regulating voltage. When a comparison was performed between conventional machines and superconducting machines, it was found that the latter exhibited novel advantages such as efficiency, compactness, lightweight and significant stable operation in power systems [120].

In 2007 proposed an eccentricity topology with a promise to enhance the power density and made use of it in the design, development, and testing on an eight-pole superconducting rotating machine. Further, the study discussed the results retrieved from the magnetic scalar potential from a Coulomb formulation by Markov Chain Monte Carlo (MCMC) method. Additionally, the flux density was calculated using derivation from the regularization method. With the purpose of reducing the computation time, the MCMC method was deployed which in turn perform the magnetic scalar calculations in specific regions of discrete geometry. By using YBaCuO high-temperature superconducting (HTS) bulk plates and low temperature superconducting NbTi wires, a high magnetic field was generated. In order to increase the cooling operation, there is a stationary superconducting inductor and a rotating armature coiled with copper wires present in the superconducting machine [121].

A detailed differentiation study was conducted [122] (2012) on the differences in development and settlement of active materials for transversal-flux machines from radial and axial ones. Lower stator copper losses were gained by increased windings space in the absence of any impact from the available space for flux in the transversal flux. As the electromagnetic structure is sophisticated, the transversal-flux machines seemed to be costly [122].

A novel low-fare methodology was proposed in 2012 to develop wind turbine electric generators from the generator from the burnt-out squirrel cage induction motors. The author first detailed the list of properties generally required for a wind turbine generator following which the methodology described the PMG, workability, multi-pole, and low-speed. The study conducted a cost comparative analysis and performance comparative analysis based on the test results achieved from a 500 W generator run at 900 RPM and a 1500 W generator at 650 RPM [123].

The efficiency of an air-cored PMSG was estimated in the study conducted [124] (2011) using finite elements and equivalent circuit modelling. The emerging trends showcase that the air-cored machines are predominantly used in wind energy systems. Instead of iron, the magnets which are captivated between the mild-steel-based rotors are present. At zero-load, the two-sided, axial-flux, air-cored machine's flux path can be seen as a stable magnetic flux that crosses axially from a magnet on one rotor to the opposite rotor which is a facing magnet. Further, the study stated that the coil is held by the stator on a plane in the middle of two magnetic sets [124].

In 2012 [125], an alternative viable solution was proposed the traditional PMSGs at MW level in direct-drive wind turbine applications via a Halbach array. It is a must to optimize the machine dimensions in order to achieve the maximum benefit of the Halbach array. This research article provides an overview of calculating the Halbach array application using analytical equations which are prevalent in the studies published earlier. The study recorded extraordinary performance by making few modifications in the existing PMSG design in which a constant magnet volume is maintained. When compared, the conventional array seemed to be more valued than the Halbach array at the time of considering the critical rotor radius. When the number of poles were increased, the critical radius got shifted to larger sizes and thus it allowed a positive leverage of the Halbach array at MW level. The analytical equation findings were verified using FEA simulation [125].

In 2008 [126], Halbach magnet array with the help of the numerical optimization method, which in turn relied upon finite element analysis. The magnetization direction of every element was designated as the design variable. In order to enhance the repulsive, attractive, and tangential magnetic forces present between the magnetic layers, the researcher investigated the optimal magnet arrays composed of two and three linear magnet layers. Two and three magnet rings altogether are present in a torsional spring and it receives the tangential force maximized by the magnet array. In this study, the researcher employed few optimization techniques such as adjoint variable methods and sequential linear programming in 2-D finite element analysis [126].

In 2005 developed a theoretical study about the magnetic circuit for a longitudinal flux PM synchronous linear generator. In order to assess the machine performance, the researcher used a coupled field and circuit model which was solved using the time-stepping finite-element technique [127]. In 2008 [128], and 2010 [129], conducted a comparison of different configurations in an axial-flux nine-phase concentrated-winding PMSG for a direct-drive wind turbine.

Various prototypes where investigated by [130] (2012) in which one of the prototypes demonstrated that the active mass of a PMG unit in a SS-PMG curtailed in a considerable fashion. For different slip-PMG concepts, the evaluation was also performed. To be specific, it is feasible to have a notable amount of minimization in active and PM mass for the new brushless-DC winding slip-PMG in comparison to existing non-overlap winding configurations. Further, it can be projected that the copper can be replaced by aluminum and there is no need to increase the mass of slip-PMG without changing the machine cost performance [130].

A low-speed three-phase generator was considered in 2014 with high induced voltage, low harmonic distortion as well as high generator efficiency, optimal generator parameters such as pole-arc to pole-pitch ratio and stator-slot-shoes dimension topology for investigation. For the purpose of obtaining sinusoidal induced voltages in stator windings, the researcher arranged the PMs in rotor structure and adopted the magnetization direction in an appropriate manner [131]. An insight was published (2006) about the basis behind the development of PMSG, a novel hybrid in Hybrid Excitation Permanent Magnet Synchronous Generator (HEPMSG). It was developed through the insertion of an exciting winding in rotor or stator [132].

In 2008, [133] developed the Flux Reversal Machine (FRM) coupled with a doubly salient stator permanent magnet machine in addition to flux linkage reversal present in the stator concentrated winding. The study conducted a comparative analysis on Full Pitch Winding Flux Reversal Machine (FPFRM) and Conventional Concentrated Stator Pole Winding FRM (CSPFRM) on the design. The results revealed that FPFRM exhibited high power density than CSPFRM [133].

In order to shuffle the standard claw pole alternator in the place of automobile application, a single-phase FRM was introduced. It has few advantages, such as it has a simple construction process, expresses high-power density, low in inertia etc. Reference [134] (2010) investigated and proposed a distributed winding for FRM (Flux Removal Machine). A high-power density is provided by FPFRM and it enhances the efficiency as well. Being a doubly-salient Permanent Magnet machine with concentrated windings, FRM has advantages of both Switched Reluctance Machine and Permanent Magnet (PM) machines. FEM analysis was carried out in order to achieve the induced EMF, winding inductances, and flux linkages. The winding function strategy received the inductance of both the machines and it was compared with FEM results. On the basis of fabricated 'electrical gear', the power densities of both CSPFRM and FPFRM with PMSM were compared. The gear ratios were provided from various FRM configurations. As the design of CSPFRM, FPFRM, and PMSM are similar with respect to outer dimensions, the volume of the magnet, and the rotor speed, the comparison of those three machines were graphically represented in Figure 18. From the graphical representation, it is observed that the machine FPFRM requires very high compensating kVAr, when compared to CSPFRM and PMSM. However, as far as active weight/kVA is concerned the FPFRM is less when compared with other two [134].

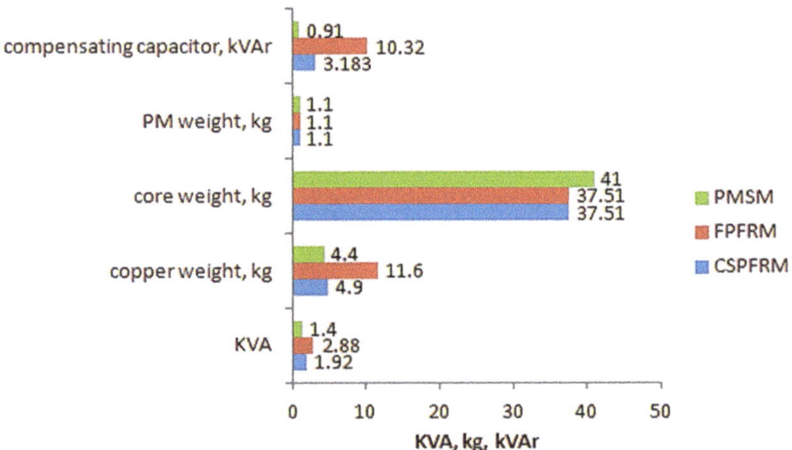

Figure 18. Comparison of Conventional Concentrated Stator Pole Winding Flux Reversal Machine (CSPFRM), Full Pitch Winding Flux Reversal Machine (FPFRM), and PMSM [134].

In 2007, [135] developed low-revolution magneto-electric generators which are custom designed for wind power engineering applications. The best and efficient way to diminish the own drag torque is to incorporate a magnetic rake so that the EMF do not exhibit a significant decrease and the adaptability of the magneto electric machine design is preserved as per the manufacturing. Among the available ones, the best alternative is the one, which is found in the magnetic rakes situated outside and inside of the rotor inductors, which is equivalent to the width of the spline way slots that were found inside and outside of the stator.

An optimal design method was proposed in 2009 [136], in which a double-layer permanent magnet (PM) Dual Mechanical Port (DMP) machine was present for wind power application with random low-wind turbine speed input and stable steep synchronous speed output. The torque was compared between the outer-rotor and inner-rotor. Further, they also compared the THD variations with a pole arc co-efficient for inner-rotor and stator winding [136].

With the purpose of overcoming the potential barriers of dimension, cost, and reliability, in 2011 [137], a multi-generator architecture was recommended. They suggested that a total of two PMSGs should be shared with one turbine-driven shaft. The outputs need to be recorded from the two PMSGs, and then rectified in order to be connected in series with an intermediate DC chopper, whereas the back-end inverter is provided with similar option [137].

In 2012, [138] developed an investigation about the novel form of transvers and axial-flux magnetic fields of the PMSG. With novel machine configuration such as rotation, the flow of the main flux would be in the transverse direction. A novel Outer Rotor-Permanent Magnet (PM) Vernier (OR-PMV) machine was introduced [139] (2010) for direct-driven wind power generation that comes packed with low speed. This is because the wind power can be easily capture, and it triggers the high-speed rotating field design in order to enhance the power density [139].

In 2003, [140] developed an operating principle called Consequent-Pole Permanent-Magnet Machine. In addition to the finite element analysis and sizing analysis, the experimental results were achieved for the prototype machine. There are many advantages associated with CPPM machine, one of which is the control on air-gap flux level excluding the demagnetization risk from magnetic pieces. In terms of low-reluctance iron poles, it is possible to execute the control action. In addition to the low field AT requirement, a wide range of air-gap flux control was also yield and this could be leveraged to either increase or diminish the air-gap flux.

In 2012, [141] sought the winding functional theory as the basis and detailed the inductance of a multi-phase synchronous machine supported with a PM or a wound field rotor. Due to the three magneto-static simulation results produced from simple machine-based geometric models, it is easy to determine the permeance function. For the inductance of stator phase, inductance of phase-to-field, and PM flux linkage calculations, the existing method was used. In order to have an accurate incorporation in numerical machine models that pertain to dynamic simulations, the study proposed the machine inductances, which are relevant to Fourier-series expansion.

A novel interpolating strategy was proposed in 2011 [142] for air-gaps by antiperiodic boundary condition when applied to AFPMSG. With the help of coupled-circuit, element analysis, and time-stepping, the performance of AFPMSG in case of isolated load was investigated. The investigation was also conducted upon the performance of short circuit. In order to produce the results of the accurate analysis, the researcher used second-order serendipity quadrilateral elements [142].

A novel three-phase 12/8-pole doubly salient permanent magnet (DSPM) machine was investigated in 2006 [143] so that it can be used in wind power generation. This in turn can be utilized to design and study the recommended DSPM generator i.e., a novel machine structure design which yields high efficiency, high power density, and high robustness in the device of system operation. In order to obtain static characteristics of the proposed generated, the researcher used FEM [143].

In 2017, [144] executed a complete model, design, and development of a novel slip-synchronous Permanent Magnet (PM) wind generator for direct-drive direct-grid connection. There is variation present in the proposed generator with that of the traditional PM induction generators mentioned in the literature. The non-overlap winding was used in the proposed model for the very first time. For the generator design to be effective, a mixture of analytical, finite element calculation and optimized design methods were employed by the researcher. The researcher minimized the critical design parameters such as load torque ripple and no-load cogging torque to the best possible minimum level in the design optimization stage. The model was verified, and the design was completed with measurements of wind generator system prototype.

Under linear condition (2007) [145] compared the predictions of the two methods listed herewith in the calculation of electro-magnetic torque with inductance of a synchronous reluctance machine.

1.   Winding function analysis (WFA);
2.   Finite-element analysis (FEA).

In the methods mentioned in the literature, the stator winding connections, stator slot effects, as well as the rotor geometry, were considered. The WFA simulation results were contrasted with that of the 2-D FEA whereas the results were the same. In case of magnetic linear condition, the winding function method seems to be quick, incur less computational costs, and quite simple.

In 2009, [146] analyzed the Hybrid Exciting Synchronous Generator (HESG) in addition to unique operating principles and structure. The upkeep required for the main output is generally taken out by PM generator and, however, the terminal voltage is regulated by a homo-planar inductor alternator. In order to execute the computation of EMF and analyze the performance of HESG, 3-D FEM was utilized.

12. Control Mechanism for WECSs

Numerous control mechanisms are employed in WECS, when designing a generator, it is important to consider the vital parameters such as aerodynamic efficacy, statistical wind distribution, and control system, since these decisive factors can be used in performance evaluation. A study considered high overload capacity generators for this specific application. It was concluded that the optimization of generator is a must to diminish the losses and achieve the highest overload capability. The wind power generation systems act to achieve the sole aim of harnessing the maximum amount of wind energy and consequently converting it to electrical energy. One can achieve this easily through the help of a control structure which allows the operation range as well as the ideal algorithm of stable system with MPPT (Maximum Power Point Tracking). The objective of MPPT presents the harnessing of maximum energy by making few changes in operating point of the system in order to tap the full energy from wind. A control structure was proposed [147] (2013) with specific reference of wind energy systems on the bases of PMSG. With the purpose of enhancing the reliability and the robustness, the study determined the optimum structure using speed and torque control. This retrieved from the analysis of conventional control structures, which used variable speed, and fixed pitch wind energy generation systems [147].

The time-stepping FEA was offered (2006) [148] pertaining to a variable speed synchronous generator in addition to the rectifier. The bi-directional alternator speeds are maintained by this model, the application is a linear generator in terms of the ocean wave energy conversion [148].

In the study conducted in 2014 [149], the authors described the output power fluctuations faced in a wind farm and the relevant problems created in the power system. The study compared the fluctuations that occurred in the output power of conventional schemes with the proposed methods. However, this study's proposed scheme had the tracking of optimal rotational speed in such a manner that the output power was smoothened. A fuzzy PID controller used instead of traditional vector control, which resulted in the tracking of the turbine's optimal rotational speed and the smoothening of wind farm's output power [149].

In 2009, [150] presented a system design model and its control approaches for a 2 MW direct-driven PMSG fed through parallel-connected full power back-to-back PWM converters. Both the electromagnetic FE analysis as well as the optimal generator design was executed in this application in terms of wind generation [150].

An elaborate design and an experimental approach proposed in 2010 [151] for a completely passive wind turbine system without the active electronic part (power and control). The efficiency of the devices predominantly relied upon the condition where the system's design parameters were reciprocally adapted using an integrated optimal design methodology. This methodology ensures simultaneous optimization of wind power extraction and losses because of the global system in a specific wind speed profile. This way, the weight of the wind turbine generator was decreased. Based on the approaches discussed in earlier studies, optimal PMSG was obtain for critical features of passive wind turbines like geometric and energetic features [151]. Figure 19 illustrates the general representation of a typical variable-speed direct-driven PMSG wind turbine connected to the grid distribution.

The study conducted by [152] developed a holistic modelling of direct-driven PMSG-based grid-connected wind turbines in addition to control schemes for the interface converters. There were two distinguishable control schemes designed in this configuration for generator and grid-side converters.

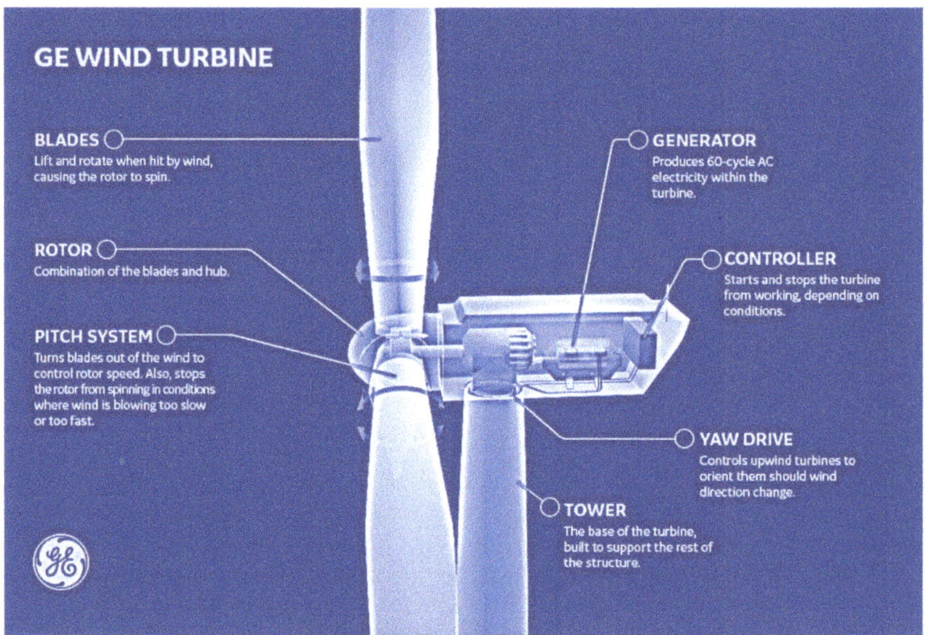

Figure 19. Typical variable-speed direct-driven PMSG wind turbine. Source: GE [153]

13. Conclusions

This review article developed a conceptual framework with an overview of research challenges, using which the proposed work of analysis, suitability, design, and control of PMSGs for Wind Energy Conversion Systems (WECS) to be carried out in future needs and development in the wind sector. The predicted influence and the preliminary results will further the progress beyond the threshold level set by the state-of-the-art envisioned research. In the literature, the WECSs and its classification as per wind turbine and generator schemes, the types of PMSG technologies where discussed elaborately. In addition, the study also suggested the solution found for optimization problems using the field computation technique. This study related to WECS development provided an advanced inter-disciplinary approach on technical parts, and compared with the pros cons of as previous studies. Information provided in this article can also be helpful in improving the WECS. It also reviewed the soft computing (SC) techniques that where applied for the optimal design methodologies of the PMSGs. When exploring the literature, unraveled mysteries and unexplored areas from the developmental perspective to take forward in future.

Author Contributions: All authors involved equally, and intensively investigated to produce the comprehensive research survey article in current form for renewable energy system for Wind Energy and its machine dynamics.

Funding: This research received no external funding.

Conflicts of Interest: The authors declare no conflict of interest.

References

1. Şahin, A.D. Progress and recent trends in wind energy. *Prog. Energy Combust. Sci.* **2004**, *30*, 501–543. [CrossRef]
2. Padmanathan, K.; Govindarajan, U.; Ramachandaramurthy, V.K.; Rajagopalan, A.; Pachaivannan, N.; Sowmmiya, U.; Padmanaban, S.; Holm-Nielsen, J.B.; Xavier, S.; Periasamy, S.K. A Sociocultural Study on Solar Photovoltaic Energy System in India: Stratification and Policy Implication. *J. Clean. Prod.* **2019**, *216*, 461–481. [CrossRef]
3. Fatih, B.; Darbari, S. The Global Energy Outlook and the Increasing Role of India. In *International Energy Agency Memorial Lecture*; IEA: Paris, France, 29 August 2018.
4. Global Wind Energy Council (GWEC). *Global Wind Report 2018*; Global Wind Energy Council (GWEC): Brussels, Belgium, 2019.
5. *Global Energy Transformation: A Roadmap to 2050 (2019 Edition)*; International Renewable Energy Agency (IRENA): Masdar City, UAE, 2019; ISBN 978-92-9260-121-8.
6. Michaelson, H.B. Techniques of Editorial Research in Electrical Engineering. *J. Franklin Inst.* **1949**, *247*, 245–253. [CrossRef]
7. Jianfeng, D.; Yi, T.; Jun, Y. Adaptive Gains Control Scheme for PMSG-Based Wind Power Plant to Provide Voltage Regulation Service. *Energies J.* **2019**, *12*, 753.
8. Ramakumar, R.; Slootweg, J.G.; Wozniak, L. Guest Editorial Introduction to the Special Issue on Wind Power. *IEEE Trans. Energy Convers.* **2007**, *22*, 1–3. [CrossRef]
9. Shafiullah, G.M.; Oo, A.M.; Ali, A.S.; Wolfs, P. Potential challenges of integrating large-scale wind energy into the power grid—A review. *Renew. Sustain. Energy Rev.* **2013**, *20*, 306–321. [CrossRef]
10. Global Wind Energy Council (GWEC). *Global Wind Energy Outlook 2014*; Global Wind Energy Council (GWEC): Brussels, Belgium, 2014.
11. Paulsen, U.S.; Vita, L.; Madsen, H.A.; Hattel, J.; Ritchie, E.; Leban, K.M.; Berthelsen, P.A.; Carstensen, S. 1st DeepWind 5 MW baseline design. *Energy Procedia* **2012**, *24*, 27–35. [CrossRef]
12. Global Wind Energy Council (GWEC). *India Wind Energy Outlook 2012*; Global Wind Energy Council (GWEC): Brussels, Belgium, 2012.
13. Mohanpurkar, M.; Ramakumar, R.G. Probability Density Functions for Power Output of Wind Electric Conversion Systems. In Proceedings of the IEEE PES General Meeting, Providence, RI, USA, 25–29 July 2010. [CrossRef]
14. Herbert, G.J.; Iniyan, S.; Sreevalsan, E.; Rajapandian, S. A review of wind energy technologies. *Renew. Sustain. Energy Rev.* **2007**, *11*, 1117–1145. [CrossRef]
15. Herbert, G.J.; Iniyan, S.; Amutha, D. A review of technical issues on the development of wind farms. *Renew. Sustain. Energy Rev.* **2014**, *32*, 619–641. [CrossRef]
16. Sareni, B.; Abdelli, A.; Roboam, X.; Tran, D.H. Model simplification and optimization of a passive wind turbine generator. *Renew. Energy* **2009**, *34*, 2640–2650. [CrossRef]
17. Islam, M.R.; Guo, Y.; Zhu, J. A review of offshore wind turbine nacelle: Technical challenges, and research and developmental trends. *Renew. Sustain. Energy Rev.* **2014**, *33*, 161–176. [CrossRef]
18. Zhong, D. Finite Element Analysis of Synchronous Machines. Ph.D. Thesis, The Pennsylvania State University, State College, PA, USA, May 2010.
19. Sethuraman, L.; Venugopal, V.; Zavvos, A.; Mueller, M. Structural integrity of a direct-drive generator for a floating wind turbine. *Renew. Energy* **2014**, *63*, 597–616. [CrossRef]
20. Chen, Z.; Blaabjerg, F. Wind farm—A power source in future power systems. *Renew. Sustain. Energy Rev.* **2009**, *13*, 1288–1300. [CrossRef]
21. Domínguez-García, J.L.; Gomis-Bellmunt, O.; Bianchi, F.D.; Sumper, A. Power oscillation damping supported by wind power: A review. *Renew. Sustain. Energy Rev.* **2012**, *16*, 4994–5006. [CrossRef]
22. Boldea, I. *The Electric Generators Hand Book—Synchronous Generators*; CRC Press, Taylor & Francis Group: Boca Raton, FL, USA, 2006.
23. Nerg, J.; Ruuskanen, V. Lumped-parameter-based thermal analysis of a doubly radial forced-air-cooled direct-driven permanent magnet wind generator. *Math. Comput. Simul.* **2012**, *90*. [CrossRef]
24. Kumar, A.; Marwaha, S.; Singh, A.; Marwaha, A. Performance investigation of a permanent magnet generator. *Simul. Model. Pract. Theory* **2009**, *17*, 1548–1554. [CrossRef]

25. Li, H.; Chen, Z. Design optimization and site matching of direct-drive permanent magnet wind power generator systems. *Renew. Energy* **2009**, *34*, 1175–1184. [CrossRef]

26. Aleksashkin, A.; Mikkola, A. *Literature review on Permanent Magnet Generators Design and Dynamic Behavior*; Lappenranta University of Technology: Lappeenranta, Finland, 2008; ISBN 978-952-214-708-0.

27. Gündoğdu, T.; Kömürgöz, G. Technological and economical analysis of salient pole and permanent magnet synchronous machines designed for wind turbines. *J. Magn. Magn. Mater.* **2012**, *324*, 2679–2686. [CrossRef]

28. Muetze, A. Deterministic global optimization of electromechanical energy converters. *IEEE Ind. Appl. Mag.* **2008**, *14*.

29. Kömürgöz, G.; Gündogdu, T. Comparison of Salient Pole and Permanent Magnet Synchronous Machines Designed for Wind Turbines. In Proceedings of the 2012 IEEE Power Electronics and Machines in Wind Applications, Denver, CO, USA, 16–18 July 2012.

30. Li, H.; Chen, Z. Design Optimization and valuation of Different Wind Generator Systems. In Proceedings of the 2008 International Conference on Electrical Machines and Systems, Wuhan, China, 17–20 October 2008; pp. 2396–2401.

31. Li, H.; Chen, Z. Overview of different wind generator systems and their comparisons. *IET Renew. Power Gener.* **2008**, *2*, 123–138. [CrossRef]

32. Li, H.; Chen, Z.; Polinder, H. Optimization of Multibrid Permanent-Magnet Wind Generator Systems. *IEEE Trans. Energy Convers.* **2009**, *24*, 82–92. [CrossRef]

33. Eriksson, S.; Bernhoff, H. Loss evaluation and design optimisation for direct driven permanent magnet synchronous generators for wind power. *Appl. Energy* **2011**, *88*, 265–271. [CrossRef]

34. Alnasir, Z.; Kazerani, M. An analytical literature review of stand-alone wind energy conversion systems from generator viewpoint. *Renew. Sustain. Energy Rev.* **2013**, *28*, 597–615. [CrossRef]

35. Rossa, R.; Król, E. Finite Element Analysis -Based Calculation of Load Characteristics of Stand-Alone PM Generators with Inner or Outer Rotor Constructions. In Proceedings of the XIX International Conference on Electrical Machines—ICEM 2010, Rome, Italy, 6–8 September 2010.

36. Park, Y.S.; Jang, S.M.; Choi, J.Y.; Choi, J.H.; You, D.J. Influence of AC-DC-DC Converter on Radial/Axial Flux Permanent Magnet Wind Power Generators with Mechanical Energy Storage System. In Proceedings of the 2013 IEEE International Conference on Industrial Technology (ICIT), Cape Town, South Africa, 25–28 February 2013.

37. Tapia, J.A.; Pyrhonen, J.; Puranen, J.; Lindh, P.; Nyman, S. Optimal Design of Large Permanent Magnet Synchronous Generators. *IEEE Trans. Magn.* **2013**, *49*, 642–650. [CrossRef]

38. Park, J.; Kim, J.; Shin, Y.; Lee, J.; Park, J. 3 MW class offshore wind turbine development. *Curr. Appl. Phys.* **2010**, *10*, S307–S310. [CrossRef]

39. Fei, W. Permanent Magnet Synchronous Machines with Fractional Slot and Concentrated Winding Configurations. Power and Drive Systems Group. Ph.D. Thesis, Cranfield University, Cranfield, UK, January 2011.

40. Ko, K.J.; Jang, S.M.; Park, J.H.; Cho, H.W.; You, D.J. Electromagnetic Performance Analysis of Wind Power Generator with Outer Permanent Magnet Rotor Based on Turbine Characteristics Variation Over Nominal Wind Speed. *IEEE Trans. Magn.* **2011**, *47*, 3292–3295. [CrossRef]

41. Duan, Y. Method for Design and Optimization of Surface Mount Permanent Magnet Machines and Induction Machines. Ph.D. Thesis, Georgia Institute of Technology, Atlanta, GA, USA, December 2010.

42. Stamenkovic, I.; Milivojevic, N.; Schofield, N.; Krishnamurthy, M. Emadi, A. Design, Analysis, and Optimization of Ironless Stator Permanent Magnet Machines. *IEEE Trans. Power Electron.* **2013**, *28*, 2527–2538. [CrossRef]

43. Antipov, V.N.; Danilevich, Y.B. Analysis and Investigation of Size Spectrum of Synchronous Machines Operating as Wind–Power–Generators in the Range of Rotation Frequencies of 75–300 rpm. *Russ. Electr. Eng.* **2009**, *80*, 23–28. [CrossRef]

44. Simion, S.V.A.; Livadaru, L.; Irimia, N.D.; Lazăr, F. FEM Analysis of a Low Speed Permanent Magnet Synchronous Machine with External Rotor for a Wind Generator. In Proceedings of the 2012 13th International Conference on Optimization of Electrical and Electronic Equipment (OPTIM), Brasov, Romania, 24–26 May 2012.

45. Colli, V.D.; Marignetti, F.; Attaianese, C. Analytical and Multiphysics Approach to the Optimal Design of a 10-MW DFIG for Direct-Drive Wind Turbines. *IEEE Trans. Ind. Electr.* **2012**, *59*, 2791–2799. [CrossRef]

46. Guo, Z.; Chang, L. FEM Study on Permanent Magnet Synchronous Generators for Small Wind Turbines. In Proceedings of the Canadian Conference on Electrical and Computer Engineering, Saskatoon, SK, CA, 1–4 May 2005; pp. 641–644.

47. Hosseini, S.M.; Agha-Mirsalim, M.; Mirzaei, M. Design, Prototyping, and Analysis of a Low Cost Axial-Flux Coreless Permanent-Magnet Generator. *IEEE Trans. Magn.* **2008**, *44*, 75–80. [CrossRef]

48. Pinilla, M. Performance Improvement in a Renewable Energy Direct Drive Permanent Magnet Machine by means of Soft Magnetic Composite Interpoles. *IEEE Trans. Energy Convers.* **2012**, *27*, 440–448. [CrossRef]

49. Stegmann, J.A.; Kamper, M.J. Design Aspects of Double-Sided Rotor Radial Flux Air-Cored Permanent-Magnet Wind Generator. *IEEE Trans. Ind. Appl.* **2011**, *47*, 767–778. [CrossRef]

50. Park, Y.S.; Jang, S.M.; Choi, J.H.; Choi, J.Y.; You, D.J. Characteristic Analysis on Axial Flux Permanent Magnet Synchronous Generator Considering Wind Turbine Characteristics According to Wind Speed for Small-Scale Power Application. *IEEE Trans. Magn.* **2012**, *48*, 2937–2940. [CrossRef]

51. Chebak, A.; Viarouge, P.; Cros, J. Optimal design of a high-speed slotless permanent magnet synchronous generator with soft magnetic composite stator yoke and rectifier load. *Math. Comput. Simul.* **2010**, *81*, 239–251. [CrossRef]

52. Spooner, E.; Williamson, A.C.; Catto, G. Modular Design of Permanent Magnet Generator for Wind Turbines. *IEE Proc.-Electr. Power Appl.* **1996**, *143*, 388–395. [CrossRef]

53. Chan, T.F.; Lai, L.L. An Axial-Flux Permanent-Magnet Synchronous Generator for a Direct-Coupled Wind-Turbine System. *IEEE Trans. Energy Convers.* **2007**, *22*, 86–94. [CrossRef]

54. Caricchi, F.; Maradei, F.; De Donato, G.; Capponi, F.G. Axial-Flux Permanent-Magnet Generator for Induction Heating Gensets. *IEEE Trans. Ind. Electron.* **2010**, *57*, 128–137. [CrossRef]

55. Hanselman, D.C. *Brushless PM Motor Design*; Marcel Dekker: New York, NY, USA, 1997.

56. Hendershot, J.R.; Miller, T.J.; Miller, E. *Design of Brushless Permanent Magnet Motor*; Oxford University Press: Oxford, UK, 1994.

57. *User's Guide of Maxwell 3D*; ANSYS Inc.: Canonsburg, PA, USA, 2010.

58. Traxler-Samek, G.; Zickermann, R.; Schwery, A. Cooling Airflow, Losses, and Temperatures in Large Air-Cooled Synchronous Machines. *IEEE Trans. Ind. Electron.* **2010**, *57*, 172–180. [CrossRef]

59. Huang, Y.; Dong, J.; Jin, L.; Zhu, J.; Guo, Y. Eddy-Current Loss Prediction in the Rotor Magnets of a Permanent Magnet Synchronous Generator with Modular Winding Feeding a Rectifier Load. *IEEE Trans. Magn.* **2011**, *47*, 4203–4206. [CrossRef]

60. Spooner, E.; Williamson, A.C. Parasitic Losses in Modular Permanent-Magnet Generators. *IEE Proc. Electr. Power Appl.* **1997**. [CrossRef]

61. Martin, F.; Zaïm, M.E.H.; Tounzi, A.; Bernard, N. Improved Analytical Determination of Eddy Current Losses in Surface Mounted Permanent Magnets of Synchronous Machine. *IEEE Trans. Magn.* **2014**, *50*, 1–9. [CrossRef]

62. Ruschetti, C.; Verucchi, C.; Bossio, G.; De Angelo, C.; García, G. Rotor demagnetization effects on permanent magnet synchronous machines. *Energy Convers. Manag.* **2013**, *74*, 1–8. [CrossRef]

63. Rodrigues, R.B.; Mendes, V.M.F.; Catalão, J.P.D.S. Protection of interconnected wind turbines against lightning effects: Over voltages and electromagnetic transients study. *Renew. Energy* **2012**, *46*, 232–240. [CrossRef]

64. Jin, Y. Fault Analysis and Protection for Wind Power Generation Systems. Ph.D. Thesis, University of Glasgow, Glasgow, UK, March 2011. Available online: http://theses.gla.ac.uk (accessed on 18 May 2015).

65. Kalokiris, G.K.; Kladas, A.G.; Hatzilau, I.K.; Cofinas, S.; Gyparis, I.K. Advances in magnetic materials and their impact on electric machine design. *J. Mater. Process. Technol.* **2007**, *181*, 148–152. [CrossRef]

66. Jung, J.W.; Lee, B.H.; Kim, D.J.; Hong, J.P.; Kim, J.Y.; Jeon, S.M.; Song, D.H. Mechanical Stress Reduction of Rotor Core of Interior Permanent Magnet Synchronous Motor. *IEEE Trans. Magn.* **2012**, *48*, 911–914. [CrossRef]

67. Geng, H.; Xu, D. Stability Analysis and Improvements for Variable-Speed Multipole Permanent Magnet Synchronous Generator-Based Wind Energy Conversion System. *IEEE Trans. Sustain. Energy* **2011**, *2*, 459–467. [CrossRef]

68. Xia, S.; Zhang, Q.; Hussain, S.; Hong, B.; Zou, W. Impacts of Integration of Wind Farms on Power System Transient Stability. *Appl. Sci.* **2018**, *8*, 1289. [CrossRef]

69. Frank, N.W.; Pakdelian, S.; Toliyat, H.A. Passive Suppression of Transient Oscillations in the Concentric Planetary Magnetic Gear. *IEEE Trans. Energy Convers.* **2011**, *26*, 933–939. [CrossRef]

70. Westlake, A.J.G.; Bumby, J.R.; Spooner, E. Damping the power-angle oscillations of a permanent-magnet synchronous generator with particular reference to wind turbine applications. *IEE Proc. Electr. Power Appl.* **1996**, *143*, 269–280. [CrossRef]

71. Klontz, K.W.; Miller, T.J.; McGilp, M.I.; Karmaker, H.; Zhong, P. Short-Circuit Analysis of Permanent- Magnet Generators. *IEEE Trans. Appl.* **2011**, *47*, 1670–1680. [CrossRef]

72. Alshibani, S.; Agelidis, V.G.; Dutta, R. Lifetime Cost Assessment of Permanent Magnet Synchronous Generators for MW Level Wind Turbines. *IEEE Trans. Sustain. Energy* **2014**, *5*, 10–17. [CrossRef]

73. Giurgea, S.; Zire, H.S.; Miraoui, A. Two-Stage Surrogate Model for Finite-Element-Based Optimization of Permanent-Magnet Synchronous Motor. *IEEE Trans. Magn.* **2007**, *43*, 3607–3613. [CrossRef]

74. Kumar, A.; Marwaha, S.; Singh, A.; Marwaha, A. Comparative leakage field analysis of electromagnetic devices using finite element and fuzzy methods. *Expert Syst. Appl.* **2010**, *37*, 3827–3834. [CrossRef]

75. Roy, R.; Hinduja, S.; Teti, R. Recent advances in engineering design optimisation: Challenges and future trends. *CIRP Ann. Manuf. Technol.* **2008**, *57*, 697–715. [CrossRef]

76. Tsekouras, G.; Kiartzis, S.; Kladas, A.G.; Tegopoulos, J.A. Neural Network Approach Compared to Sensitivity Analysis Based on Finite Element Technique for Optimization of Permanent Magnet Generators. *IEEE Trans. Magn.* **2001**, *37*, 3618–3621. [CrossRef]

77. Pinilla, M.; Martinez, S. Optimal design of permanent-magnet direct-drive generator for wind energy considering the cost uncertainty in raw materials. *Renew. Energy* **2012**, *41*, 267–276. [CrossRef]

78. Fontchastagner, J.; Messine, F.; Lefèvre, Y. Design of Electrical Rotating Machines by Associating Deterministic Global Optimization Algorithm with Combinatorial Analytical and Numerical Models. *IEEE Trans. Magn.* **2007**, *43*, 3411–3419. [CrossRef]

79. Liu, C.; Chau, K.T.; Jiang, J.Z.; Jian, L. Design of a New Outer-Rotor Permanent Magnet Hybrid Machine for Wind Power Generation. *IEEE Trans. Magn.* **2008**, *44*, 1494–1497.

80. Leung, Y.W.; Wang, Y. An Orthogonal Genetic Algorithm with Quantization for Global Numerical Optimization. *IEEE Trans. Evolut. Comput.* **2001**, *5*, 41–53. [CrossRef]

81. Kicinger, R.; Arciszewski, T.; De Jong, K. Evolutionary computation and structural design: A survey of the state-of-the-art. *Comput. Struct.* **2005**, *83*, 1943–1978. [CrossRef]

82. Saridakis, K.M.; Dentsoras, A.J. Soft computing in engineering design—A review. *Adv. Eng. Inform.* **2008**, *22*, 202–221. [CrossRef]

83. Yusup, N.; Zain, A.M.; Hashim, S.Z.M. Evolutionary techniques in optimizing machining parameters: Review and recent applications (2007–2011). *Expert Syst. Appl.* **2012**, *39*, 9909–9927. [CrossRef]

84. Zhong, W.; Liu, J.; Xue, M.; Jiao, L. A Multiagent Genetic Algorithm for Global Numerical Optimization. *IEEE Trans. Syst. Cybern. Part B Cybern.* **2004**, *34*, 1128–1141. [CrossRef]

85. Rahnamayan, S.; Tizhoosh, H.R.; Salama, M.M. Opposition versus randomness in soft computing techniques. *Appl. Soft Comput.* **2008**, *8*, 906–918. [CrossRef]

86. Banos, R.; Manzano-Agugliaro, F.; Montoya, F.G.; Gil, C.; Alcayde, A.; Gómez, J. Optimization methods applied to renewable and sustainable energy: A review. *Renew. Sustain. Energy Rev.* **2011**, *15*, 1753–1766. [CrossRef]

87. Ahn, Y.; Park, J.; Lee, C.G.; Kim, J.W.; Jung, S.Y. Novel Memetic Algorithm implemented With GA (Genetic Algorithm) and MADS (Mesh Adaptive Direct Search) for Optimal Design of Electromagnetic System. *IEEE Trans. Magn.* **2010**, *46*, 1982–1985. [CrossRef]

88. Mukherjee, I.; Ray, P.K. A review of optimization techniques in metal cutting processes. *Comput. Ind. Eng. J.* **2006**, *50*, 15–34. [CrossRef]

89. Dennis, B.H.; Dulikravich, G.S. Optimization of magneto-hydrodynamic control of diffuser flows using micro-genetic algorithms and least-squares finite elements. *Finite Elem. Anal. Des.* **2001**, *37*, 349–363. [CrossRef]

90. Padmanathan, K.; Uma, G.; Vigna, K.R.; Sudar, O.S.T. Multiple Criteria Decision Making (MCDM) Based Economic Analysis of Solar PV System with Respect to Performance Investigation for Indian Market'. *Sustainability J.* **2017**, *9*, 820.

91. Dias, A.H.F.; de Vasconcelos, J.A. Multiobjective genetic algorithms applied to solve optimization problems. *IEEE Trans. Magn.* **2002**, *38*, 1133–1136. [CrossRef]

92. Kiartzis, S.; Kladas, A. Deterministic and artificial intelligence approaches in optimizing permanent magnet generators for wind power applications. *J. Mater. Process. Technol.* **2001**, *108*, 232–236. [CrossRef]

93. Das, S.; Abraham, A.; Konar, A. Particle Swarm Optimization and Differential Evolution Algorithms: Technical Analysis, Applications and Hybridization Perspectives. In *Advances of Computational Intelligence in Industrial Systems. Studies in Computational Intelligence*; Springer: Berlin/Heidelberg, Germany, 2008; Volume 116, pp. 1–38.

94. Bi, X.; Wang, Y. An Improved Artificial Bee Colony Algorithm. In Proceedings of the 2011 IEEE Computer Research and Development (ICCRD), Shanghai, China, 11–13 March 2011.

95. Babayigit, B.; Ozdemir, R. A modified artificial bee colony algorithm for numerical function optimization. In Proceedings of the 2012 IEEE Symposium on Computers and Communications (ISCC), Cappadocia, Turkey, 1–4 July 2012; pp. 245–249.

96. Rostami, N.; Feyzi, M.R.; Pyrhonen, J.; Parviainen, A.; Behjat, V. Genetic Algorithm Approach for Improved Design of a Variable Speed Axial-Flux Permanent-Magnet Synchronous Generator. *IEEE Trans. Magn.* **2012**, *48*, 4860–4865. [CrossRef]

97. Song, C.; Ye, J.; Liu, D.; Kang, Q. Receding Horizon Control of Fuzzy Systems Based on Numerical Optimization Algorithm. *IEEE Trans. Fuzzy Syst.* **2009**, *17*, 1336–1352. [CrossRef]

98. Chaturvedi, D.K. *Soft Computing Techniques and Its Applications in Electrical Engineering*; Springer: Berlin/Heidelberg, Germany, 2008.

99. Peng, F.; Tang, K.; Chen, G.; Yao, X. Population-Based Algorithm Portfolios for Numerical Optimization. *IEEE Trans. Evolut. Comput.* **2010**, *14*, 782–800. [CrossRef]

100. Bastos Filho, C.J.; de Lima Neto, F.B.; Lins, A.J.; Nascimento, A.I.; Lima, M.P. A Novel Search Algorithm based on Fish School Behavior. In Proceedings of the 2008 IEEE International Conference on Systems, Man and Cybernetics (SMC 2008), Suntec, Singapore, 12–15 October 2008; pp. 2646–2651.

101. Ho, S.L.; Yang, S.; Ni, G.; Wong, H.C.C. A Particle Swarm Optimization Method with Enhanced Global Search Ability for Design Optimizations of Electromagnetic Devices. *IEEE Trans. Magn.* **2006**, *42*, 1107–1110. [CrossRef]

102. Matić, D.; Kulić, F.; Pineda-Sánchez, M.; Kamenko, I. Support vector machine classifier for diagnosis in electrical machines: Application to broken bar. *Expert Syst. Appl.* **2012**, *39*, 8681–8689. [CrossRef]

103. Ho, S.L.; Yang, S.; Ni, G.; Wong, H.C.C. A Tabu Method to Find the Pareto Solutions of Multiobjective Optimal Design Problems in Electromagnetics. *IEEE Trans. Magn.* **2002**, *38*, 1013–1016. [CrossRef]

104. Wadhwa, R.S.; Lien, T. Electromagnet Shape Optimization using Improved Discrete Particle Swarm Optimization (IDPSO). In Proceedings of the 2011 COMSOL Conference, Stuttgart, Germany, 26–28 October 2011.

105. Arkadan, A.A.; Mneimneh, M.; Al-Aawar, N. R-FL-C Model for Design Optimizationof PM Generators. *IEEE Trans. Magn.* **2007**, *43*, 1649–1652. [CrossRef]

106. Li, G.; Niu, P.; Xiao, X. Development and investigation of efficient artificial bee colony algorithm for numerical function optimization. *Appl. Soft Comput.* **2012**, *12*, 320–332. [CrossRef]

107. Hahn, I. Heuristic Structural Optimization of the Permanent Magnets Used in a Surface Mounted Permanent-Magnet Synchronous Machine. *IEEE Trans. Magn.* **2012**, *48*, 118–127. [CrossRef]

108. Sha, H.; Mei, F.; Zhang, C.; Pan, Y.; Zheng, J. Identification Method for Voltage Sags Based on K-means-Singular Value Decomposition and Least Squares Support Vector Machine. *Energies* **2019**, *12*, 1137. [CrossRef]

109. Ho, S.L.; Yang, S.; Ni, G.; Wong, H.C.C. An Improved Tabu Search for the Global Optimizations of Electromagnetic Devices. *IEEE Trans. Magn.* **2001**, *37*, 3570–3574. [CrossRef]

110. Liu, L.; Liu, W.; Cartes, D.A. Particle swarm optimization-based parameter identification applied to permanent magnet synchronous motors. *Eng. Appl. Artif. Intell.* **2008**, *21*, 1092–1100. [CrossRef]

111. Yang, S.; Machado, J.M.; Ni, G.; Ho, S.L.; Zhou, P. A Self-Learning Simulated Annealing Algorithm for Global Optimizations of Electromagnetic Devices. *IEEE Trans. Magn.* **2000**, *36*, 1004–1008.

112. Sudhoff, S.D.; Cale, J.; Cassimere, B.; Swinney, M. Genetic Algorithm Based Design of a Permanent Magnet Synchronous Machine. In Proceedings of the IEEE International Conference on Electric Machines and Drives, 2015, San Antonio, TX, USA, 15 May 2005.

113. Ho, S.L.; Yang, S.; Wong, H.C.C.; Cheng, K.W.E.; Ni, G. An Improved Ant Colony Optimization Algorithm and Its Application to Electromagnetic Devices Designs. *IEEE Trans. Magn.* **2005**, *41*, 1764–2004. [CrossRef]

114. Karaboga, D.; Basturk, B. A powerful and efficient algorithm for numerical function optimization: artificial bee colony (ABC) algorithm. *J. Glob. Optim.* **2007**, *39*, 459–471. [CrossRef]

115. Karaboga, D.; Basturk, B. A comparative study of Artificial Bee Colony algorithm. *Appl. Math. Comput.* **2009**, *214*, 108–132. [CrossRef]

116. Hasanien, H.M.; Muyeen, S.M. Design Optimization of Controller Parameters Used in Variable Speed Wind Energy Conversion System by Genetic Algorithms. *IEEE Trans. Sustain. Energy* **2012**, *3*, 200–207. [CrossRef]

117. Ho, S.L.; Yang, S.; Wong, H.C.C.; Ni, G. A Meshless Collocation Method Based on Radial Basis Functions and Wavelets. *IEEE Trans. Magn.* **2004**, *40*, 1021–1024. [CrossRef]

118. Im, C.H.; Jung, H.K.; Kim, Y.J. Hybrid Genetic Algorithm for Electromagnetic Topology Optimization. *IEEE Trans. Magn.* **2003**, *39*, 2163–2169.

119. Potgieter, J.H.; Kamper, M.J. Design of New Concept Direct Grid-Connected Slip-Synchronous Permanent-Magnet Wind Generator. *IEEE Trans. Ind. Appl.* **2012**, *48*, 913–922. [CrossRef]

120. Chen, Y.; Wang, D. FEM Analysis of Surface-Inserted Permanent Magnets Synchronous Generator. In Proceedings of the 2011 International Conference on Electrical and Control Engineering, Yichang, China, 16–18 September 2011; pp. 665–668.

121. Ailam, E.H.; Netter, D.; Leveque, J.; Douine, B.; Masson, P.J.; Rezzoug, A. Design and Testing of a Superconducting Rotating Machine. *IEEE Trans. Appl. Supercond.* **2007**, *17*, 27–33. [CrossRef]

122. Widyan, M.S.; Hanitsch, R.E. High-power density radial-flux permanent-magnet sinusoidal three-phase three-slot four-pole electrical generator. *Electr. Power Energy Syst.* **2012**, *43*, 1221–1227. [CrossRef]

123. Farooqui, S.Z. Conversion of squirrel cage induction motors to wind turbine PMG. *Renew. Energy* **2012**, *41*, 345–349. [CrossRef]

124. Subiabre, E.E.; Mueller, M.A. Efficiency Estimation of an Air-Cored Permanent Magnet Synchronous Generator Using Finite Elements and Equivalent Circuit Modelling. In Proceedings of the 2011 IEEE International Electric Machines & Drives Conference (IEMDC), Niagara Falls, ON, CA, 16–18 September 2011; pp. 377–382.

125. Alshibani, S.; Dutta, R.; Agelidis, V.G. An Investigation of the Use of a Halbach Array in MW Level Permanent Magnet Synchronous Generators. In Proceedings of the 2012 XX International Conference on Electrical Machines, Marseille, France, 2–5 September 2012; pp. 59–65, ISBN 978-1-4673-0142-8. [CrossRef]

126. Choi, J.S.; Yoo, J. Design of a Halbach Magnet Array Based on Optimization Techniques. *IEEE Trans. Magn.* **2008**, *44*, 2361–2366. [CrossRef]

127. Danielsson, O.; Leijon, M.; Sjostedt, E. Detailed Study of the Magnetic Circuit in a Longitudinal Flux Permanent-Magnet Synchronous Linear Generator. *IEEE Trans. Magn.* **2005**, *41*, 2490–2495. [CrossRef]

128. Brisset, S.; Vizireanu, D.; Brochet, P. Design and Optimization of a Nine-Phase Axial-Flux PM Synchronous Generator With Concentrated Winding for Direct-Drive Wind Turbine. *IEEE Trans. Ind. Appl.* **2008**, *44*, 707–715. [CrossRef]

129. Legranger, J.; Friedrich, G.; Vivier, S.; Mipo, J.C. Combination of Finite-Element and Analytical Models in the Optimal Multi domain Design of Machines: Application to an Interior Permanent-Magnet Starter Generator. *IEEE Trans. Ind. Appl.* **2010**, *46*, 232–239. [CrossRef]

130. Potgieter, J.H.; Kamper, M.J. Optimum Design and Technology Evaluation of Slip Permanent Magnet Generators for Wind Energy Applications. In Proceedings of the 2012 IEEE Energy Conversion Congress and Exposition (ECCE), Raleigh, NC, USA, 15–20 September 2012; pp. 2342–2349.

131. Hsiao, C.; Yeh, S.; Hwang, J. Design of High Performance Permanent-Magnet Synchronous Wind Generators. *Energies* **2014**, *7*, 7105–7124. [CrossRef]

132. Jin, W.B.; Zhang, D.; An, Z.L.; Tan, R.Y. Analysis and Design of Hybrid Excitation Permanent Magnet Synchronous Generators. *Front. Electr. Electron. Eng. China* **2006**, *1*, 63–66. [CrossRef]

133. More, D.S.; Kalluru, H.; Fernandes, B.G. Design and analysis of full pitch winding and concentrated stator pole winding three-phase flux reversal machine for low speed application. *Sādhanā* **2008**, *33*, 671–687. [CrossRef]

134. More, D.S.; Fernandes, B.G. Power density improvement of three phase flux reversal machine with distributed winding. *IET Electr. Power Appl.* **2010**, *4*, 109–120. [CrossRef]

135. Zakharenko, A.B.; Semenchukov, G.A. Research on the Synchronous Electric Machine with a Rake of Permanent Magnets. *Russ. Electr. Eng.* **2007**, *78*, 100–105. [CrossRef]

136. Sun, X.; Cheng, M.; Hua, W.; Xu, L. Design of Double-Layer Permanent Magnet Dual Mechanical Port Machine for Wind Power Application. *IEEE Trans. Magn.* **2009**, *45*, 4613–4616.

137. Zhang, S.; Tseng, K.J.; Nguyen, T.D. PMSG based Multi-Generator Architecture for Wind Generation Application. In Proceedings of the 2011 6th IEEE Conference on Industrial Electronics and Applications, Beijing, China, 21–23 June 2011; pp. 335–340.
138. Chan, T.F.; Wang, W.; Lai, L.L. Field in a Transverse- and Axial-Flux Permanent Magnet Synchronous Generator From 3-D Finite Element Analysis. *IEEE Trans. Magn.* **2012**, *48*, 1055–1058. [CrossRef]
139. Li, J.; Chau, K.T.; Jiang, J.Z.; Liu, C.; Li, W. A New Efficient Permanent-Magnet Vernier Machine for Wind Power Generation. *IEEE Trans. Magn.* **2010**, *46*, 1475–1478. [CrossRef]
140. Tapia, J.A.; Leonardi, F.; Lipo, T.A. Consequent-Pole Permanent-Magnet Machine with Extended Field-Weakening Capability. *IEEE Trans. Ind. Appl.* **2003**, *39*, 1704–1709. [CrossRef]
141. Tessarolo, A. Accurate Computation of Multiphase Synchronous Machine Inductances Based on Winding Function Theory. *IEEE Trans. Energy Convers.* **2012**, *27*, 895–904. [CrossRef]
142. Wang, W.; Cheng, K.W.E.; Ding, K.; Meng, L.C. A Novel Approach to the Analysis of the Axial-Flux Permanent-Magnet Generator with Coreless Stator Supplying a Rectifier Load. *IEEE Trans. Magn.* **2011**, *47*, 2391–2394. [CrossRef]
143. Fan, Y.; Chau, K.T.; Cheng, M. A New Three-Phase Doubly Salient Permanent Magnet Machine for Wind Power Generation. *IEEE Trans. Ind. Appl.* **2006**, *42*, 53–60.
144. Tiwari, R.; Padmanaban, S.; Neelakandan, R.B. Coordinated Control Strategies for a Permanent Magnet Synchronous Generator Based Wind Energy Conversion System. *Energies* **2017**, *10*, 1493. [CrossRef]
145. Lubin, T.; Hamiti, T.; Razik, H.; Rezzoug, A. Comparison Between Finite-Element Analysis and Winding Function Theory for Inductances and Torque Calculation of a Synchronous Reluctance Machine. *IEEE Trans. Magn.* **2007**, *43*, 3406–3410. [CrossRef]
146. Xinghe, F.; Jibin, Z. Numerical Analysis on the Magnetic Field of Hybrid Exciting Synchronous Generator. *IEEE Trans. Magn.* **2009**, *45*, 4590–4593. [CrossRef]
147. Carranza, O.; Figueres, E.; Garcerá, G.; Gonzalez-Medina, R. Analysis of the control structure of wind energy generation systems based on a permanent magnet synchronous generator. *Appl. Energy* **2013**, *103*, 522–538. [CrossRef]
148. Thorburn, K.; Karlsson, K.E.; Wolfbrandt, A.; Eriksson, M.; Leijon, M. Time stepping Finite Element Analysis of a variable speed synchronous generator with rectifier. *Appl. Energy* **2006**, *83*, 371–386. [CrossRef]
149. Varzaneh, S.G.; Gharehpetian, G.B.; Abedi, M. Output power smoothing of variable speed wind farms using rotor-inertia. *Electric Power Syst. Res.* **2014**, *116*, 208–217. [CrossRef]
150. Huang, K.; Zhang, Y.; Huang, S.; Lu, J.; Gao, J.; Cai, L. Some Practical Consideration of a 2MW Direct-drive Permanent-magnet Wind-power Generation System. In Proceedings of the 2009 International Conference on Energy and Environment Technology, Guilin, China, 16–18 October 2009; pp. 819–828.
151. Tran, D.H.; Sareni, B.; Roboam, X.; Espanet, C. Integrated Optimal Design of a Passive Wind Turbine System: An Experimental Validation. *IEEE Trans. Sustain. Energy* **2010**, *1*, 48–56. [CrossRef]
152. Alaboudy, A.H.K.; Daoud, A.A.; Desouky, S.S.; Salem, A.A. Converter controls and flicker study of PMSG-based grid connected wind turbines. *Ain Shams Eng. J.* **2012**, *4*, 75–91. [CrossRef]
153. Available online: https://www.ge.com/renewableenergy/wind-energy/onshore-wind/turbines (accessed on 14 June 2019).

 *energies*

Article

Multi-Channel-Based Microgrid for Reliable Operation and Load Sharing

Ali Elrayyah and Sertac Bayhan *

Qatar Environment and Energy Research Institute, Hamad Bin Khalifa University, Doha 34110, Qatar;
aelrayyah@hbku.edu.qa
* Correspondence: sbayhan@hbku.edu.qa; Tel.: +974-4454-7188

Received: 6 May 2019; Accepted: 28 May 2019; Published: 30 May 2019

Abstract: This paper presents a novel approach to distribute available power among critical and non-critical loads in microgrids. The approach is based on supplying power over a number of channels with distinguishable frequencies where loads could be served by these channels according to their level of importance. The multi-channel scheme not only offers flexibility to supply loads but also to share power among adjacent microgrids. The control system, which can deal with multi-channel scheme, is presented and different applications that can be offered whereby are discussed. The number of channels that can be supplied by any inverter is determined based on the parameters of the used filter. Moreover, the power exchange efficiencies over the active channels at various power levels are determined and approximated formulas for quick evaluation are presented. To verify the proposed solution performance, simulation and experimental studies were performed. The obtained results demonstrate the effectiveness of using multi-channel scheme for power exchange in microgrid and also confirm the accuracy of the provided formula related to power exchange efficiencies.

Keywords: multi-channel power exchange; microgrid; power electronics-based source/loads; distributed generators; critical loads; uninterruptable power supply; power exchange efficiency

---

1. Introduction

The development of the microgrid concept has become an important element in the of the future utility grids and a priority in many countries due to its considerable environmental, economic, and social benefits. An increase of microgrid deployment rate will play an important role to meet future electricity demands without significant investment in new power plants. The main advantage of the microgrid is that it operates either in islanded mode or grid-connected mode. On the other hand, the power generation capacity of the distributed generators (DGs) is limited and it is often not sufficient to meet load demand in the islanded mode of operation. Therefore, a control system should be designed to ensure the continuity of the energy at critical loads (hospitals, data centers, etc.) and special attention should be paid for such loads in the islanded operation [1].

There has been extensive research and development in technologies, methods, and systems to secure a reliable power supply for various electric loads (critical and non-critical loads) [2]. In some aspects of their operation, microgrid systems still rely on the same operating principles that were established over a century ago [3]. Considering the advancement in power electronics technologies, loads and sources which are based on power electronics systems are expected to have significant share within power systems in near future [4–6]. Power electronics-based sources and loads can thus provide new means to revolutionize power exchange in microgrids.

One principle that remains rooted in microgrids is the non-differentiable sharing of power among loads. In all power systems, sources set up a single channel voltage and loads draw power from that channel. This indicates that critical as well as non-critical loads extract their need using the same approach. It would be very useful if there was an ability to enable sources to limit the supplied power to serve critical loads when there is a lack of sufficient power supply. However, sources cannot discriminate among the loads in these cases. There are solutions to address this problem which are based on establishing centralized energy management systems (EMSs) that monitor the whole system and adjust the loads accordingly [7,8]. However, EMSs could be challenging to establish especially when addressing large systems [9]. Moreover, the dynamics in the future inverter-based system could be too fast to be accommodated by such centralized-based systems [10]. The power packet presented in [11] proposes a scheme for power switching from one source to a specific load similar to the concept of circuit switching used in old telephone networks. This approach could be very effective but it requires a number of parallel lines and it needs high synchronization between switches.

In this paper, multi-channel-based microgrid is proposed to enhance system reliability and power sharing flexibility. The concept of using more than one frequency channel in microgrid is used in [12]. However, that concept is limited to superimposing a small AC signal driven by droop relation on a DC microgrid to ensure proper sharing among sources. The idea is further extended in [13] to manage power sharing among interlinking converters that transfer power between AC and DC sides of hybrid power systems. The approach proposed in this paper aims to use multi-channel not as communication mean, but rather to achieve an intelligent power exchange within/among microgrids.

The paper is organized as follows. The structure, operation, and control of multi-channel-based microgrid are discussed in Section 2 while the control of the sources and loads within that system are covered in Section 3. Section 4 provides theoretical analysis for the number of channels that can be used within a microgrid and estimate for the associated losses. The effectiveness of the proposed concept is demonstrated through simulation and experimental studies covered in Section 5. Finally, the paper is concluded in Section 6.

2. The Proposed Multi-Channel-Based Microgrid Structure and Operation

Figure 1 illustrates a conceptual diagram of the proposed multi-channel-based microgrid. Power sources—renewable-based or energy storage—can generate power at different frequency channels (50, 100, and 200 Hz in this case) through power electronics converters. The loads, either non-critical or critical, can then tune their power reception to one of more of these channels. The power sources can limit their supply to channels devoted to critical loads such that their power supply service is maintained uninterrupted when overloading occurs. The ability to exchange power over number of channels is not limited to individual sources and loads, but rather it can take place between subsystems such as microgrids.

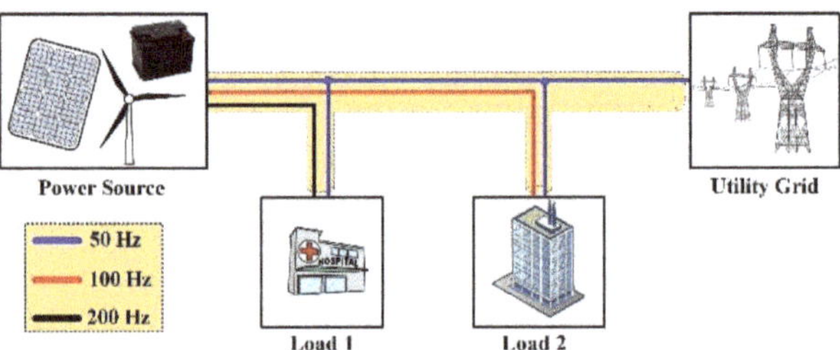

Figure 1. The conceptual diagram of the proposed multi-channel-based power system.

Fortunately, power electronics converters can support multi-channel-based power exchange with merely changes in their control algorithms and no hardware modification. Although the generated voltage signals contain harmonics, these are injected deliberately for power distribution in this application. To block the harmonic components, serial compensators (SCs), which are inverters connected in series with line to apply compensating voltages, are employed. The SCs can also be used in exchanging power among adjacent power systems. The operation of SCs is similar to the electric springs proposed in [14]. Electric springs regulate voltage and perform demand response by adding voltage components in series with load branches [15]. However, the operational scope of SCs is wider than that of electric springs as it deals with several channels.

The single-line diagram of the multi-channel microgrids is depicted in Figure 2. The system contains a number of multi-channel sources/loads as well as conventional ones. The power system is divided into subsystems. Grid and conventional loads are operated by a single frequency (50 Hz) while subsystems A and B can exchange power in multi-channel mode. In Figure 2, subsystem A can supply power in three channels; grid, local critical loads, and power exchange with subsystem B. It is well known that sharing power among interconnected subsystems in coordinated manner can provide several benefits for efficient and reliable system operation [16]. As shown in Figure 2, subsystem A can secure sufficient supply to its critical loads and also designate some power to subsystem B that does not get interfered by other loads within the system. The connections between the subsystems are done through SCs which are operated to block the current flow of certain channels. A communication system can be used to set the various frequencies and to manage the power flow, but it is not critical for the system reliability.

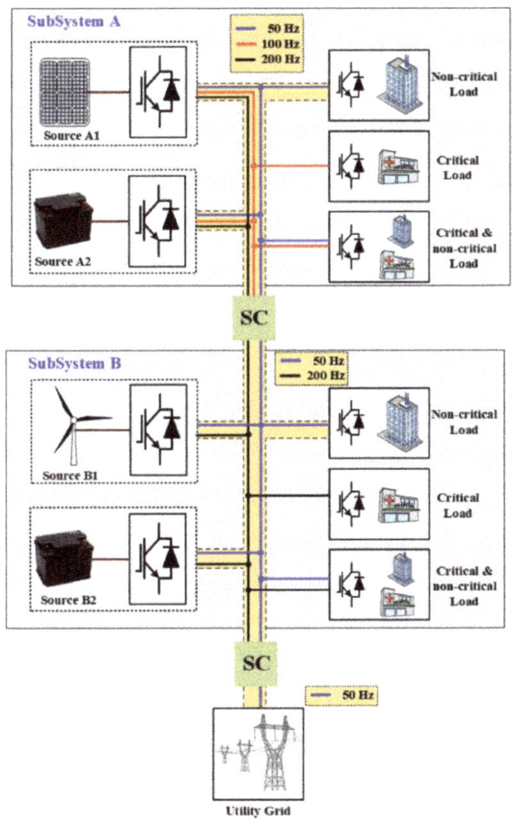

Figure 2. Single-line diagram of the multi-channel-based microgrids.

The power sources and loads in this system are connected to the network through power electronic converters. The line voltage in this power system composes of a number of frequency channels (harmonics). Each source can contribute power in any of these channels and likewise loads can receive power from any of these channels. Each source acts on parts of these channels as a voltage source where droop control can be applied. For the remaining channels, the source acts as a current source to feed some or zero current on each one of them.

Multi-channel-based microgrids shown in Figures 1 and 2 require changes in power system structure and the associated cost with these changes needs to be considered to evaluate the effectiveness of such approach. Fortunately, power electronics-based loads and sources are expected to have a larger share within microgrids in the future. As the following section shows, only changes in the inverter control logic are required to apply a multi-channel power exchange which does not impose any extra hardware cost. The need for SCs, on the other hand, might require an investment in additional hardware. However, the deployment of series compensating devices is expected to increase within distribution systems to accommodate for renewable energy sources integration [17]. Accordingly, the hardware components needed to apply the multi-power exchange are expected to be available within microgrids in the near future which opens the doors for this scheme to be utilized once the appropriate control logic is developed.

3. Control of Multi-Channel Sources and Loads

The control system of a multi-channel source is shown in Figure 3. Central controllers manage the various sources by informing them about the active channels and the amount of power that needs to be supplied in each channel. The phase locked loop block (PLL) in each source monitors the magnitude and angle of each channel to:

- Synchronize output voltage with active channels;
- Perform voltage/current control tasks and;
- Detect an activation/deactivation action in the system.

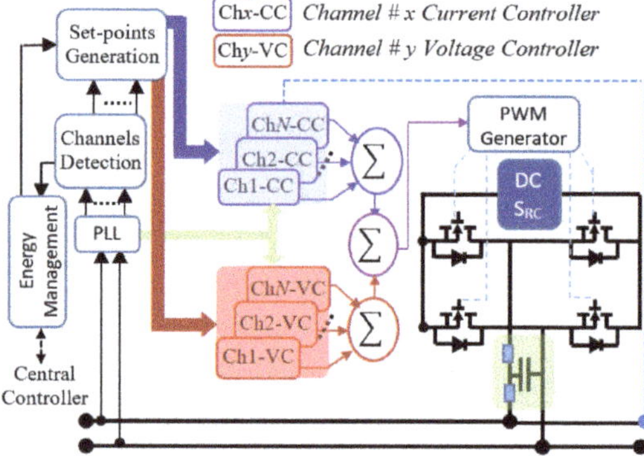

Figure 3. The block diagram of the control system of a source in a multi-channel-based power system.

Based on the applied operation scheme, the channels angles detected by the PLL are sent to current and/or voltage controllers to supply power through those channels. The outputs of these controllers are then combined to drive the inverter switches through the power width modulation (PWM) generator.

The multi-channel loads on the other hand operate as current sinks. In general, some channels could be defined for critical services and the other are for non-critical ones. Any load can then consume the power it needs for its essential functions from the critical channels, while it consumes the remaining demand from the other ones. The SC control structure is shown in Figure 4. This system works merely as a current source inverter where it adjusts the current at all channels that need to be blocked to zero. However, it does not contribute any voltage at the channel allowed to pass.

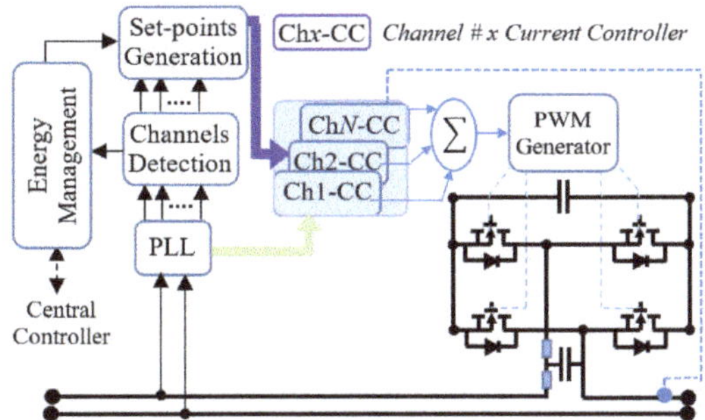

Figure 4. The block diagram of serial compensator control system.

The flowchart of multi-channel operation is given in Figure 5. The cycle starts by updating the list of active channels based on commands sent from central controller. The controller then identifies active channels in the line through the PLL block. The importance of this step is of two folds. First, it is needed to identify the frequency and phase of each channel. Second, the PLL can detect channels' activation/deactivation immediately as it does not have to wait for command from the central controller. The control system then updates the power flow in various channels. Generally, every source needs to maintain a certain power reserve to accommodate for demand increase. As the sources supply more power, this reserve might not be maintained. To recover the required reserve level, sources can perform the following steps:

- If the source acts as a current source in some channels, it can reduce the supply to them till the reserve is maintained;
- If zero power is reached in current fed channels, sources can turn low-priority voltage-supplied channels into the current fed type and limit the current to a value that maintains the required reserve;
- If only high priority channels are active, sources can send a notification to central management unit to drop some of these channels;
- Voltage control loops and current control loops will then be updated based on their respective set-points and their combined output will be sent to the PWM controller that drives the switches.

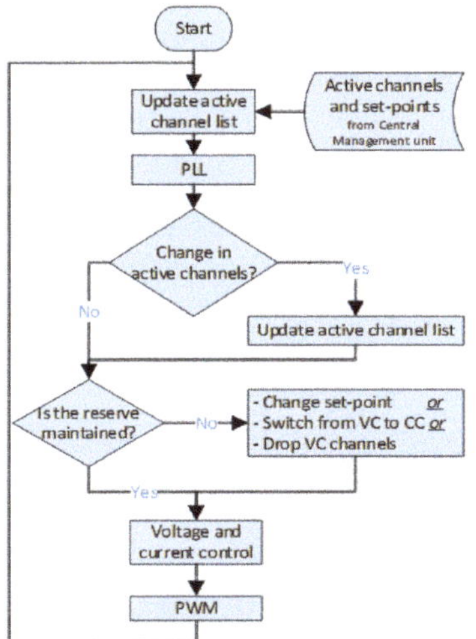

Figure 5. The flowchart of multi-channel operation scheme.

The SC operates similar to the sources without the part related to maintaining the reserve. After detecting the active channels, the currents at blocked channels are set to zero through their current controller operation.

4. Determination of Number and Frequencies of Usable Channels

Activated frequencies in multi-channel systems cannot be arbitrarily selected. For effective operation, there must be clear separations between their frequencies, otherwise the PLL may not be able to identify them accurately. Moreover, even if the channel frequencies are selected with reasonable difference among them, their higher-order harmonics might not be distinguishable from one another. In this paper frequencies that meet these requirements are selected as:

$$f_k = 2^{k-1} f_L, \quad k = 1, 2, \ldots, k_{mx} \tag{1}$$

where f_L is the frequency of the fundamental channel, f_k is the frequency of the kth channel, and k_{mx} is the number of channels that can be activated in the power zone. Since merely odd harmonics are experienced in power systems [18], there is at least 50 Hz separation between the harmonics components of the channels indicated in Equation (1).

Different types of filters could be used for multi-channel inverters. However, LCL is considered in this paper since in requires relatively low inductors. Having a low inductor is very important to reduce its associated voltage drop. As indicated in [19], the resonance frequency of LCL filter needs to satisfy

$$10 f_L < f_{res} < \frac{1}{2} f_{sw} \, , f_{res} = \frac{1}{2\pi} \sqrt{\frac{L_1 + L_2}{L_1 L_2 C}} \tag{2}$$

where f_{res} is the resonance frequency of the LCL filter and f_{sw} is the switching frequency. In the case of a multi-channel system, Equation (2) implies

$$10 \times 2^{k_{mx} - 1} f_L < f_{res} < \frac{1}{2} f_{sw} \tag{3}$$

Equation (3) leads to the following relation

$$2^{k_{mx}-1} < \frac{f_{sw}}{20 f_L} \rightarrow k_{mx} < 3.32 \left(\log \left(\frac{f_{sw}}{f_L} \right) - 1 \right) \tag{4}$$

Accordingly, for a 50 Hz system with a 20 kHz switching frequency, $k_{mx} \leq 5$.

To investigate the other constraints for acceptable value of k_{mx}, the following information needs to be provided for each channel: frequency (f_k), rated power ($P_{mx,k}$), and voltage magnitude ($V_{mx,k}$). Consider a multi-channel inverter system shown in Figure 6 where power is transferred over the kth channels from a source to a load. As multi-channel systems are based on power electronics sources and loads, it is more effective to transfer only real power as the reactive power needed by loads could be provided by their local inverters. Accordingly, the line voltage $V_{o,k}$ and output current $i_{o,k}$ are assumed to be in phase. The voltage at the capacitor $V_{C,k}$ of the source inverter is given by

$$V_{C,k} = V_{o,k} + j\omega_k L_2 i_{o,k} \rightarrow |V_{C,k}| = \sqrt{V_{o,k}^2 + \left(\omega_k L_2 i_{o,k} \right)^2} \approx V_{o,k} \tag{5}$$

The approximation in Equation (5) is done to simplify the analysis and since the voltage drop $\omega_k L_2 i_{o,k}$ should to be very small in comparison with $V_{o,k}$ for reasonable power transfer. The reactive power supplied to the capacitor $\left(Q_{C,k} \right)$ can then be approximated by

$$Q_{C,k} \approx -\omega_k C V_{o,k}^2 \tag{6}$$

The current that flows over the inductor L_1 ($i_{i,k}$) accordingly becomes:

$$i_{i,k} \approx i_{o,k} + j\omega_k C V_{o,k} \tag{7}$$

By summing the reactive power supplied to L_1, L_2, and C total reactive power Q_k supplied by the inverter can then be written as:

$$\begin{aligned}
Q_k &\approx -\omega_k C V_{o,k}^2 + \omega_k \left(\omega_k C V_{o,k}\right)^2 L_1 + \omega_k i_{o,k}^2 (L_1 + L_2) \\
&= \omega_k\left(-C V_{o,k}^2\left(1 - \omega_k^2 L_1 C\right) + P_{o,k}^2 / V_{o,k}^2 (L_1 + L_2)\right)
\end{aligned} \tag{8}$$

Assuming R_{L1}, R_{L2}, and R_C are the resistances of L_1, L_2, and C the power transmission efficiency (η_k) from source to load could be estimated using the relation:

$$\eta_k = \frac{P_{o,k}}{P_{o,k} + 2\left(\frac{P_{o,k}}{V_{o,k}}\right)^2 (R_{L1} + R_{L2}) + 2\left(Q_{C,k}/V_{o,k}\right)^2 (R_{L1} + R_C)} \tag{9}$$

Equations (8) and (9) can be used to determine whether certain power could be transferred over a specific power depending on the associated reactive power and losses.

Another constraint for multi-channel inverter is related to its DC bus voltage (V_{DC}). The DC bus voltage needs to be high enough to generate the required output voltage at all channels. The appropriate value for the DC bus voltage depends on the phase angle of the various channels. When the phase angles of the channels are allowed to be arbitrary set, the DC bus voltage needs to satisfy the following relation

$$\sqrt{2} \sum_{k=1}^{k_{mx}} V_{mx,k} < V_{DC} \tag{10}$$

Allowing the channels to have any phase is very important for certain applications. For example, when droop control is implemented over the various channels, the exact channel frequency varies slightly around the nominal channel frequency and so does the phase angle. In this case Equation (10) needs to be satisfied. However, in another application the master source could be responsible to set the line voltage (V_L). In this case, the channel frequencies and phases can be maintained at fixed values set by the master source as:

$$V_L = \sqrt{2} \sum_{k=1}^{k_{mx}} V_{mx,k} \sin\left(2^{k-1} \pi f_L t + \phi_k\right) \tag{11}$$

If ϕ_k of the various channels do not have the same value, the positive and negative half cycles will have different shapes. It is therefore preferred that all channels have the same value of ϕ_k which can be taken as zero. To determine the peak voltage of V_L in this case, let $\pi f_L t$ be defined as θ, then the following equation needs to be solved for θ in the range $[0, \pi]$

$$\frac{dV_L}{d\theta} = \sqrt{2} \sum_{k=1}^{k_{mx}} 2^{k-1} V_{mx,k} \cos 2^{k-1}\theta = 0 \tag{12}$$

The values of θ that solves Equation (12) can then be used to determine the peak voltage. For example, consider the case of three channels with RMS values of $V_{mx,1}$, $V_{mx,2}$, and $V_{mx,3}$ given by V_m, $0.5V_m$, and $0.5V_m$, respectively. Solving Equation (12) yields a peak voltage for V_L as $1.37\sqrt{2}V_m$. The DC bus voltage in this case can be taken as $1.37\sqrt{2}V_m$ representing 68% of the value set by Equation (10). This indicates that, depending on the intended application, selecting the right setting for the channel frequencies and phases can have significant impact in components' sizing and design.

Besides the constraints related to switching/resonant frequencies, reactive power supply, and DC bus voltage, ripple in inverter current and voltage drop over the inductors are usually considered while designing LCL filters. The ripple in the inverter output current Δi_{L1} in unipolar PWM switching is given by [20]:

$$\Delta i_{L1} = \frac{V_{DC}L_1}{8f_{sw}} \tag{13}$$

L_1 must then be selected to maintain Δi_{L1} below the required limit. This value of L_1 can be used to estimate the voltage drop of the LCL filter. The voltage drop over the filter inductors is usually needed to be kept below a certain value for proper voltage regulation. For the kth channel, the voltage drop over the filter inductors can be written as:

$$\Delta V_{o,k} = V_{o,k} - m_{a,k}V_{DC} \approx j\omega_k L_1\left(i_{C,k} + i_{o,k}\right) + j\omega_k L_2 i_{o,k} = \omega_k^2 L_1 C V_{o,k} + j\omega_k(L_2 + L_1)i_{o,k} \tag{14}$$

$m_{a,k}$ is the inverter modulation index. The term $V_{o,k}\omega_k^2 L_1 C$ has a 180 degree phase shift from $V_{o,k}$ and it is added to the inverter output voltage while the term $j\omega_k i_{o,k}(L_1 + L_2)$ has a 90 degree phase shift from $V_{o,k}$. Based on any requirement for the line voltage drop, the relation in Equation (14) can be used to check the possibility of activating a certain channel with the power system under consideration.

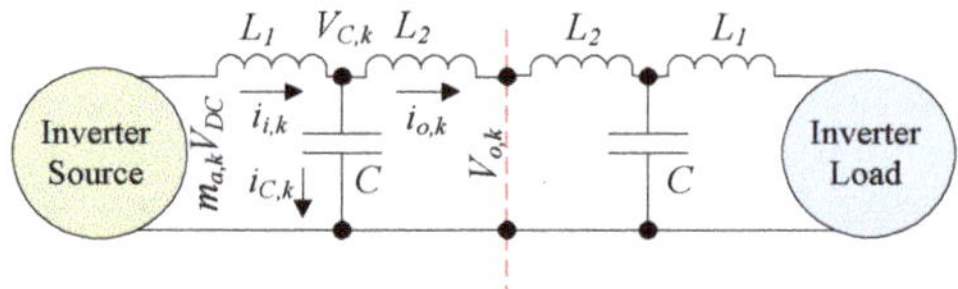

Figure 6. LCL filter for multi-channel inverter.

5. Simulation and Experimental Studies

To verify the effectiveness of the proposed multi-channel-based power systems, simulation and experimental studies have been conducted. In the experimental setup, a Chroma regenerative grid simulator is used to set up the three channels. Since the grid simulator can absorb power it is used to set the line voltage and to emulate loads in the system. Two sources are then used to supply power over the various channels. Each of the sources is supplied by Magna-Power power supply and it uses IAP inverter and controlled by TMS320F28335 DSP. The system is also simulated using Matlab to analyze its efficiency at different power levels. The considered diagram for the microgrid is shown in Figure 6 where the source and load inverters have identical parameters which are listed in Table 1. The resistors R_{L1}, R_{L2}, and R_C are assumed to be connected in series with L_1, L_2, and C, respectively.

Table 1. Inverter parameter of the considered system.

Parameter	Value	Parameter	Value
C	22.5 μF	R_C	1.0 Ω
L_1	0.5 mH	R_{L1}	0.5 Ω
L_2	0.5 mH	R_{L2}	0.5 Ω
Voltage	120 V	f_L	50 Hz

From Table 1, $f_{res} = 2.1\ kHz$ indicating that k_{mx} can at most be 3. For source inverter, Figure 7a compares the simulated per unit reactive power and the one provide by Equation (8). For the small value of $P_{o,k}$ the term $-\omega_k C V_{o,k}^2$ is dominant, especially for higher-order channels, and the inverter supplies more reactive power. However, as $P_{o,k}$ increases, the value of $\omega_k P_{o,k}^2 / V_{o,k}^2 (L_1 + L_2)$ causes the supplied reactive power to decrease. Note that the power ranges that correspond to high amounts of reactive power need to be avoided to minimize operating losses. This can be observed clearly in Figure 7b, since when channel 3 supplies less than 0.5 kW, the transmission efficiency is relatively low. When the transmitted amount of power increases, the losses due to the term $\left(Q_k/V_{o,k}\right)^2\left(R_{L1}\right)$ decrease, while that of $\left(P_{o,k}/V_{o,k}\right)^2\left(R_{L1}+R_{L2}\right)$ becomes dominant and hence the efficiencies of all channels become comparable and they all decrease by the same rate.

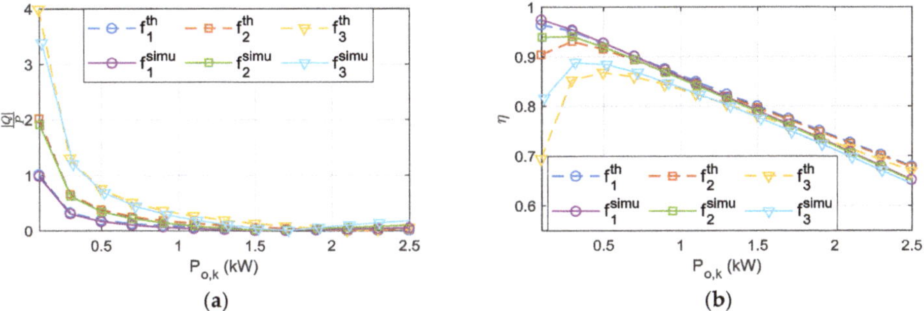

(a)   (b)

Figure 7. Reactive power supplied by multi-channel inverter to LCL components over the various channels: (**a**) Reactive power supply using Equation (8) and simulation study; (**b**) power exchange efficiency over various channels.

The microgrid system in Figure 8 is considered for experimental studies. Two sources (Src1 and Src2) feed power as current sources to loads in three channels ch1, ch2, and ch3 which have the frequencies 50, 100, and 200 Hz, respectively. The line voltage in the system is given by:

$$v_L = V_1 \sin \omega_1 t + V_2 \sin \omega_2 t + V_3 \sin \omega_3 t \tag{15}$$

where $V_1 = V_2 = V_3 = 83$. Though all channels are generated by Chroma power simulator, ch1 is taken to represent the grid. Two case studies are considered where the sources feed power over the channels differently and the event of grid disconnection are analyzed.

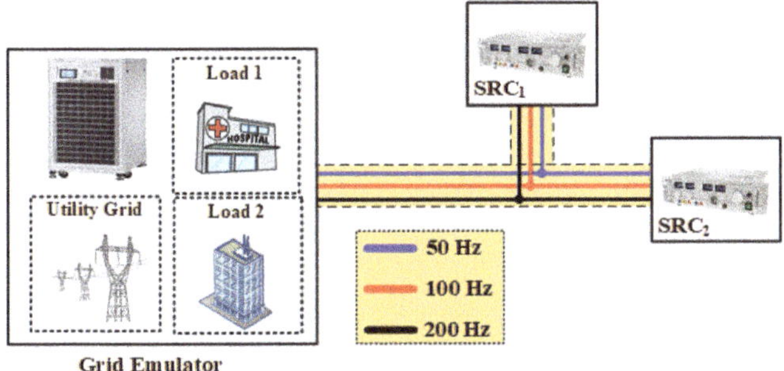

Figure 8. Considered microgrid for experimental study.

5.1. Case Study I

In the first case study, Src1 supplied 7 A at ch1 only while Src2 supplied 7 A in each of the three channels. This can be used in applications where the supply of Src1 is unreliable and thus is used to supply uncritical loads over ch1, while Src2 supplies critical loads that are served by ch2 and ch3 as well as uncritical ones in ch1. Figure 9a shows the line voltage and current supplied by the two sources.

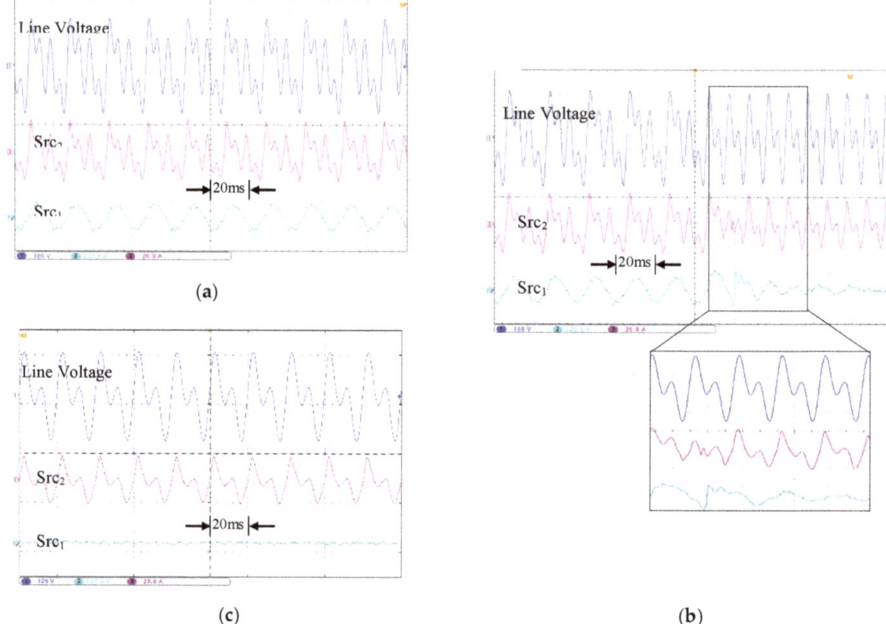

Figure 9. Experimental study sources' behavior in multi-channel power system after grid disconnection where one source supplies power to the grid channel only while the other feeds power to all channels; (a) Src1 feeds power to ch1 and Src2 feeds power over all channels; (b) grid disconnection causing elimination of ch1; (c) Src1 and Src2 output currents after grid disconnection.

At t = 4.5 s, the grid disconnection was imposed to analyze the system performance. The controllers of Src1 and Src2 could detect this event locally and eliminate any power supply over ch1. The line voltage and sources currents waveforms reflected that as shown in Figure 9b. Eventually, the line voltage and sources current were adjusted to feed the loads that were served by ch2 and ch3 as shown in Figure 9c.

The power and energy fed by the two sources for case study I are demonstrated in Figure 10. Before t = 4.5 s, Src2 fed three times the power of Src1 as it was active over three channels. However, after the grid disconnection, the power supplied over ch1 was dropped, which represented all the power supplied by Src1 and one third of Src2 supplied power. This behavior could be very useful to enable sources to play different roles in supplying load-demand with fast and reliable responses to system changes and without explicit inter-controllers communication.

5.2. Case Study II

Case study II represents an application where certain critical loads need not to experience any disruption in their power supply. In this case, sources were configured such that one source (Src2) was exclusively used to serve critical loads in ch2 and ch3 while Src1 participated a guaranteed amount of power to critical loads, with any extra power being fed to ch1. Originally, as Figure 11a shows, Src1 supplied 7 A over the three channels and Src2 supplied 7 A in each of ch2 and ch3. At t = 4.5 s, grid disconnection was imposed which eliminated ch1 from the line voltage. As shown in Figure 11b, Src1 could respond to this event by adjusting their current at ch1 to zero without any major disturbance in the sources current supply over ch2 and ch3. The system then maintained the operation under the new condition highlighted in Figure 11c.

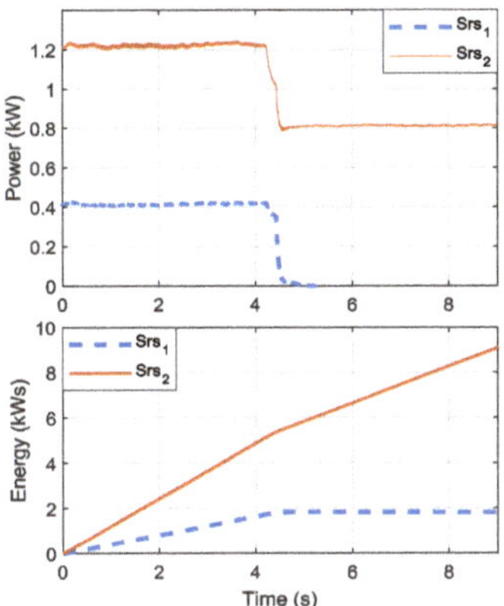

Figure 10. Power and energy supplied by Src1 and Src2 in case study I.

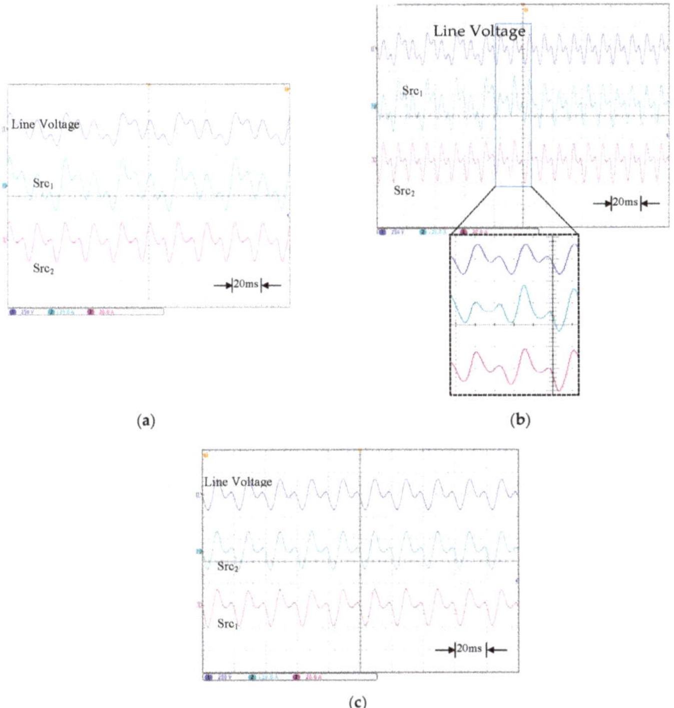

Figure 11. Experimental study sources behavior in multi-channel power system after grid disconnection where one sources supply power to all channels while the other supplies power merely to critical loads on non-grid supplied channels; (**a**) Src1 feeding power in ch2 and ch3 and Src2 feeding power over all three channels; (**b**) Grid disconnection casuing elimination of ch1, (**c**) Src1 and Src2 output currents after grid disconnection.

The power and energy fed by the two sources are demonstrated in Figure 12. The power supply of Src2 was maintained at the same value throughout the experiment duration as it goes in its entirety to critical loads. On the other hand, one third of the Src1 supplied power was withdrawn after t = 4.5 since that power was used to feed uncritical loads or to be fed back to the grid.

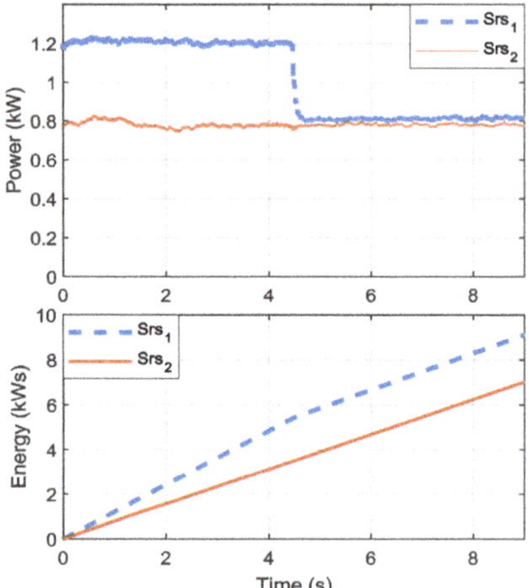

Figure 12. Power and energy supplied by Src1 and Src2 in case study II.

6. Conclusions

The use of sources that supply power over a number of frequencies in multi-channel microgrids provides many benefits to power systems. A multi-channel-based power supply helps maintain an uninterruptable power supply to critical loads, enhances system reliability, and allows more liberalized power trading among adjacent microgrids. The paper shows the viability of supplying power over more than one channel through the proper configuration of voltage and current controllers. The parameters of an inverter filter determine the maximum number of channels that can be supplied by that inverter. Moreover, supplying low power at higher-order channels has poor efficiency as a significant amount of reactive power is supplied in these cases to the reactive components of the filter. The formulas provided in the paper for the allowed number of channels and power exchange efficiencies assist in determining the effectiveness of applying the concept of a multi-channel scheme in any microgrid. The paper focuses on introducing the concept of a multi-channel power exchange, its implementation scheme, some of the related advantages, and discussion of its limitations. In future publications, we plan to cover and analyze in detail subjects such as system stability, controller design, and interactions among various channels.

Author Contributions: The research presented in this paper was a collaborative effort among both authors. A.E. and S.B. conceived, implemented, and got the results along with the paper write-up. A.E. and S.B. wrote the paper and discussed the results and revised the manuscript critically.

Funding: The publication of this article was funded by the Qatar National Library.

Acknowledgments: The publication of this article was funded by the Qatar National Library.

Conflicts of Interest: The authors declare no conflict of interest.

References

1. Li, J.; Liu, Y.; Wu, L. Optimal Operation for Community-Based Multi-Party Microgrid in Grid-Connected and Islanded Modes. *IEEE Trans. Smart Grid* **2018**, *9*, 756–765. [CrossRef]
2. Liu, M.; Li, W.; Wang, C.; Polis, M.P.; Wang, l.; Li, J. Reliability Evaluation of Large Scale Battery Energy Storage Systems. *IEEE Trans. Smart Grid* **2017**, *8*, 2733–2743. [CrossRef]
3. Tan, D.; Novosel, D. Energy Challenge, Power Electronics & Systems (PEAS) Technology and Grid Modernization. *CPSS Trans. Power Electron. Appl.* **2017**, *2*, 3–11.
4. Huber, J.E.; Kolar, J.W. Applicability of Solid-State Transformers in Today's and Future Distribution Grids. *IEEE Trans. Smart Grid* **2018**, *10*, 317–326. [CrossRef]

5. Xu, S.; Huang, A.Q.; Burgos, R. Review of Solid-state Transformer Technologies and Their Application in Power Distribution Systems. *IEEE Trans. Emerg. Sel. Top. Power Electron.* **2013**, *1*, 186–198.

6. Che, L.; Shahidehpour, M. DC Microgrids: Economic Operation and Enhancement of Resilience by Hierarchical Control. *IEEE Trans. Smart Grid* **2014**, *5*, 2517–2526.

7. Farzin, H.; Fotuhi-Firuzabad, M.; Moeini-Aghtaie, M. Role of Outage Management Strategy in Reliability Performance of Multi-Microgrid Distribution Systems. *IEEE Trans. Power Syst.* **2018**, *33*, 2359–2369. [CrossRef]

8. Lopes, J.A.P.; Moreira, C.L.; Madureira, A.G. Defining Control Strategies for Microgrids Islanded Operation. *IEEE Trans. Power Syst.* **2006**, *21*, 916–924. [CrossRef]

9. Han, H.; Hou, X.; Yang, J.; Wu, J.; Su, M.; Guerrero, J.M. Review of Power Sharing Control Strategies for Islanding Operation of AC Microgrids. *IEEE Trans. Smart Grid* **2016**, *7*, 200–215. [CrossRef]

10. Olivares, D.E.; Mehrizi-Sani, A.; Etemadi, A.H.; Cañizares, C.A.; Iravani, R.; Kazerani, M.; Hajimiragha, A.H.; Gomis-Bellmunt, O.; Saeedifard, M.; Palma-Behnke, R.; et al. Trends in Microgrid Control. *IEEE Trans. Smart Grid* **2014**, *5*, 1905–1919. [CrossRef]

11. Takahashi, R.; Tashiro, K.; Hikihara, T. Router for Power Packet Distribution Network: Design and Experimental Verification. *IEEE Trans. Smart Grid* **2015**, *6*, 618–626. [CrossRef]

12. Peyghami, S.; Mokhtari, H.; Blaabjerg, F. Autonomous Operation of a Hybrid AC/DC Microgrid with Multiple Interlinking Converters. *IEEE Trans. Smart Grid* **2018**, *9*, 6480–6488. [CrossRef]

13. Peyghami, S.; Mokhtari, H.; Loh, P.C.; Davari, P.; Blaabjerg, F. Distributed Primary and Secondary Power Sharing in a Droop-Controlled LVDC Microgrid with Merged AC and DC Characteristics. *IEEE Trans. Smart Grid* **2018**, *9*, 2284–2294. [CrossRef]

14. Hui, S.Y.R.; Lee, C.K.; Wu, F.F. Electric springs-A New Smart Grid Technology. *IEEE Trans. Smart Grid* **2012**, *3*, 1552–1561. [CrossRef]

15. Chen, X.; Hou, Y.; Hui, S.Y.R. Distributed Control of Multiple Electric Springs for Voltage Control in Microgrid. *IEEE Trans. Smart Grid* **2017**, *8*, 1350–1359. [CrossRef]

16. Wang, H.; Huang, J. Incentivizing Energy Trading for Interconnected Microgrids. *IEEE Trans. Smart Grid* **2018**, *9*, 2647–2657. [CrossRef]

17. Wang, J.; Wang, Z.; Xu, L.; Wang, Z. A Summary of Applications of D-FACTS on Microgrid. In Proceedings of the 2012 Asia-Pacific Power and Energy Engineering Conference, Shanghai, China, 27–29 March 2012; pp. 1–6.

18. Bidram, A.; Davoudi, A. Hierarchical Structure of Microgrids Control System. *IEEE Trans. Smart Grid* **2012**, *3*, 1963–1976. [CrossRef]

19. Reznik, A.; Simoes, M.G.; Al-Durra, A.; Muyeen, S.M. LCL Filter Design and Performance Analysis for Grid-Interconnected Systems. *IEEE Trans. Ind. Appl.* **2014**, *50*, 1225–1232. [CrossRef]

20. Ruan, X.; Wang, X.; Pan, D.; Yang, D.; Li, W.; Bao, C. *Control Techniques for LCL-type Grid-connected Inverters*; Springer: Berlin, Germany, 2018; pp. 31–61.

Article

A Novel Integrated Topology to Interface Electric Vehicles and Renewable Energies with the Grid

Alfredo Alvarez-Diazcomas [1,†], Héctor López [1,†], Roberto V. Carrillo-Serrano [2,†], Juvenal Rodríguez-Reséndiz [2,†,*], Nimrod Vázquez [1,†] and Gilberto Herrera-Ruiz [2,†]

1   Electronics Engineering Department, Instituto Tecnológico de Celaya, 38010 Celaya, Mexico;
    m1703081@itcelaya.edu.mx (A.A.-D.); hector.lopez@itcelaya.edu.mx (H.L.);
    nimrod.vazquez@itcelaya.edu.mx (N.V.)
2   Facultad de Ingeniería, Universidad Autónoma de Querétaro, 76010 Queretaro, Mexico;
    roberto.carrillo@uaq.mx (R.V.C.-S.); gherrera@uaq.mx (G.H.-R.)
*   Correspondence: juvenal@uaq.edu.mx; Tel.: +52-442-192-1200
†   These authors contributed equally to this work.

Received: 19 September 2019; Accepted: 23 October 2019; Published: 26 October 2019

Abstract: Electric Vehicles (EVs) are an alternative to internal combustion engine cars to reduce the environmental impact of transportation. It is common to use several power sources to achieve the requirements of the electric motor. A proper power converter and an accurate control strategy need to be utilized to take advantage of the characteristics of every source. In this paper is presented a novel topology of a multiple-input bidirectional DC-DC power converter to interface two or more sources of energy with different voltage levels. Furthermore, it can be used as a buck or a boost in any of the possible conversion of energy. It is also possible to independently control the extracted power in each source and any combination of the elements of the system can be used as source and destiny for a transfer. Finally, the interaction with the grid is possible. The operation, analysis and design of the converter are presented with different modes of power transfer. Simulation results are shown where the theoretical analysis of the converter is validated.

Keywords: DC-DC/DC-AC power converter; electric vehicle; multi-input converter; sliding mode control; photovoltaic module; grid; renewable energies

---

1. Introduction

Global warming is one of the greatest challenges today for humankind. The transportation sector is one of the largest contributors to the emissions of greenhouse gases. It represents 27% in the European Union in 2016 and 28% in the United States for the same year which represents the major contribution. Moreover, it is responsible for the greatest growth in emissions currently due to the growth of tourism, the globalized economy and the increase in living standards. A viable alternative to reduce emissions due to transportation is using electric vehicles (EVs), which practically behave like zero-emission cars [1].

In this type of cars, it is common to use several power sources to achieve the requirements of the electric motor such as fuel cells, batteries, ultracapacitors (UCs), and so forth. The aim of a hybrid energy storage system is to make use of the strong features of each source while eliminating their weaknesses [2]. Researchers have hybridized batteries and UCs in References [3–6] to create an ideal energy storage unit with high energy/power density, low cost/weight per unit capacity and a long cycle life. The battery can be used when the vehicle maintains a relatively constant velocity to take advantage of its high energy density characteristic. Also, the peak power transients during acceleration and regenerative braking can be avoided by the inclusion of a higher power density element such as an UC. The ability of the UC to handle higher power for a higher number of charge/discharge cycles not only increases the life span of the battery but also improves the overall system efficiency [7–9]. The active hybridization of the aforementioned energy storage system, in which the power/current in its output can be fully controlled, is only possible by means of utilizing power converters.

These converters can be isolated or non-isolated. In Reference [10] a novel multiport isolated bidirectional DC-DC converter for hybrid battery and supercapacitor applications is presented, which can achieve zero voltage switching for all switches in the whole load range. Moreover, the current ripples are greatly decreased by interleaved control, which is good for battery and supercapacitor. In Reference [11] it is proposed architecture eliminates two boost switches which are present in the two-stage counterpart. Moreover, the input inductors are operated in discontinuous conduction mode; thus, power can be shared between input sources through proper selection of input inductors. In Reference [12] a new modified LCLC series resonant circuit based dual-input single-output isolated converter is proposed for hybrid energy systems. With this novel converter topology, two different voltage sources can be decoupled completely and transfer the power from two separate dc sources to dc load simultaneously. Moreover, it consists of only two controllable switches for integrating two separate voltage sources; it can provide good voltage regulation and soft switching over a wide load range. Nevertheless, these converters use a transformer to achieve galvanic isolation between sources and output; therefore, are much more complex in terms of designing and control when compared to the non-isolated ones.

In References [13,14] is presented a simple way to build a non-isolated hybrid energy storage system, connecting one of the sources directly while linking the other utilizing a DC-DC converter; yet this method does not permit the adjustment of the DC bus voltage. Another technique is to link each of the sources with the DC bus with an individual converter as presented in References [15–18]. In this way, it is possible to manage the DC bus voltage but it is an expensive solution due to the utilization of multiple converters. In order to decrease the cost, multiple-input converters have been proposed to achieve the goals of EVs. In Reference [19] is demonstrated that the multiple-input converters are cost effective, reliable, simple and easy to control. In Reference [20], energy flow between N different sources and the DC link are discussed. In this topology, it is not possible to transfer energy directly between DC sources. In Reference [21] a Z-source converter for EV application is presented, although this topology is suitable for optimal devices and components, the number of voltage sources is limited to two and it is not possible to extend this topology for multiple-input sources. In Reference [22] a modular multiple-input converter is presented, whose input ports are connected to the DC bus via half-bridges. However, with this topology it is not possible to transfer the energy directly between the sources. In Reference [23] is presented a converter with the same characteristics as that presented in Reference [22] but with a reduced element count. In Reference [24] a flexible topology is presented that can be used as a boost or a buck in any transfer of the energy and allows the direct exchange of energy between the sources. A non-desirable characteristic of this converter is the presence of an inductor per input, due to the intrinsic weight of these elements. In Reference [25] a topology is introduced that presents a greater gain compared to other existing ones, for its use on fuel cell-based EVs. In detriment of this converter, it can be said that it is not bidirectional. In References [26] and [27] are proposed converters with multiple-input and multiple-output, utilizing only one inductor. These types of converters are very useful for its use with multilevel inverters. Nevertheless, are not bidirectional converters and that can be a limitation in this type of application. Moreover, in Reference [28] is presented an inverter for the injection of energy generated in a panel into the grid. As stated in Reference [29] it is very important to take advantage of the renewable energies and for that reason in this work the topology presented in Reference [28] was the base of the DC-DC converter proposed in order to achieve the interaction with the grid.

There are several solutions for converters to harvest the energy generated in a Photovoltaic module (PV) and store it in the battery.Reference [30] presents an isolated multiport bidirectional DC-DC converter capable of parallel power management of various renewable energy sources. The advantage of this converter is it utilizes less number of controllable switches and provides soft switching for converter primary switch. Reference [31] proposes a battery charger for an EV based on a Zeta converter. This converter has the advantage of an output current without ripple and it is possible to buck or boost the input voltage. On the other hand, in Reference [32] it is utilized a cascaded buck-boost converter for the application. Such converters are typical of PV-battery systems for its simplicity, its bidirectional capability and the versatility to buck or boost the input voltage. Another solution is presented in Reference [33]—a non-isolated three-port switching boost converter. By controlling three degrees of freedom, the ports have boost, buck and buck-boost characteristics. Nevertheless, as stated in Reference [34], the boost converters provide the lower cost and higher efficiency of the non-isolated converters for PV systems.

In order to minimize the size, weight and cost of the traditional on-board chargers, integrated chargers have been proposed, some topologies and techniques are reviewed in Reference [35]. One concept of integration was proposed by Rippel and Cocconi in References [36,37], consisting in the use of the existent inverter and motor windings for the charging operation. Since the traction operation and charging the battery are not simultaneous, using the drivetrain components could reduce the size and cost of the on-board chargers. This modified structure supports the charging/discharging process of the battery. Some examples based on an induction machine are presented in References [38–40], based on PMSMs in References [41,42], based on windings rearrangement in Reference [43] and finally based on multiphase machines in Reference [44]. Another technique consists of combining a modified DC-DC converter with the AC-DC bidirectional rectifier. In Reference [45] the integrated converter is able to function as an AC-DC battery charger and to transfer electrical energy between the battery pack and the high-voltage bus of the electric traction system. The converter has a reduced number of high-current inductors and current transducers and presents fault-current tolerance in PHEV conversion. In Reference [46] a single-stage integrated converter is proposed based on direct AC-DC conversion theory. The proposed converter eliminates the full-bridge rectifier, reduces the number of semiconductor switches and high current inductors and improves the conversion efficiency. In Reference [47] is presented a bidirectional converter that not only enables beneficial vehicle-to-grid (V2G) interactions but also ensures that all power delivered to and from the grid has good power factor and near zero current harmonics. To accomplish this task, a multi-level bidirectional AC-DC converter is combined with an integrated bidirectional DC-DC converter. The proposed converter has four different modes of operation that allows it to supply power to or from the battery to either the grid or the high voltage bus of the EV. As stated in Reference [48], the increase of the renewable energies can affect the power system efficacy, the power quality, the security, among other problems; therefore, these converters need to have the possibility of power factor correction (PFC) when charging the battery and low total harmonic distortion (THD) when injecting current into the grid to have an interaction with the grid without disturbing the quality of the energy.

The aim of this work is to propose an integrated topology that combines a multiple-input DC-DC converter and a bidirectional rectifier for the interaction of the storage devices with the grid. The proposed converter topology has all the considered advantages from the architectures presented in the literature, such as allowing the bidirectional power flow, the possibility of directly transferring the energy between sources, that for every transfer it is possible to boost or buck the input voltage, that any storage element of the system can be the source or destination for a transfer and has only one inductor, which means less weight, and finally, that it permits the interaction of the storage devices with the grid.

For this purpose, in the present investigation three main sources are considered—a PV, an UC and a battery. Furthermore, a pedal was thought to be mechanically connected to the engine to generate energy, through the electric machine when the vehicle is parked. In addition, when the automobile is in motion, it can help the engine with its mechanical load. All these elements can be seen in Figure 1, which shows a diagram of the considered EV. The focus of this paper is the DC-DC converter.

The operation and steady-state analysis of the proposed converter topology, with all its cases, is explained in Section 2. The size of the elements of the circuit is presented in Section 3. The control strategy used to regulate the converter is shown in Section 4. Finally, the simulation results for all cases are presented in Section 5.

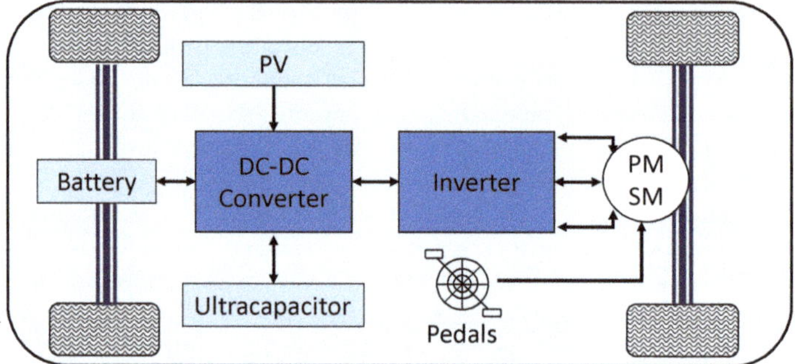

Figure 1. Diagram of the electric vehicle (EV) considered in this investigation.

2. Operation and Analysis of the Converter

The proposed power converter topology is displayed in Figure 2. It can be built by connecting each bidirectional source/output through two switches to the inductor L_2 while the unidirectional sources/outputs only need one switch. For a better understanding and analysis of the converter, this is divided into several stages in the present work. This can be done because they work independently. One of these stages is dedicated to the maximum power point tracking (MPPT) of the PV, shown in Figure 3. The other stage is the multiple-input converter shown in Figure 4. In case that an interaction with the grid is required to charge the UC or the battery, the diagram is shown in Figure 5. Furthermore, the injection of current into the grid was considered starting from the energy generated in the PV. In this way, when the vehicle is parked in daylight the energy can be harvested. Figure 6 indicates the circuit for this case. This article presents a detailed analysis of the operation of these converters, as well as simulation results for each of the possible cases.

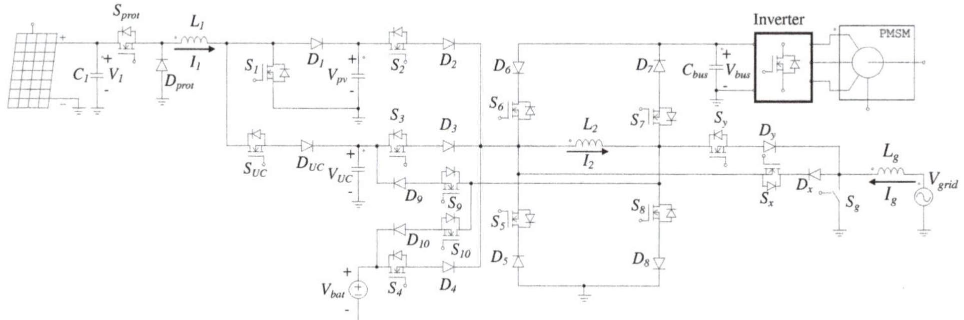

Figure 2. Proposed topology of the power converter.

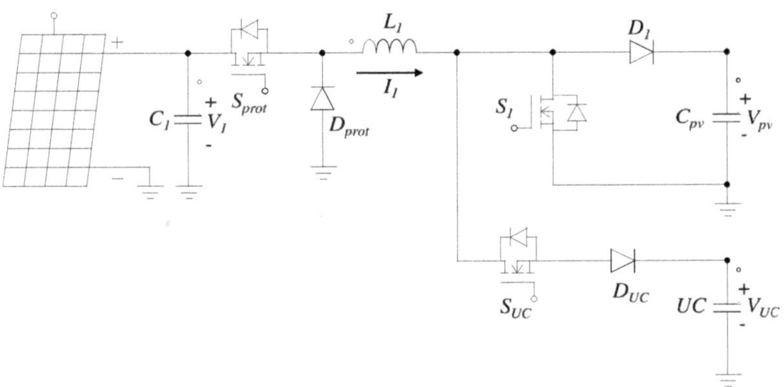

Figure 3. Converter associated with the PV.

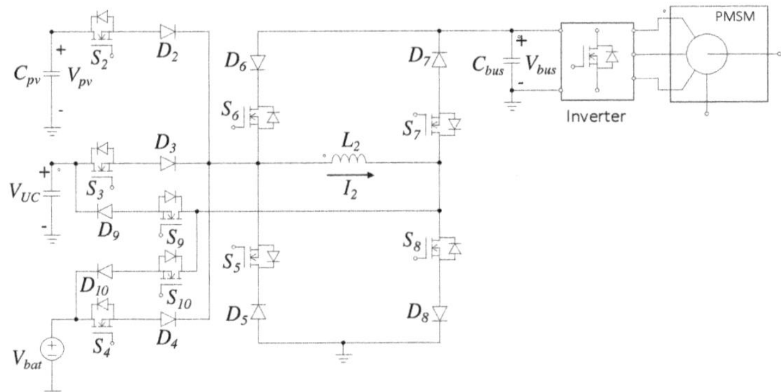

Figure 4. Multiple-input converter.

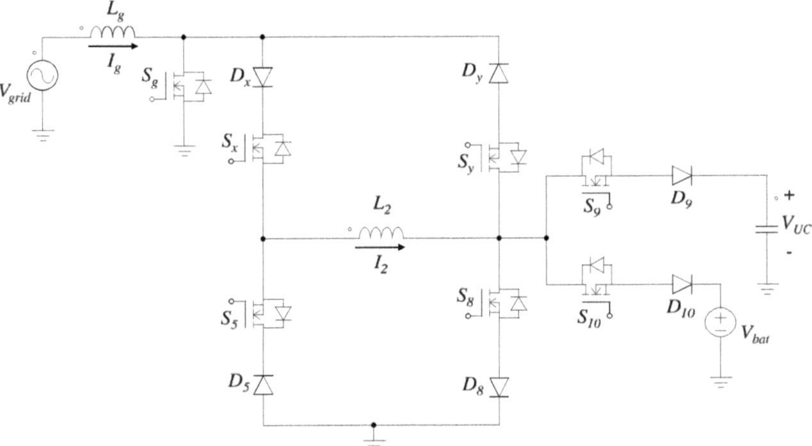

Figure 5. Circuit of the converter when it is needed the charge of the ultracapacitor (UC) or the battery.

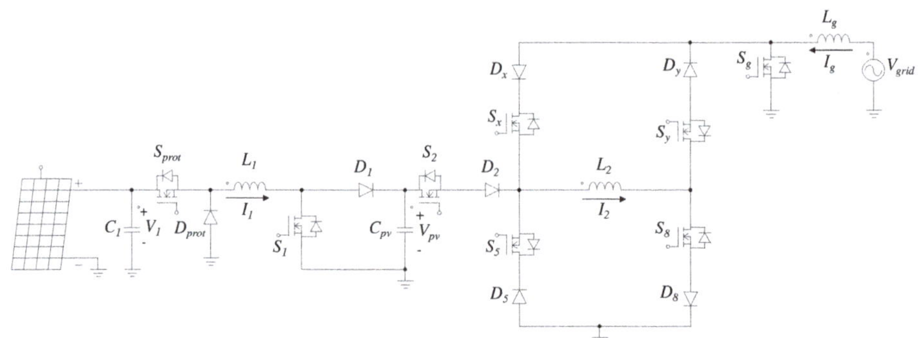

Figure 6. Circuit of the converter when the vehicle is parked in the sunlight.

In Table 1, various topologies are compared, attending several parameters of interest. Compared to the converters presented in the literature, the proposed topology presents more switches to its operation. Nevertheless, this is justified because it gathers all the advantages identified in the other converters.

Table 1. Comparison of the proposed converter with the existing ones.

Topology	Bidirectional	Transfer Directly between Sources	Extension to Multiport	Operation in Buck/Boost Mode	Interaction with the Grid	Number of Devices (n is Number of Inputs)
[22]	Yes	No	Yes	No	No	n inductors 2n switches
[23]	Yes	No	Yes	No	No	n inductors 2+n switches
[24]	Yes	Yes	Yes	Yes	No	n inductors 2n+2 switches
[25]	No	Yes	No	No	No	2 inductors 4 switches
Proposed converter	Yes	Yes	Yes	Yes	Yes	1 inductor 2n+4 switches

Now, the converter associated with the PV will be analyzed. When it is dedicated a converter to achieve the MPPT it is easiest to control the system. Some modifications were done to the boost converter such as the addition of the elements S_{prot} and D_{prot} for the protection of the system. It allows to isolate the PV in case no more energy is required without damaging the inductor L_1. Finally, the S_{UC} switch is added to provide greater versatility to the circuit.

In order to make better use of the energy the UC voltage is monitored to define the switch on which to act. In case that the $V_{UC} > V_1$ the energy of the PV is transferred to the UC and the capacitor C_{pv}. Switch S_1 is first turned on and the voltage V_1 appears across L_1, resulting in the increase of the current in this device with a slope of V_1/L_1. In the next interval, this switch is turned off and switch S_{UC} and diode D_1 are turned on. In this way, the energy stored in the inductor is transferred to the UC and the capacitor C_{pv}. It was decided to not include a switch in the position of the diode D_1 because it is acceptable for the application that the voltage in capacitor C_{pv} matches the voltage in the UC. Figure 7 shows the two stages needed to achieve the transfer of the energy.

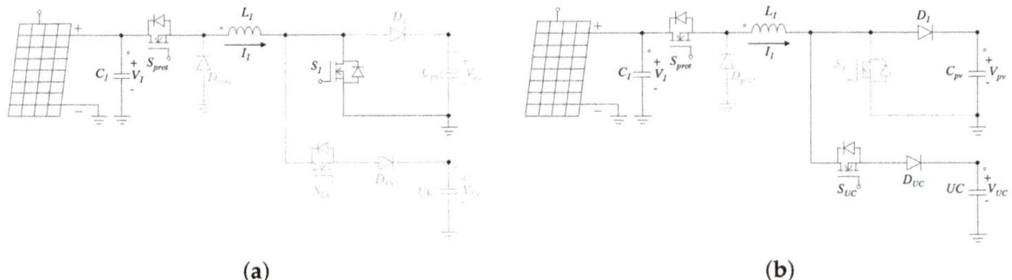

Figure 7. Operating modes of the converter associated with the PV when $V_{UC} > V_1$, (**a**) charge of the inductor, (**b**) discharge of the inductor.

In Table 2 are represented the operating modes of the converter associated with the PV. If $V_1 > V_{UC}$, the inductor is charged through the S_{UC} switch in order to emulate the behavior of the boost converter. In both cases, the point of operation in steady-state is giving by Equation (1). It is necessary to point out that in case that V_{UC} is greater than V_1, then $V_{pv} = V_{UC}$.

$$V_{pv} = \frac{1}{1 - D} V_1 \tag{1}$$

where D is the duty cycle giving by Equation (2).

$$D = \frac{t_{1A}}{t_{1A} + t_{1B}} \tag{2}$$

where t_{1A} is the length of time of the mode $1A$ of Table 2 and t_{1B} is the length of time of the mode $1B$ of Table 2.

Table 2. Operating modes of the converter associated with the PV.

Transfer	$V_{UC} > V_1$	Mode	Switches on	L_1	C_{pv}	UC
1. PV $\rightarrow C_{pv}$,UC	Yes	1A	S_1	$+$ [a]	$=$ [b]	$=$
		1B	S_{UC}, D_1	$-$ [c]	$+$	$+$
2. PV $\rightarrow C_{pv}$,UC	No	2A	S_{UC}	$+$	$=$	$+$
		2B	D_1	$-$	$+$	$=$

[a] Increase the state of charge; [b] The same state of charge was kept; [c] Decrease the state of charge.

Now, the multiple-input converter will be analyzed. The operating modes used in the multiple-input converter emulates the behavior of a buck converter or a boost one. Figure 8 shows the operating modes for a transfer from the battery to the capacitor C_{bus}. Switches S_4 and S_8 are first turned on and the voltage V_{bat} appear across inductor L_2, resulting in the increase of the current in this device with a slope of V_{bat}/L_2. In the next interval, switch S_8 is turned off and switch S_7 is turned on to complete the transfer of energy to the bus. Therefore, S_4 stay on throughout the transfer and switches S_7 and S_8 work in a complementary manner. As mentioned before, the proposed converter is very flexible, allowing that any element of the system can be source or destiny for a transfer and any transfer can be performed by boosting or reducing the input voltage. Hence, the aforementioned transfer can be realized reducing the input voltage with the correct combination of switches. Nevertheless, for this application, only the transfers present in Table 3 were considered.

One of the main concerns when using the number of switches utilized is the efficiency of the system. However, this is not a serious problem since only three switches are used to perform a transfer of energy; one that stays on and two that work in a complementary way, as can be appreciated in Table 3.

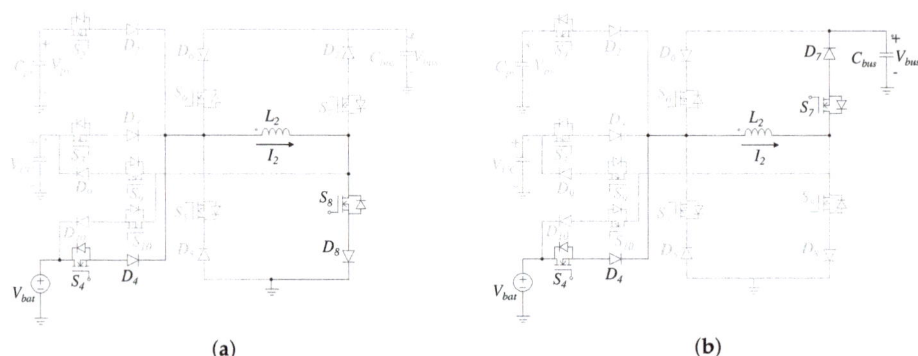

(a)   (b)

Figure 8. Operating modes of the multiple-input converter, (**a**) charge of the inductor, (**b**) discharge of the inductor.

Table 3. Operating modes of the multiple-input converter.

Transfer	Behavior	Mode	Switches on	L_2	C_{pv}	UC	Bat	Bus	Steady-State
3. Bat→Bus	Boost	3A	S_4, S_8	+	=	=	-	=	$V_{bus} = \dfrac{1}{1-D} V_{bat}$
		3B	S_4, S_7	-	=	=	-	+	
4. UC→Bus	Boost	4A	S_3, S_8	+	=	-	=	=	$V_{bus} = \dfrac{1}{1-D} V_{UC}$
		4B	S_3, S_7	-	=	-	=	+	
5. PV→Bus	Boost	5A	S_2, S_8	+	-	=	=	=	$V_{bus} = \dfrac{1}{1-D} V_{pv}$
		5B	S_2, S_7	-	-	=	=	+	
6. PV→UC	Boost	6A	S_2, S_8	+	-	=	=	=	$V_{UC} = \dfrac{1}{1-D} V_{pv}$
		6B	S_2, S_9	-	-	+	=	=	
7. PV→Bat	Boost	7A	S_2, S_8	+	-	=	=	=	$V_{bat} = \dfrac{1}{1-D} V_{pv}$
		7B	S_2, S_{10}	-	-	=	+	=	
8. Bat→UC	Buck	8A	S_4, S_9	+	=	+	-	=	$V_{UC} = D V_{bat}$
		8B	S_5, S_9	-	=	+	=	=	
9. Bus→Bat	Buck	9A	S_6, S_{10}	+	=	=	=	-	$V_{bat} = D V_{bus}$
		9B	S_5, S_{10}	-	=	=	+	=	
10. Bus→UC	Buck	10A	S_6, S_9	+	=	+	=	-	$V_{UC} = D V_{bus}$
		10B	S_5, S_9	-	=	+	=	=	

The losses analysis of the proposed topology can be done with the circuit presented in Figure 9. The power dissipated by the devices S_4, D_4 and L_2 is mainly by means of conduction due to the current is flowing in these devices at all time and is given by the Equation (3). The conduction power loss of the switches S_7 and S_8 are given by the Equation (4), while the switching power loss is given by Equation (5). Finally, total power dissipation due to all the elements is given by the Equation (6). Where I_2 is the current through the inductor L_2, D is the duty cycle, f_{sw} is the switching frequency, t_r is the rise time of the Metal-Oxide-Semiconductor Field-effect Transistors (MOSFETs), t_f is the fall time of the MOSFETs and R_{SL}, $R_{DS(on)S4}$, $R_{DS(on)D4}$, $R_{DS(on)S8}$, $R_{DS(on)D8}$, $R_{DS(on)S7}$, $R_{DS(on)D7}$ are the series resistance of the inductor L_2, switch S_4, diode D_4, switch S_8, diode D_8, switch S_7 and diode D_7 respectively. From this analysis, it can be appreciated that it is not a major issue the power loss compared to topologies that exist in the literature and the use of all the switches can be justified by the advantages gathered in the proposed topology.

$$P_{L_1} = I_2^2 \left(R_{SL} + R_{DS(on)S4} + R_{DS(on)D4} \right) \tag{3}$$

$$P_{L_2} = I_2^2 D(R_{DS(on)S8} + R_{DS(on)D8}) + I_2^2(1 - D)(R_{DS(on)S7} + R_{DS(on)D7}) \tag{4}$$

$$P_{L_3} = V_{in} I_2 f_{sw}(t_r + t_f) \tag{5}$$

$$P_{L\_Total} = P_{L_1} + P_{L_2} + P_{L_3} \tag{6}$$

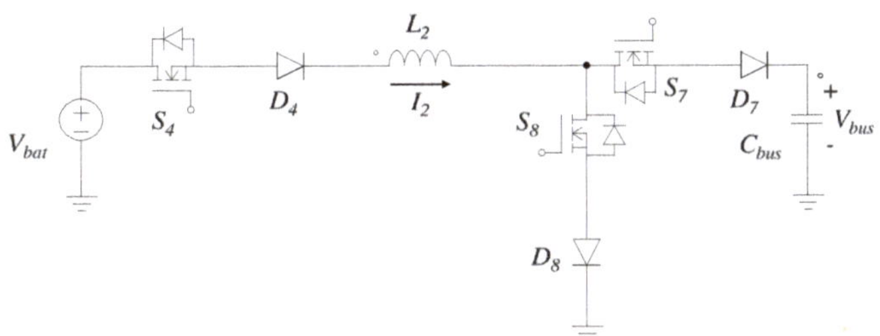

Figure 9. Circuit considered for the analysis of the losses.

For the efficiency calculation it can be assumed that the on-resistance of the MOSFETs and diodes is of 0.5 Ω, a rise time (t_r) of 28 *ns* and a decrease time (t_f) of 44 *ns*. Finally, the series resistance of the inductor is estimated to 1 Ω. The transfer that causes a greater loss is from UC to the bus and for that reason this conversion was considered for the power loss calculation. In this conversion the current through I_2 is of 10 A and the voltage in the UC is of 125 V. Substituting the corresponding values in Equations (3)–(6) it is obtained that the total power loss is of 309 V. The efficiency is given by Equation (7), which for an output power of 1.8 kW the efficiency is of 85.35%.

$$\eta = \frac{P_O}{P_O + P_{L\_Total}} 100 \tag{7}$$

3. Design of the Converter

In this section, the analysis for the sizing of the inductors L_1 and L_2 will be explained, as well as the capacitors C_{pv} and C_{bus}. First, the calculation of the components associated with the PV converter were performed. For the calculation of these elements, the methodology described in Reference [49] was followed. In this way, the capacitor was calculated as described in Equation (8), while the inductor was obtained as set out in Equation (9).

$$C_{pv} = \frac{DP_O}{V_O \Delta V_O f} \tag{8}$$

$$L_1 = \frac{D(1 - D)^2 V_O^2}{2fP_O} \tag{9}$$

where D is the duty cycle, P_O is the output power, V_O is the output voltage and f is the switching frequency.

To make the calculation was necessary to take into account some design considerations with respect to the PV and the desired operation for the converter. In this application was considered a PV that has an open circuit voltage of 50 V, short circuit current of 10 A and with a maximum power point of energy transfer around 25 V and 4–5 A (100–125 W). As proposed in [31], a $D = 0.5$ is used, the switching frequency was set in 50 kHz, while the typical output power was fixed in 150 W. Finally, the variation in the output voltage was defined as 10 V. For this data, the capacitor value needs to be greater than 3 μF. Finally, the inductor value was calculated and determined as 21 μH.

Next, the values of the inductor and the capacitor of the multiple-input converter were calculated. As explained in the previous section, this converter always behaves as a buck or as a boost. For this reason, the strategy followed to obtain these parameters was to calculate the necessary values for each case and take the critical values. The equations used to dimension the power elements in a buck converter are described in Reference [50] and are presented in (10) and (11). While the equations used for the sizing of the power elements in a boost converter are described in Reference [51] and are presented in (12) and (13).

$$C = \frac{\Delta I_L}{8f\Delta V_O} \tag{10}$$

$$L = \frac{V_O(V_{in} - V_O)}{\Delta I_L f V_{in}} \tag{11}$$

$$C = \frac{I_O D}{f\Delta V_O} \tag{12}$$

$$L = \frac{V_{in}(V_O - V_{in})}{\Delta I_L f V_O} \tag{13}$$

where I_L is the current through the inductor, f is the switching frequency, V_O is the output voltage, V_{in} is the input voltage, I_O is the output current and D is the duty cycle.

Taking into account these equations, and the behavior that was highlighted in the previous section for each state, Table 4 was constructed. In this table are presented the nominal values of the parameters needed for the sizing of the power elements. In this way, was possible to calculate the minimum inductance and capacitance allowable for the proper operation of the converter. It is important to highlight that the transfers concerning the PV were not included because this source of energy is considered as auxiliary. It can be concluded after analyzing this table that the value needed for the inductor L_2 must be higher than 3.86 mH and for the capacitor C_{bus} it must be greater than 144 µF.

Table 4. Sizing of the power elements of the multiple-input converter.

Transfer	Behavior	V_O [V]	V_{in} [V]	ΔI_L [A]	ΔV_O [V]	I_O [A]	D	f [kHz]	L_2 [mH]	C_{bus} [µF]
3. Bat→Bus	Boost	450	140	0.5	1	10	0.69	50	**3.86**	138
4. UC→Bus	Boost	450	125	0.5	1	10	0.72	50	3.61	**144**
8. Bat→UC	Buck	125	140	1	-	-	-	50	0.27	-
9. Bus→Bat	Buck	140	450	1	-	-	-	50	1.93	-
10. Bus→UC	Buck	125	450	1	-	-	-	50	1.81	-

4. Control Strategy

The control diagram of the system is shown in Figure 10.

It can be appreciated that there are several levels of hierarchy. The purpose of the lower level is to achieve the regulation of the state variables of the converter. Moreover, are present the blocks of MPPT, which are responsible for the extraction of the maximum power. Finally, there is an energy management system (EMS) that dictates the proper transfer at every moment. There are several strategies to control the state variables in a switching power converter as the ones presented in Reference [52]. Nevertheless, in this work, the current through L_2 was regulated utilizing a sliding mode control (SMC) because this technique is naturally suited for the regulation of switched controlled systems such as DC-DC power converters [53] and for its robustness, as stated in References [54–56]. Moreover, the control parameter can be the same to regulate the current in a boost and a buck converter as demonstrated in Table 5. In this table, V_{in}, v_C and i_L are the variables of the respective converter, while u is the input signal (for this case is the state of the switch), σ is the sliding surface and S is the control parameter of the SMC.

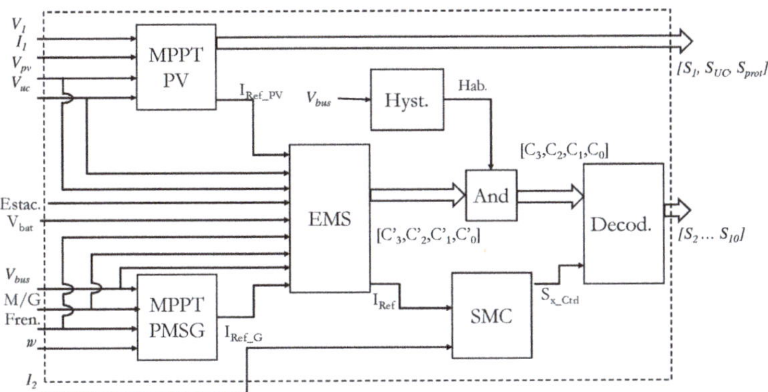

Figure 10. Block diagram of the controller of the system.

Table 5. Demonstration of the homogeneity of the control parameter for the regulation of the current with SMC in a buck and a boost converter.

Action	Buck Converter	Boost Converter
Model Obtention	$\begin{bmatrix} \dot{i_L} \\ \dot{v_C} \end{bmatrix} = \begin{bmatrix} 0 & -\frac{1}{L} \\ \frac{1}{C} & -\frac{1}{RC} \end{bmatrix} \begin{bmatrix} i_L \\ v_C \end{bmatrix} + \begin{bmatrix} \frac{u}{L} \\ 0 \end{bmatrix} V_{in}$	$\begin{bmatrix} \dot{i_L} \\ \dot{v_C} \end{bmatrix} = \begin{bmatrix} 0 & \frac{u-1}{L} \\ \frac{1-u}{C} & -\frac{1}{RC} \end{bmatrix} \begin{bmatrix} i_L \\ v_C \end{bmatrix} + \begin{bmatrix} \frac{1}{L} \\ 0 \end{bmatrix} V_{in}$
Proposing the sliding surface	$\sigma = S(i_L\text{-}i_{REF})$ $u = \begin{pmatrix} 1 & si & \sigma < 0 \\ 0 & si & \sigma > 0 \end{pmatrix}$	$\sigma = S(i_L\text{-}i_{REF})$ $u = \begin{pmatrix} 1 & si & \sigma < 0 \\ 0 & si & \sigma > 0 \end{pmatrix}$
Obtaining the derivative of the sliding surface	$\dot{\sigma} = S(-\frac{v_C}{L} + \frac{uV_{in}}{L})$	$\dot{\sigma} = S(\frac{(u-1)v_C}{L} + \frac{V_{in}}{L})$
Verify the existence condition ($\sigma > 0$)	$\dot{\sigma} = S(-\frac{v_C}{L}) = S * Negative\ number$	$\dot{\sigma} = S(\frac{V_{in}-v_C}{L}) = S * Negative\ number$
Verify the existence condition ($\sigma < 0$)	$\dot{\sigma} = S(\frac{V_{in}-v_C}{L}) = S * Positive\ number$	$\dot{\sigma} = S(\frac{V_{in}}{L}) = S * Positive\ number$
Define the sign of the parameter $S(\sigma\ \dot{\sigma}<0)$	S must be positive	S must be positive

The other state variable is the bus voltage (V_{bus}). This voltage was regulated with an on-off controller, because it is not necessarily an exact value in the input of the inverter and because of its simplicity. The range of the voltage was defined from 400 V to 500 V. This controller enable/disable the input current to the capacitor C_{bus}.

In order to achieve the MPPT of the PMSG there are several techniques like that presented in Reference [57] and is used in this work due to its simplicity. This algorithm is based on the principle that if the voltage in the capacitor C_{bus} is kept constant, then the maximum possible energy is being extracted. On the other hand, the MPPT applied for the PV was based on a Perturb and Observe (P&O) algorithm. This algorithm was implemented with the PV as the active input in the multiple-input converter and manipulating the current through the inductor L_2. Nevertheless, this is not always possible as the multiple-input converter can be needed for other functions. If this were the case, then the algorithm manipulates the state of the switches S_{UC} and S_1.

Finally, the EMS is the element that establishes the proper transfer in every moment. It is based on if-else rules to achieve its purpose. Figure 11 shows the main cases treated in this controller. If the vehicle is parked, then the source generating energy transfer is the same as the storage device that needs it. In this case, the UC is prioritized because it is desirable that the demanded energy at the start of the motion was supplied by this device. If the vehicle is in motion with positive acceleration and there is energy in both power sources, the flowchart depicted in Figure 12 is followed. If, on the contrary, there is only one power source with energy, then this device supplies the power during the driving cycle. Lastly, when the vehicle is in regenerative braking the energy generated is stored in the capacitor C_{bus}. For each case, the EMS establishes a code and the output decoder with this information and the control signal from the SMC sends the correct signal for each switch.

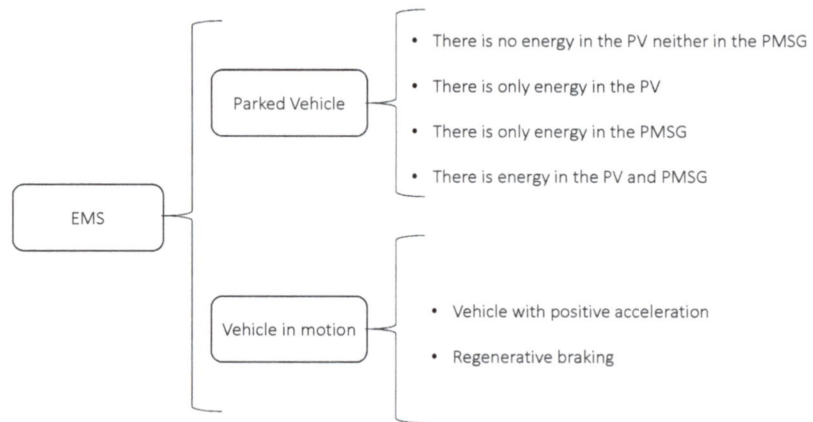

Figure 11. Cases taking into account for the implementation of the EMS.

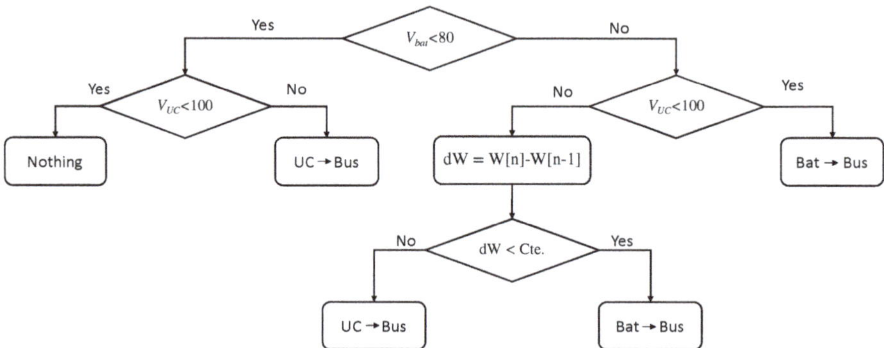

Figure 12. Flowchart followed by the EMS when the vehicle is in motion and there is energy in both power sources.

The control strategy previously presented requires a complement when it is required to interact with the grid. For this purpose, there are two well differentiated cases; charging the battery or the UC from the grid or injecting the energy generated in the PV into the grid. Figure 13 shows the block diagram for the charge of the storage elements. The objective is to regulate the current in the inductor L_2 and send it to the device that requires it. In Reference [58], it was proved that the SMC technique is adequate for achieving the control of an inverter interacting with the grid. For that reason this strategy was used and, in addition, it was necessary to know if the voltage grid is in the positive or negative half cycle to act on the necessary switch. Table 6 shows the behavior of the charge decoder. The switches that are not considered in Table 6 are kept off all the time.

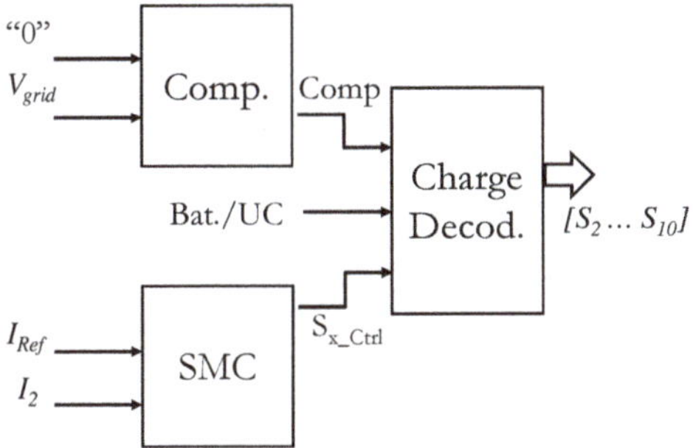

Figure 13. Block diagram of the control strategy of the charge from the grid.

Table 6. Behavior of the charge decoder.

Bat./UC	Comp	S_{x_Ctrl}	S_5	S_x	S_y	S_8	S_9	S_{10}
0	0	0	0	1	0	0	1	1
0	0	1	1	0	0	0	1	1
0	1	0	0	0	1	0	1	0
0	1	1	0	0	1	0	1	0
1	0	0	0	0	1	1	0	1
1	0	1	1	0	0	0	1	1
1	1	0	0	1	0	1	0	0
1	1	1	0	1	0	1	0	0

On the other hand, Figure 14 shows the block diagram for the injection of current into the grid. One of the objectives is to create a hysteresis behavior in the voltage V_{pv} acting on the current I_2 reference. If the voltage increases above a set value, the reference is augmented in order to decrease the voltage below a permissible value where is increased the current reference again. It is necessary to maintain the voltage V_{pv} above 170 V to guaranteed a satisfactory injection of the current into the grid. Moreover, in order to achieve an interaction with low THD at the beginning and at the end of the injection are executed in the zero-crossing. Finally, two separated SMC are implemented for each inductor (I_2 and I_g). With this information and the half-cycle in which the grid is operating, the injection decoder establishes the corresponding signal in each inductor. Table 7 shows the behavior of the injection decoder. The switches that are not considered in Table 7 are kept off all the time.

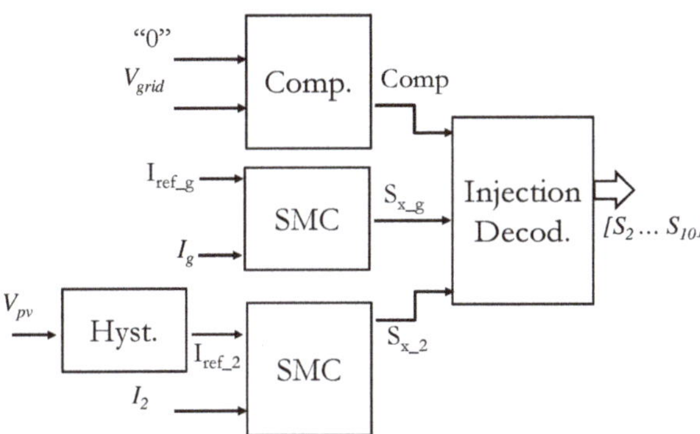

Figure 14. Block diagram of the control strategy of the injection of current into grid.

Table 7. Behavior of the injection decoder.

Comp	S_{x_g}	S_{x_2}	S_4	S_5	S_x	S_y	S_8	S_g
0	0	0	0	1	0	0	1	1
0	0	1	1	0	0	0	1	1
0	1	0	0	0	1	0	1	0
0	1	1	0	0	1	0	1	0
1	0	0	0	0	1	1	0	1
1	0	1	1	0	0	0	1	1
1	1	0	0	1	0	1	0	0
1	1	1	0	1	0	1	0	0

5. Simulation Results

The validation of the proposed converter and the control strategy was realized by the simulation of the system in the PSIM software (version 9.0.3). The main considerations for the simulation are presented in Table 8.

Table 8. Considerations for the simulation of the multiple-input converter.

Power Source	Sizing
Battery	100–144 V, 15 Ah
UC	125 V, 5 F
PV	100–125 W
PMSM	2.5 kW
Power converter	3.5 kW

The validation of the blocks of MPPT was possible in a separate way. Figure 15a shows the MPPT achieved for the PV utilizing the multiple-input converter for this purpose. It can be appreciated that the system has a good response with this method. The settling time is less than 0.1 s for a power variation of 100 W. Moreover, in steady-state, the ripple current is less than 0.25 A and the ripple power is less than 1 W. Otherwise, the second method is manipulating the switches S_{UC} and S_1, obtaining the response shown in Figure 15b. It is realized that this is a response of lesser quality. The settling time is of 0.02 s and in steady-state, there is a current ripple of 15 A and a power ripple of 8 W. Despite the lesser quality response obtained with this method, it has the advantage that the multiple-input converter can be dedicated to other tasks.

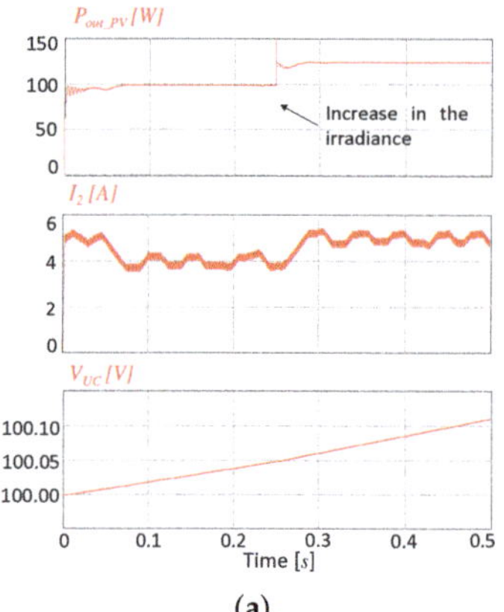

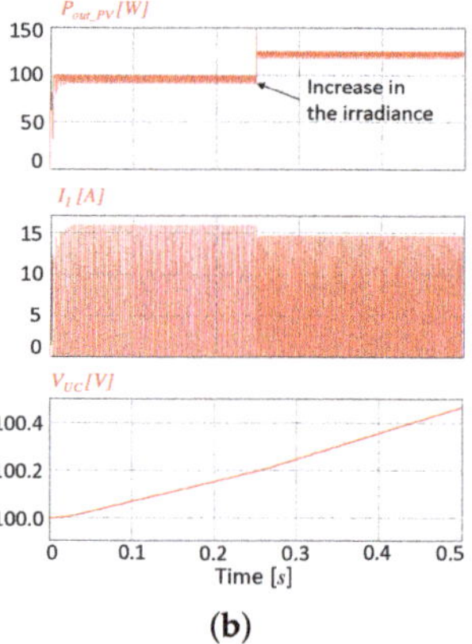

Figure 15. Methods implemented to achieve the MPPT of the PV, (**a**) utilizing the multiple-input converter and manipulating the reference of the current I_2, (**b**) manipulating the switches S_{UC} and S_1.

The PMSG also needs an MPPT algorithm for the harvesting of the energy. Figure 16 shows the simulation results for the implemented algorithm. It can be appreciated that by maintaining the voltage V_{bus} constant the MPPT is achieved. With a previous characterization of the electric machine it can be concluded that 92% and 97% of the possible total were extracted, respectively, for the conditions of the test.

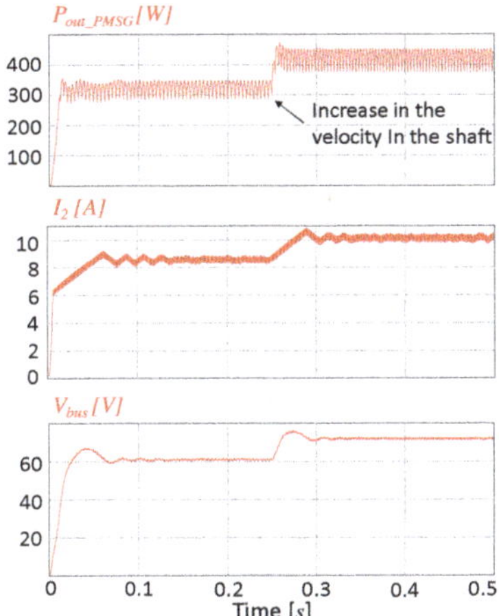

Figure 16. MPPT achieved in the PMSG by keeping the voltage in the capacitor C_{bus} constant.

For the validation of the EMS and the performance of the converter, the driving cycle ECE-15, which was legislated in the European Union in 1970, was utilized. Figure 17 shows the behavior of the state variables of the converter during the cycle. It can be concluded that the regulation of these variables was successful, with a ripple current of 0.8 A in steady-state and the desired hysteresis of 100 V for the voltage V_{bus}. Figure 18 shows the performance of the power sources. It is important to point out that the UC supplies the energy during the velocity transitions and when the velocity remains relatively constant the demanded energy is supplied by the battery as desired. Finally, the behavior of the PV and the PMSG is shown in Figure 19. The PV is generating energy during the entire duration of the cycle and verifies the proper switch between the two methods implemented. The generation of energy from the pedals was simulated in the beginning of the cycle, obtaining the desired behavior of the system.

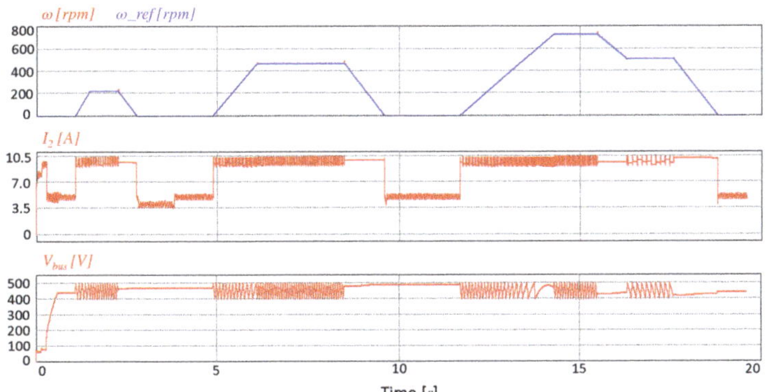

Figure 17. Behavior of the state variables through the ECE-15 cycle. The speed of the motor following the ECE-15 cycle; the current through inductor L_2, I_2 and the voltage in the capacitor C_{bus}, V_{bus}.

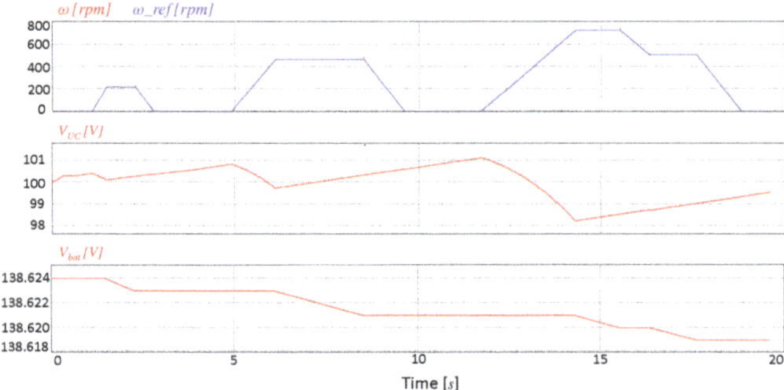

Figure 18. Behavior of the energy storage devices through the ECE-15 cycle. The speed of the motor following the ECE-15 cycle; the voltage in the ultracapacitor V_{UC} and the voltage in the battery V_{bat}.

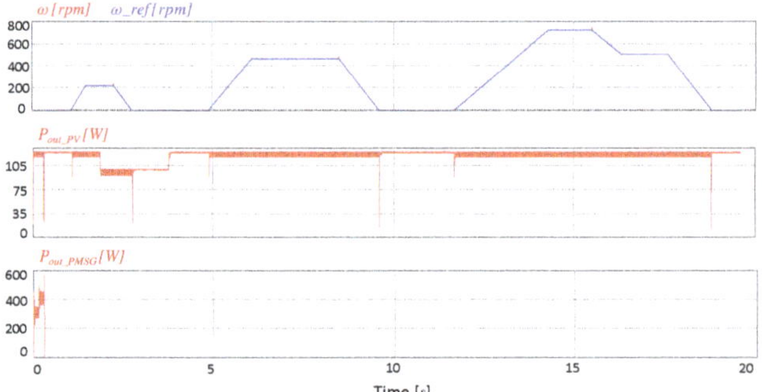

Figure 19. Behavior of the energy sources through the ECE-15 cycle. The speed of the motor following the ECE-15 cycle; the power harvested from the solar panel P_{out_PV} and the power harvested from the PMSG P_{out_PMSG}.

Finally, the interaction with the grid was validated. Figure 20 shows the charge of the battery. It can be appreciated that the battery has been charged and the current I_2 is regulated in 10 A with a ripple of 1 A. Despite the main objective being achieved, the system presents a deficient power quality. In order to achieve a better response, a PFC strategy need to be implemented.

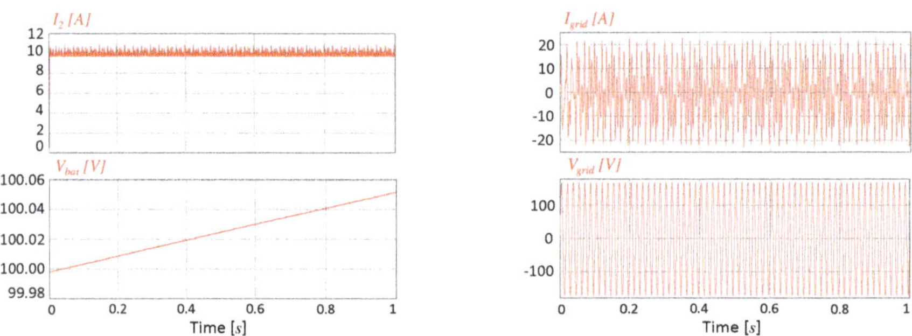

Figure 20. Charge of the battery from the grid: Current through the inductor L_2, I_2; voltage in the battery V_{bat}; current in the grid I_{grid} and the voltage in the grid V_{grid}.

Figure 21 demonstrates the injection of current into the grid. It can be appreciated that the MPPT is working properly as is the hysteresis of the voltage V_{pv}. This behavior of the voltage is permissible because the only restriction is that needs to be greater than 170 V and it is achieved by modifying the current amplitude injected into the grid. Figure 22 shows a zoom of the injected current where it can be appreciated that it is in phase with the normalized voltage grid and it is also notable that the start of the injection coincides with the zero-crossing.

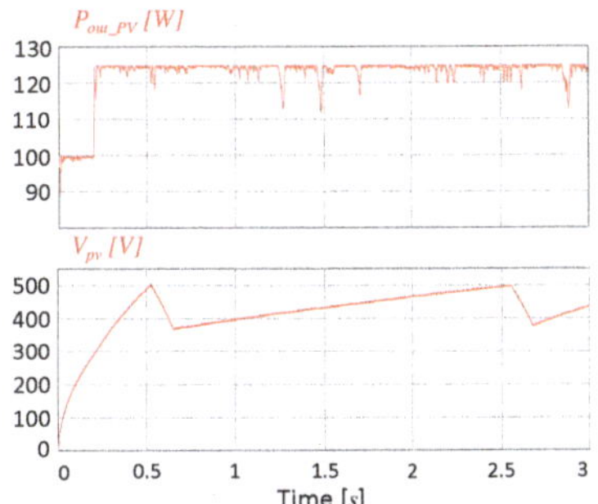
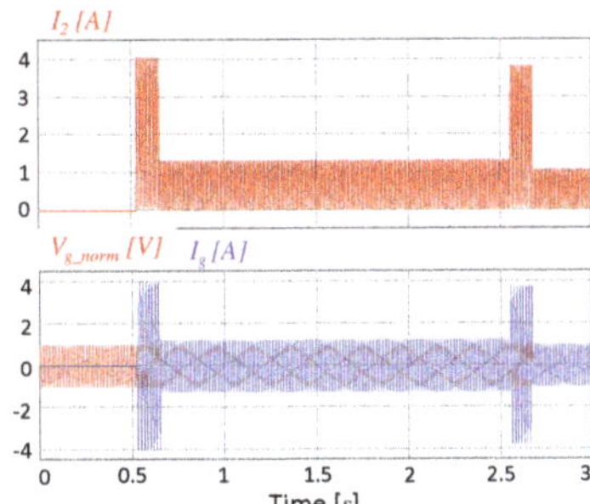

Figure 21. Injection of current into the grid—the power harvested from the solar panel P_{out_PV}; the voltage in the capacitor C_{pv}, V_{pv}; the current in the inductor L_2, I_2 and the voltage and current of the grid V_{grid} and I_{grid} respectively.

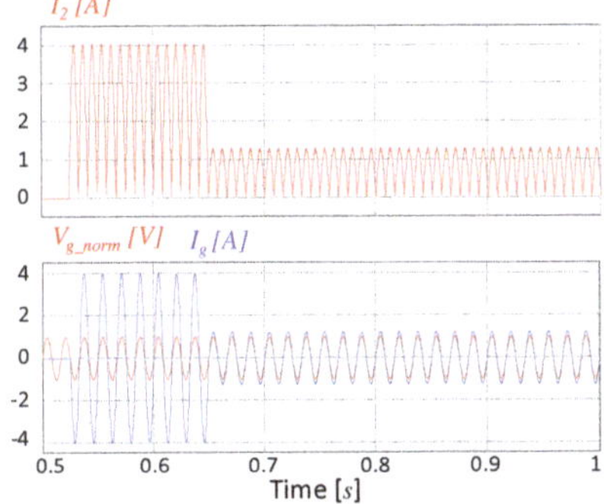

Figure 22. Zoom of the currents in the interval 0.5 s to 1 s—the current in the inductor L_2, I_2 and the voltage and current of the grid V_{grid} and I_{grid} respectively.

6. Conclusions

In this work, a bidirectional multiple-input DC-DC converter topology is proposed for electric vehicles. A very flexible topology has been achieved, which gathers all the advantages identified in the converters presented in the literature. This converter allows the interaction of two or more energy sources with a wide range of voltage in its inputs with a direct current bus. It can be operated in buck or boost mode for each possible transfer. In addition, each element of the system can be a source or destination in a transfer. Moreover, it permits the direct transfer of energy between sources and it only presents a single inductor which will impact the weight of the vehicle. Finally, due to the arrangement of the switches, it can be said that it allows the interaction with the grid to charge the battery and UC and to deliver energy from these devices.

Likewise, a detailed analysis of the operation of the converter was presented in the operating modes of interest for the application. A control strategy based on different levels of hierarchy was implemented to achieve the proper performance of the system. The regulation of the current of the converter was achieved by an SMC, while the voltage was regulated with an on-off controller. The MPPT was realized in the PV using two different converters, based on the P&O algorithm. By using two converters it simplifies the control system. Meanwhile, the MPPT in the PMSG was accomplished by controlling the voltage on the capacitor of the direct current bus. In addition, an EMS was developed to decide the corresponding transfer at each time.

Because of its versatility, this converter is not only limited to its application in electric or hybrid vehicles. It can also be used in smart grids, microgrids, as a resource in distributed energy systems, battery charging management systems and so forth, in any application where the interaction of two or more sources is needed with the possibility of bidirectional power flow.

For future works the overall control system will be redesigned; an EMS based on fuzzy logic, which will increase the efficiency of the system and it will be more robust in situations that were not considered and a stability analysis will be developed. Moreover, a PFC strategy needs to be implemented for the charging of the power sources, and a control strategy that permit the boosting of the input voltage needs to be designed for the injection of current into the grid in order to be able to use the battery. Finally, experimental results that validate the system need to be performed, where it is assumed that the main constraints will be associated with the switching frequency in the current control loop of the L_2 inductor and in the transitions between the EMS states.

Author Contributions: Conceptualization, A.A.-D. and H.L.; methodology, A.A.-D., H.L. and R.V.C.-S.; software, A.A.-D., H.L. and N.V.; validation, A.A.-D., H.L., R.V.C.-S., J.R.-R. and N.V.; formal analysis, H.L., R.V.C.-S., J.R.-R. and N.V.; writing—original draft preparation, A.A.-D., H.L., R.V.C.-S. and J.R.-R.; writing—review and editing, G.H.-R.; supervision, R.V.C.-S., J.R.-R. and N.V.

Funding: This research was funded by the "Consejo Nacional de Ciencia y Tecnología (CONACYT)".

Conflicts of Interest: The authors declare no conflict of interest.

References

1. Solaymani, S. CO$_2$ emissions patterns in 7 top carbon emitter economies: The case of transport sector. *Energy* **2019**, *168*, 989–1001. [CrossRef]
2. Khaligh, A.; Li, Z. Battery, ultracapacitor, fuel cell, and hybrid energy storage systems for electric, hybrid electric, fuel cell, and plug-in hybrid electric vehicles: State of the art. *IEEE Trans. Veh. Technol.* **2010**, *59*, 2806–2814. [CrossRef]
3. Dusmez, S.; Hasanzadeh, A.; Khaligh, A. Comparative analysis of bidirectional three-level DC-DC converter for automotive applications. *IEEE Trans. Ind. Electron.* **2014**, *62*, 3305–3315. [CrossRef]
4. Shen, J.; Dusmez, S.; Khaligh, A. Optimization of sizing and battery cycle life in battery/ultracapacitor hybrid energy storage systems for electric vehicle applications. *IEEE Trans. Ind. Inform.* **2014**, *10*, 2112–2121. [CrossRef]
5. Lu, S.; Corzine, K.A.; Ferdowsi, M. A new battery/ultracapacitor energy storage system design and its motor drive integration for hybrid electric vehicles. *IEEE Trans. Veh. Technol.* **2007**, *56*, 1516–1523. [CrossRef]
6. Dusmez, S.; Khaligh, A. A supervisory power-splitting approach for a new ultracapacitor–battery vehicle deploying two propulsion machines. *IEEE Trans. Ind. Inform.* **2014**, *10*, 1960–1971. [CrossRef]
7. Emadi, A.; Williamson, S. Fuel cell vehicles: Opportunities and challenges. In Proceedings of the IEEE Power Engineering Society General Meeting, Denver, CO, USA, 6–10 June 2004; pp. 1640–1645.
8. Aso, S.; Kizaki, M.; Nonobe, Y. Development of fuel cell hybrid vehicles in Toyota. In Proceedings of the 2007 Power Conversion Conference-Nagoya, Nagoya, Japan, 2–5 April 2007; pp. 1606–1611.
9. Rajashekara, K. Present status and future trends in electric vehicle propulsion technologies. *IEEE J. Emerg. Sel. Top. Power Electron.* **2013**, *1*, 3–10. [CrossRef]
10. Ding, Z.; Yang, C.; Zhang, Z.; Wang, C.; Xie, S. A novel soft-switching multiport bidirectional DC-DC converter for hybrid energy storage system. *IEEE Trans. Power Electron.* **2013**, *29*, 1595–1609. [CrossRef]
11. Dusmez, S.; Li, X.; Akin, B. A new multiinput three-level DC-DC converter. *IEEE Trans. Power Electron.* **2015**, *31*, 1230–1240. [CrossRef]
12. Reddi, N.; Ramteke, M.; Suryawanshi, H. Dual-Input Single-Output Isolated Resonant Converter with Zero Voltage Switching. *Electronics* **2018**, *7*, 96. [CrossRef]
13. Lu, S.; Corzine, K.A.; Ferdowsi, M. A unique ultracapacitor direct integration scheme in multilevel motor drives for large vehicle propulsion. *IEEE Trans. Veh. Technol.* **2007**, *56*, 1506–1515. [CrossRef]

14. Camara, M.B.; Gualous, H.; Gustin, F.; Berthon, A.; Dakyo, B. DC-DC converter design for supercapacitor and battery power management in hybrid vehicle applications—Polynomial control strategy. *IEEE Trans. Ind. Electron.* **2009**, *57*, 587–597. [CrossRef]

15. Kollimalla, S.K.; Mishra, M.K.; Narasamma, N.L. Design and analysis of novel control strategy for battery and supercapacitor storage system. *IEEE Trans. Sustain. Energy* **2014**, *5*, 1137–1144. [CrossRef]

16. Tani, A.; Camara, M.B.; Dakyo, B. Energy management based on frequency approach for hybrid electric vehicle applications: Fuel-cell/lithium-battery and ultracapacitors. *IEEE Trans. Veh. Technol.* **2012**, *61*, 3375–3386. [CrossRef]

17. Zandi, M.; Payman, A.; Martin, J.P.; Pierfederici, S.; Davat, B.; Meibody-Tabar, F. Energy management of a fuel cell/supercapacitor/battery power source for electric vehicular applications. *IEEE Trans. Veh. Technol.* **2010**, *60*, 433–443. [CrossRef]

18. Payman, A.; Pierfederici, S.; Meibody-Tabar, F.; Davat, B. An adapted control strategy to minimize DC-bus capacitors of a parallel fuel cell/ultracapacitor hybrid system. *IEEE Trans. Power Electron.* **2009**, *26*, 3843–3852. [CrossRef]

19. Nejabatkhah, F.; Danyali, S.; Hosseini, S.H.; Sabahi, M.; Niapour, S.M. Modeling and control of a new three-input DC-DC boost converter for hybrid PV/FC/battery power system. *IEEE Trans. Power Electron.* **2011**, *27*, 2309–2324. [CrossRef]

20. Khaligh, A. A multiple-input DC-DC positive buck-boost converter topology. In Proceedings of the 2008 Twenty-Third Annual IEEE Applied Power Electronics Conference and Exposition, Austin, TX, USA, 24–28 February 2008; pp. 1522–1526.

21. Peng, F.Z.; Shen, M.; Holland, K. Application of Z-source inverter for traction drive of fuel cell—Battery hybrid electric vehicles. *IEEE Trans. Power Electron.* **2007**, *22*, 1054–1061. [CrossRef]

22. Solero, L.; Lidozzi, A.; Pomilio, J.A. Design of multiple-input power converter for hybrid vehicles. *IEEE Trans. Power Electron.* **2005**, *20*, 1007–1016. [CrossRef]

23. Akar, F.; Tavlasoglu, Y.; Ugur, E.; Vural, B.; Aksoy, I. A bidirectional nonisolated multi-input DC-DC converter for hybrid energy storage systems in electric vehicles. *IEEE Trans. Veh. Technol.* **2015**, *65*, 7944–7955. [CrossRef]

24. Hintz, A.; Prasanna, U.R.; Rajashekara, K. Novel modular multiple-input bidirectional DC-DC power converter (MIPC) for HEV/FCV application. *IEEE Trans. Ind. Electron.* **2014**, *62*, 3163–3172. [CrossRef]

25. Ahrabi, R.R.; Ardi, H.; Elmi, M.; Ajami, A. A novel step-up multiinput DC-DC converter for hybrid electric vehicles application. *IEEE Trans. Power Electron.* **2016**, *32*, 3549–3561. [CrossRef]

26. Babaei, E.; Abbasi, O. Structure for multi-input multi-output DC-DC boost converter. *IET Power Electron.* **2016**, *9*, 9–19. [CrossRef]

27. Nahavandi, A.; Hagh, M.T.; Sharifian, M.B.B.; Danyali, S. A nonisolated multiinput multioutput DC-DC boost converter for electric vehicle applications. *IEEE Trans. Power Electron.* **2014**, *30*, 1818–1835. [CrossRef]

28. López, H.; Rodríguez-Reséndiz, J.; Guo, X.; Vázquez, N.; Carrillo-Serrano, R.V. Transformerless common-mode current-source inverter grid-connected for PV applications. *IEEE Access* **2018**, *6*, 62944–62953. [CrossRef]

29. Zsiborács, H.; Hegedűsné Baranyai, N.; Csányi, S.; Vincze, A.; Pintér, G. Economic Analysis of Grid-Connected PV System Regulations: A Hungarian Case Study. *Electronics* **2019**, *8*, 149. [CrossRef]

30. Mali, V.D.; Thorat, A.; Sawant, N.K. An Isolated Multiport Bidirectional DC-DC Converter for PV Battery System. In Proceedings of the 2018 Second International Conference on Inventive Communication and Computational Technologies (ICICCT), Coimbatore, India, 20–21 April 2018; pp. 302–306.

31. Khatab, A.M.; Marei, M.I.; Elhelw, H.M. An Electric Vehicle Battery Charger Based on Zeta Converter Fed from a PV Array. In Proceedings of the 2018 IEEE International Conference on Environment and Electrical Engineering and 2018 IEEE Industrial and Commercial Power Systems Europe (EEEIC/I&CPS Europe), Palermo, Italy, 12–15 June 2018; pp. 1–5.

32. Chen, X.; Pise, A.A.; Elmes, J.; Batarseh, I. Ultra-Highly Efficient Low Power Bi-directional Cascaded-Buck-Boost Converter for Portable PV-Battery-Devices Applications. *IEEE Trans. Ind. Appl.* **2019**, *55*, 3989–4000. [CrossRef]

33. Liu, S.; Gao, Y.; Yang, L. Research on Application of Non-Isolated Three-Port Switching Boost Converter in Photovoltaic Power Generation System. *Electronics* **2019**, *8*, 746. [CrossRef]

34. Chewale, M.A.; Wanjari, R.A.; Savakhande, V.B.; Sonawane, P.R. A Review on Isolated and Non-isolated DC-DC Converter for PV Application. In Proceedings of the 2018 International Conference on Control, Power, Communication and Computing Technologies (ICCPCCT), Kannur, India, 23–24 March 2018; pp. 399–404.

35. Fahem, K.; Chariag, D.E.; Sbita, L. On-board bidirectional battery chargers topologies for plug-in hybrid electric vehicles. In Proceedings of the 2017 International Conference on Green Energy Conversion Systems (GECS), Hammamet, Tunisia, 23–25 March 2017; pp. 1–6.
36. Rippel, W.E. Integrated Traction Inverter and Battery Charger Apparatus. U.S. Patent 4,920,475, 24 April 1990.
37. Rippel, W.E.; Cocconi, A.G. Integrated Motor Drive and Recharge System. U.S. Patent 5,099,186, 24 March 1992.
38. Cocconi, A.G. Combined Motor Drive and Battery Recharge System. U.S. Patent 5,341,075, 23 August 1994.
39. Sul, S.K.; Lee, S.J. An integral battery charger for four-wheel drive electric vehicle. *IEEE Trans. Ind. Appl.* **1995**, *31*, 1096–1099.
40. Hegazy, O.; Van Mierlo, J.; Lataire, P. Design and control of bidirectional DC/AC and DC/DC converters for plug-in hybrid electric vehicles. In Proceedings of the 2011 International Conference on Power Engineering, Energy and Electrical Drives, Malaga, Spain, 11–13 May 2011; pp. 1–7.
41. De Sousa, L.; Silvestre, B.; Bouchez, B. A combined multiphase electric drive and fast battery charger for electric vehicles. In Proceedings of the 2010 IEEE Vehicle Power and Propulsion Conference, Lille, France, 1–3 September 2010; pp. 1–6.
42. Lacroix, S.; Labouré, E.; Hilairet, M. An integrated fast battery charger for electric vehicle. In Proceedings of the 2010 IEEE Vehicle Power and Propulsion Conference, Lille, France, 1–3 September 2010; pp. 1–6.
43. Haghbin, S.; Lundmark, S.; Alakula, M.; Carlson, O. An isolated high-power integrated charger in electrified-vehicle applications. *IEEE Trans. Veh. Technol.* **2011**, *60*, 4115–4126. [CrossRef]
44. Subotic, I.; Levi, E. An integrated battery charger for EVs based on a symmetrical six-phase machine. In Proceedings of the 2014 IEEE 23rd International Symposium on Ind. Electron (ISIE), Istanbul, Turkey, 1–4 June 2014; pp. 2074–2079.
45. Lee, Y.J.; Khaligh, A.; Emadi, A. Advanced integrated bidirectional AC/DC and DC/DC converter for plug-in hybrid electric vehicles. *IEEE Trans. Veh. Technol.* **2009**, *58*, 3970–3980.
46. Chen, H.; Wang, X.; Khaligh, A. A single stage integrated bidirectional ac/dc and dc/dc converter for plug-in hybrid electric vehicles. In Proceedings of the 2011 IEEE Vehicle Power and Propulsion Conference, Chicago, IL, USA, 6–9 September 2011; pp. 1–6.
47. Erb, D.C.; Onar, O.C.; Khaligh, A. An integrated bi-directional power electronic converter with multi-level AC-DC/DC-AC converter and non-inverted buck-boost converter for PHEVs with minimal grid level disruptions. In Proceedings of the 2010 IEEE Vehicle Power and Propulsion Conference, Lille, France, 1–3 September 2010; pp. 1–6.
48. Kumar, G.; Sarojini, R.K.; Palanisamy, K.; Sanjeevikumar, P.; Holm-Nielsen, J.B. Large Scale Renewable Energy Integration: Issues and Solutions. *Energies* **2019**, *12*, 1996. [CrossRef]
49. Mudhol, A.; Pius, P. Design and implementation of boost converter for photovoltaic systems. *Int. J. Innov. Res. Elect. Electron. Instrum. Control Eng.* **2016**, *4*, 110–114. [CrossRef]
50. Hauke, B. *Basic Calculation of Buck Converter's Power Stages*; Technical report, Texas, Tech. Rep. SLVA477, Dec; Texas Instruments: Dallas, TX, USA, 2011.
51. Hauke, B. *Application Report: Basic Calculation of a Boost Converter's Power Stage*; Texas Instruments: Dallas, TX, USA, 2014.
52. Abbas, G.; Gu, J.; Farooq, U.; Abid, M.; Raza, A.; Asad, M.; Balas, V.; Balas, M. Optimized Digital Controllers for Switching-Mode DC-DC Step-Down Converter. *Electronics* **2018**, *7*, 412. [CrossRef]
53. Sira-Ramirez, H.J.; Silva-Ortigoza, R. *Control Design Techniques in Power Electronics Devices*; Springer-Verlag: London, UK, 2006.
54. Yang, Y.; Tan, S.C. Trends and Development of Sliding Mode Control Applications for Renewable Energy Systems. *Energies* **2019**, *12*, 2861. [CrossRef]
55. Cucuzzella, M.; Rosti, S.; Cavallo, A.; Ferrara, A. Decentralized sliding mode voltage control in DC microgrids. In Proceedings of the 2017 American Control Conference (ACC), Seattle, WA, USA, 24–26 May 2017; pp. 3445–3450.

56. Cucuzzella, M.; Lazzari, R.; Trip, S.; Rosti, S.; Sandroni, C.; Ferrara, A. Sliding mode voltage control of boost converters in DC microgrids. *Control Eng. Pract.* **2018**, *73*, 161–170. [CrossRef]
57. Kim, S.K.; Lee, K.B. Robust DC-Link Voltage Tracking Controller with Variable Control Gain for Permanent Magnet Synchronous Generators. *Electronics* **2018**, *7*, 339. [CrossRef]
58. Senthilnathan, K.; Annapoorani, I. Multi-port current source inverter for smart microgrid applications: A cyber physical paradigm. *Electronics* **2019**, *8*, 1. [CrossRef]

Article

Large Scale Renewable Energy Integration: Issues and Solutions

G. V. Brahmendra Kumar [1], Ratnam Kamala Sarojini [1], K. Palanisamy [1,*], Sanjeevikumar Padmanaban [2,*] and Jens Bo Holm-Nielsen [2]

[1]   School of Electrical Engineering, Vellore Institute of Technology, Vellore 632014, India; brahmendrakumar.g@gmail.com (G.V.B.K.); kamala.rtnm@gmail.com (R.K.S.)

[2]   Center for Bioenergy and Green Engineering, Department of Energy Technology, Aalborg University, 6700 Esbjerg, Denmark; jhn@et.aau.dk

*   Correspondence: kpalanisamy@vit.ac.in (K.P.); san@et.aau.dk (P.S.)

Received: 16 April 2019; Accepted: 21 May 2019; Published: 24 May 2019

Abstract: In recent years, many applications have been developed for the integration of renewable energy sources (RES) into the grid in order to satisfy the demand requirement of a clean and reliable electricity generation. Increasing the number of RES creates uncertainty in load and power supply generation, which also presents an additional strain on the system. These uncertainties will affect the voltage and frequency variation, stability, protection, and safety issues at fault levels. RES present non-linear characteristics, which requires effective coordination control methods. This paper presents the stability issues and solutions associated with the integration of RES within the grid.

Keywords: renewable energy sources; energy storage system; voltage; frequency; grid integration

---

1. Introduction

The majority of countries across the world are concentrating on RES, due to an increase of environmental concerns and depletion of fossil fuels [1]. For example, China is aiming for RES to represent somewhere around 35 percent of consumption by 2030. India has set an ambitious RES target of 175 GW [2]. The European Union and the United States also have targets regarding RES [3]. Some countries successfully integrated large shares of RES in the power grid in 2017, and most countries have set targets to receive their power generation from RES by 80% in 2050, as is shown in Figure 1.

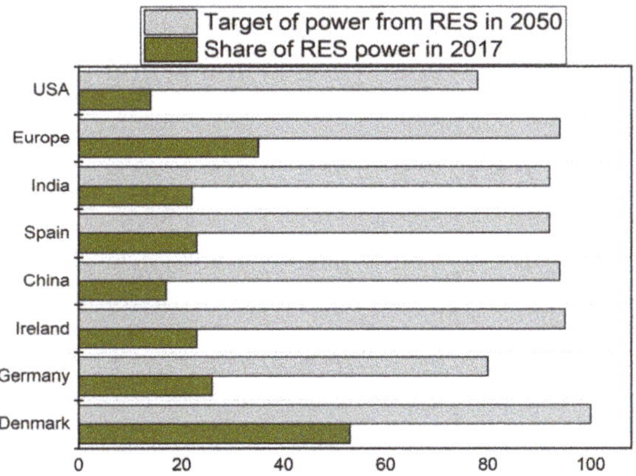

Figure 1. Renewable energy share and targets by countries in 2017 and 2050 [4].

Due to the irregularities and fluctuating features of RES, new challenges to the grid over a unified anticipated generation have been created. The RES outputs are influenced by meteorological conditions. These characteristics affect the power system efficacy, power quality, system reliability, load management, security, and safety in various ways and are also very significant factors to be viewed in the integration of RES within the grid [6]. This influence is commonly observed with solar and wind energy, but geothermal, hydropower and biomass energy resources are more anticipated and have irrelevant difficulties in their association with the grid [7]. The comparison of RES with a synchronous generator (SG) is shown in Table 1.

Table 1. The comparison of renewable energy sources (RES) with synchronous generator (SG) [5].

Characteristics	Solar	Wind	SG
Fluctuations	More	Less	No
Cost for large scale	More	Medium	Low to medium
Maintenance cost	Minimum	More	Moderate
Inertia	No inertia	Low inertia	High
Capacity factor	Very Low	Low to medium	More
Annual growth in power industry	Very High	More	More

The RES integrated with the grid is shown in Figure 2. Figure 2 is incorporated with RES, grid and electronic components. The power generation from RES to the grid is unidirectional. The RES requires the converters to be connected to the grid. These converters will achieve efficient operation and obtain the energy quality requirements related to the harmonics level. Also, this allows the integration of RES in a power inverter with high energy features and safety.

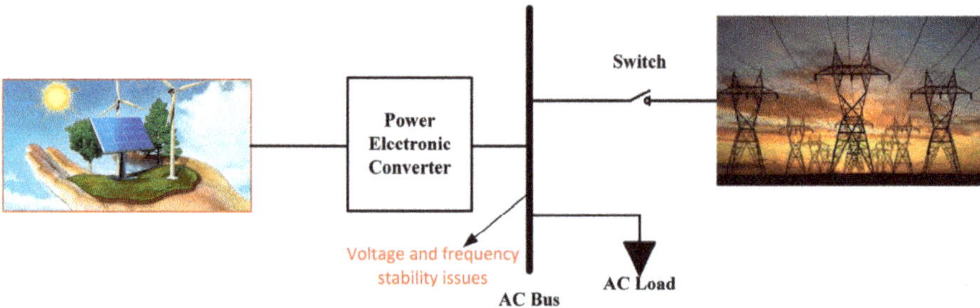

Figure 2. RES integration with grid.

The remainder of this paper is organized as follows: Section 2 considers the grid integration issues with a high share of RES, and solutions for variable RES is discussed in Section 3. The Energy Storage System (ESS) support is explained in Section 4, and followed by smart grid features in voltage control with RES in Section 5. Section 6 is the conclusion.

2. Grid Integration Issues with a High Share of RES

As the power generation from RES increases, the installed capacity of power converters increases. By utilizing the large scale of solar energy, the RES system is able to supply power to the grid. Thus, by increasing the capacity of the RES, the grid-connected RES will create a negative impact on the grid when fault or disturbance occurs. Hence, the Point of Common Coupling (PCC) voltage drop is triggered by a fault in grid power and the RES will become off-grid across a wide area range. Moreover, these fault impacts cause the grid voltage and frequency collapse, also affecting the safe, stable and reliable operation of the grid and even triggering large economic losses [8].

2.1. Impact of Large-Scale Integration of RES on Frequency

The frequency of a power system must be preserved near to its nominal value (either 50 Hz or 60 Hz based on the grid). The frequency deviations will only arise when there is a mismatch between generation and load. A stiff power system preserves the frequency subsequent to a contingency event [9].

The frequency of the power system is maintained at the nominal value only when the active power of generation and demand is balanced as indicated in Figure 3. If the demand is more than the generation, then the frequency decreases from the nominal value. In the case of surplus generation, the system frequency increases. The kinetic energy (KE) stored in the rotor of the rotating machines present in the power system contribute to inertia. The inertia in the power system regulates the frequency in demand generation imbalances. If the inertia is more in the power system, then it is less sensitive to small power imbalances. The inertia in the power system provides energy for some time to reduce the frequency deviations.

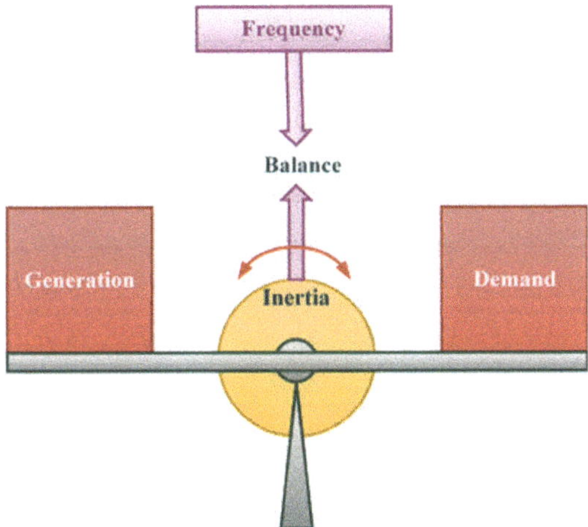

Figure 3. Power balance in a power system network.

In order to balance the power generation and demand, several control techniques are employed in a power system in multiple time frames as shown in Figure 4. The frequency response of a power system is comprised of the following steps [10]:

1. Inertial response (a few seconds).
2. Primary frequency response—Governor Response (1–10 s).
3. Secondary frequency response—Area Governor Control response (seconds to minutes).

When the power imbalance occurs, the frequency starts to fall. The Rate of Change of Frequency (ROCOF) is more at the initial stage. ROCOF is associated with how much inertia is present in the system; the inertia slows down the initial frequency deviation, and this is called the inertial frequency response. The governor response is the primary frequency technique which acts within the first few seconds (typically 10–30 s) after a frequency event and seeks to reduce the frequency deviation. As the frequency reaches a minimum value (the frequency nadir); then the governor control is activated. Hence, the active power output increases and the frequency settles at a point slightly below the nominal value [11].

As the penetration level of RES increases, the frequency deviations are more frequent. As these RES are connected to the grid through a power electronic inverter, substituting the conventional SG with power electronic inverters will decrease the inertia of the power system. To handle the frequency stability issues raised from the low inertia and reserve power, new frequency control techniques need to be employed for RES to participate in a frequency regulation process.

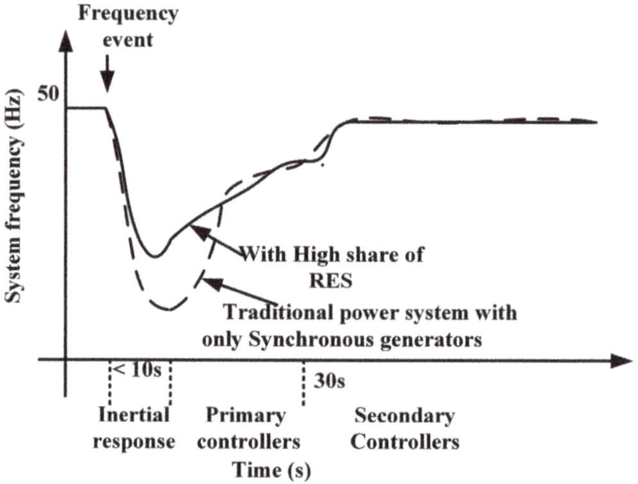

Figure 4. Frequency response in a power system network.

2.2. Voltage Rise and Fluctuation

It is considered that electricity is distributed at the consumer's terminal within the tolerable limit. The normal allowable voltage range is ±6% of the nominal value [12]. When large RES is integrated with lightly loaded feeders, the impact is unbearable. Whenever there is any change in load, the voltage fluctuation occurs at PCC. The voltage fluctuations and dips occur due to earth leakage faults and earth short-circuits located in the electrical power system (EPS). These faults weaken the voltage quality at PCC, based on fault situations. This is a significant element, particularly for solar and wind energy sources which possess irregular characteristics due to the wind speed disparity and solar irradiance that changes with time. The sensitivity of both electrical and electronic appliances that contributes to the life span deficiency of maximum devices is due to the voltage deviation [13].

Impact of RES on Voltage Drops in the Grid

Voltage is a significant factor in EPS. Hence, voltage control requirements are essential in EPS for both transmission and distribution levels. The source voltage and voltage drop at the feeder is determined at the end of the feeder [14]. The voltage drop in the feeder is due to the conductor impedance, current flow, and load. Also, the drop should not be lower when in a peak load situation, and it should not be more than the maximum voltage in a light load condition.

$$\Delta V = V_1 - V_2 = \frac{R_{LN}(P_L - P_G) + X_{LN}(Q_L - Q_G)}{V_2} \tag{1}$$

Equation (1) shows the injection of reactive power from RES within the grid; the RES will continually diminish the drop at the feeder. Where ΔV is the voltage between bus 1 & 2, P and Q are active and reactive power, and R_{LN} & X_{LN} are the line resistance and reactance, respectively. The grid is incorporated with RES and the load is shown in Figure 5. If the supply power is lower than the demand, the RES will inject power into the grid. Hence, this process is subject to RES active/reactive power that is relevant to the load active/reactive power and line X/R ratio [15].

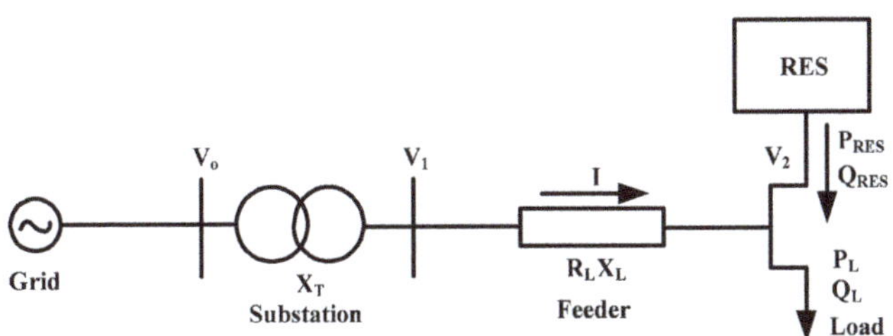

Figure 5. Grid connected system with RES and load.

3. Solutions for Variable RES

Various control techniques need to be employed to increase the power generation from RES and to decrease the negative impact of the RES on the power grid. This section gives a brief overview of the frequency and voltage control techniques for RES.

3.1. Frequency Control Techniques for RES

The frequency control techniques employed for RES are shown in Figure 6. The frequency control techniques used for both wind and solar are discussed in this section. For RES-based power plants, the frequency can be controlled by reserving the active power by using the de-load operation or by using the energy storage system (ESS). Generally, Inertia and frequency control methods for RES are categorized in two ways: control techniques to de-load the RES, and control practices for RES with ESS.

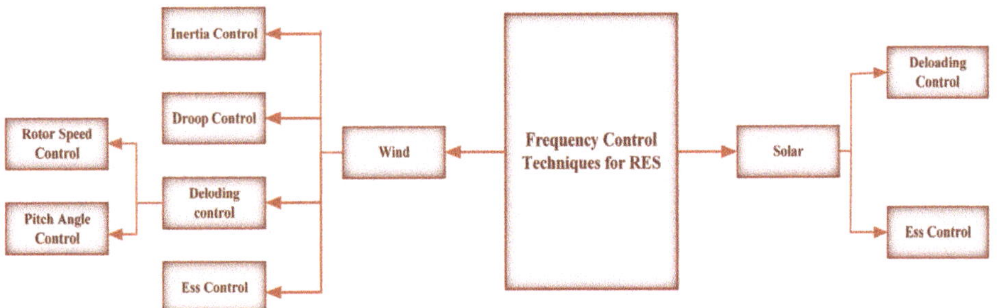

Figure 6. Frequency control techniques for RES.

3.1.1. Control techniques Used for Wind Turbines

The frequency control techniques for the wind turbine can be classified in three ways, i.e., 1. Inertia control, 2. Droop control, 3. De-loading control.

3.1.2. Inertia Control

The inertia of the wind turbines needs to be emulated to maintain the frequency stability under the high penetration of wind power generators. The inertia of the wind turbines can be emulated by either "hidden" inertia emulation or by the fast power reserve emulation.

(a)   Hidden inertia emulation

The emulation of hidden inertia present in the wind turbines is used to reduce the frequency deviation and to maintain the frequency stability. The suitable control algorithm for the power electronic converter for the wind turbines allows the wind turbine to release the KE stored in the rotating blades. The KE stored in the blades of the wind turbine helps to regulate the frequency under unbalance condition through its inertial response [16]. There are two ways to emulate the inertial response, either by considering the ROCOF alone or by incorporating both ROCOF and frequency deviation. Figure 7 shows the control diagram for the inertia emulation, considering only ROCOF to release the KE stored in the wind turbine. The inertia constant (H) is used to express the inertial characteristics. The "hidden" inertia of wind turbines can be known as,

$$H = \frac{J\omega_{nom}^2}{2S} \tag{2}$$

where J the inertia of the wind turbine is, S is the VA rating of the machine, ω_{nom} is the nominal angular frequency.

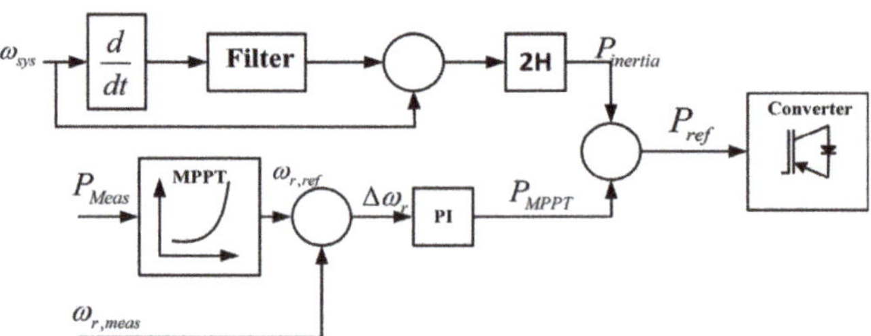

Figure 7. Hidden inertia emulation control [17].

The inertial active power control signal $P_{inertia}$ of the hidden inertial emulation control is known as,

$$P_{inertia} = 2H * \omega_{sys} * \frac{d\omega_{sys}}{dt} \tag{3}$$

where ω_{sys} the angular frequency of the system is, ω_r is the reference angular frequency.

The inertia control algorithm with both ROCOF and frequency deviation is shown in Figure 8. In the power balanced condition, the active set power is monitored by the MPPT controller. Under power imbalance/frequency disturbance conditions the control algorithm acts and produces the extra inertial power signal. In this controller, the inertial power is calculated from both the ROCOF loop and frequency deviation loop. The inertial power calculated from the Figure 8 is known as,

$$P_{inertia} = K_I \frac{df}{dt} + K_{droop}\Delta f \tag{4}$$

where K_I and K_{droop} are the inertia and drooping gains respectively.

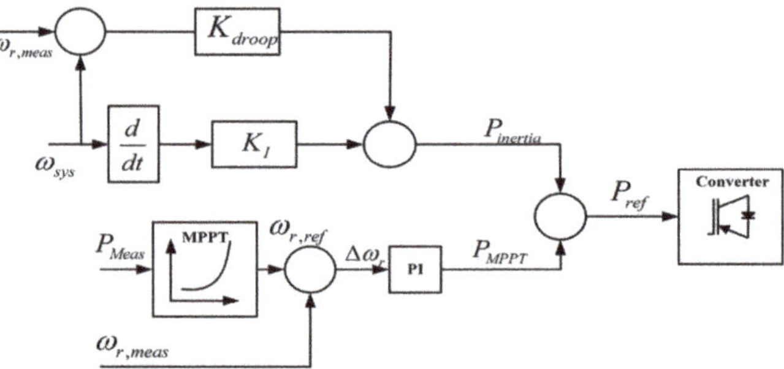

Figure 8. Inertia emulation control with droop control [18].

The wind generators can store and release KE instantly compared to the SG, due to the power electronic converter controller. The variable speed wind turbines can participate more in the frequency regulation by releasing more KE than fixed speed wind turbines and SG [19].

(b) Fast Power Reserve

Usually, the emulated inertia for the wind turbines can be determined based on the frequency deviation or ROCOF, as illustrated in the above section. Whereas the fast power reserve is defined as the constant active power support regardless of the wind speed [16]. The fast power characteristics are shown in Figure 9. The fast power reserve is the temporary power, released from the KE stored in the rotating blades of the wind turbine.

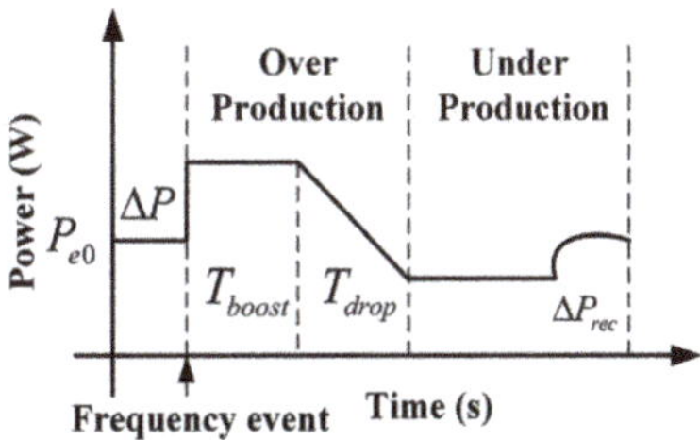

Figure 9. Fast power reserve characteristics.

The control diagram of fast power reverse is shown in Figure 10. This fast power reserve can be realized by regulating the set value of the rotor speed. This is given by,

$$P_{Const,t} = \frac{1}{2}J\omega_{r,0}^2 - \frac{1}{2}J\omega_{r,t}^2 \tag{5}$$

where t ($t < t_{max}$) is the lasting time of the fast power reserve since the beginning of the frequency event, $\omega_{r,0}$ is the rotor speed at the start and $\omega_{r,t}$ is the rotor speed at t, $P_{Const,t}$ is the constant active power available at t, hence the refence angular speed in this method can be calculated as ω_{ref}.

$$\omega_{ref} = \sqrt{\omega_{r,0}^2 - 2\frac{P_{Const,t}}{J}} \tag{6}$$

The fast power reserve control operates when the frequency deviation is more than the threshold value. This control delivers the extra power in the frequency event through the KE stored in the rotor and can be known as "over-production". To recover the KE after the event is known as "under-production". The shifting of an over-production state to an under-production state must be in a sloped manner, to avoid a sudden dip in the active power [20].

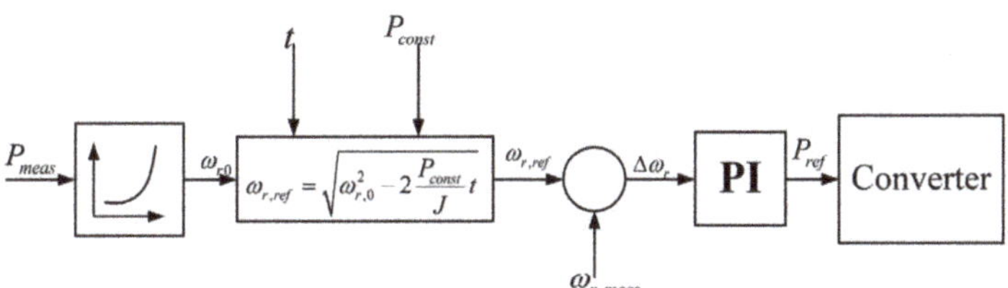

Figure 10. Fast power reserve control.

3.1.3. Droop Control for Frequency Regulation

The droop control of a wind turbine adjusts the active power response based on the frequency deviation. The droop controller shown in Figure 11 decreases the frequency nadir. Regulated active power and frequency is linearly related and it is shown in Figure 12.

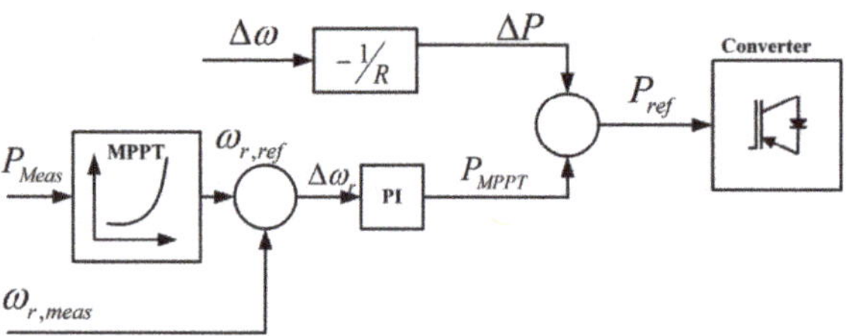

Figure 11. Droop control for wind turbine.

When the frequency drops from f_{ref} to f_{cal}, the wind generator increases the output of power from P_0 to P_1 to regulate the frequency deviation. Hence, the active power regulated by the droop control can be given as,

$$\Delta P = P_1 - P_0 = -\frac{f_{meas} - f_{nom}}{R} \tag{7}$$

where R is the droop coefficient, f_{meas} and P_1 are the measured frequency and wind turbine output power, respectively, while f_{nom} and P_0 are the initial operating points.

The analytical method for estimating the influence of inertia and droop responses from wind for frequency control was presented [21]. The droop control for wind turbine is shown in Figure 12. In order to use the wind turbine in frequency regulation under higher penetration, the adaptive gains for both the inertia, and droop controller was proposed in [22] by Yuan-Kang Wu. The inertia and droop gains are altered from time to time depending on the frequency imbalance.

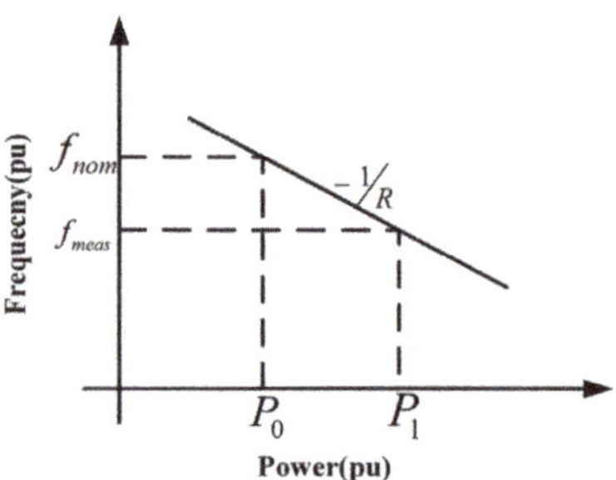

Figure 12. Droop characteristics.

3.1.4. De-Loaded Operation in Wind Turbines

The control techniques presented in this section would help to eliminate the adverse impact of the higher RES penetration level on the frequency stability. This section addresses the inertia and frequency control techniques with the de-loaded operation of RES. The reserve power in RES-based generators from de-loaded operation can be utilized for the inertial response, and primary frequency support. Generally, the wind turbines are operated with a maximum power point tracking technique and this does not contribute to frequency regulation. To maintain the stability of the power system, the wind turbines need to participate in the frequency regulation with the increasing penetration level of wind in the power system. The de-loaded control techniques enable the wind turbines in frequency regulation. The de-loaded operation of wind generators for the fast frequency reserve was initially proposed in [23]. Some of the authors [9] proposed maintaining the active power reserve for high wind speeds, and the reserve being unavailable in lower wind speeds. Another way to maintain the reserve power for the wind generators is by de-loading them in low wind speeds. Although, the reserve is not available above the wind-rated speeds [24]. The author of [25–28] proposed the de-loaded operation of a wind generator over the entire speed range, either in lower wind speeds or in higher wind speeds. The power and rotor characteristics for the de-loaded operation of wind turbine are shown in Figure 13.

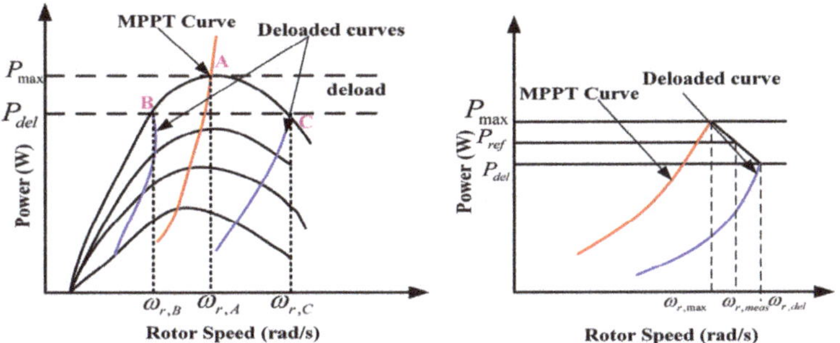

Figure 13. De-loaded operation for wind turbine.

(a) Speed Control

The de-loading operation of the wind turbine can be realized by changing the operating point from the maximum power point (P_{mpp}) to a sub-optimal power point ($P_{Suboptimal}$). The power can be varied from P_{sub} to P_{mpp} by regulating the rotor speed. The speed control method controls the tip speed ratio (λ) and it is known as,

$$\lambda = \frac{\omega_r R}{v} \tag{8}$$

where R the rotor radius and v is is the velocity of wind.

By controlling the speed ratio (λ) the operating point of the wind turbine can be altered and is shown in Figure 14.

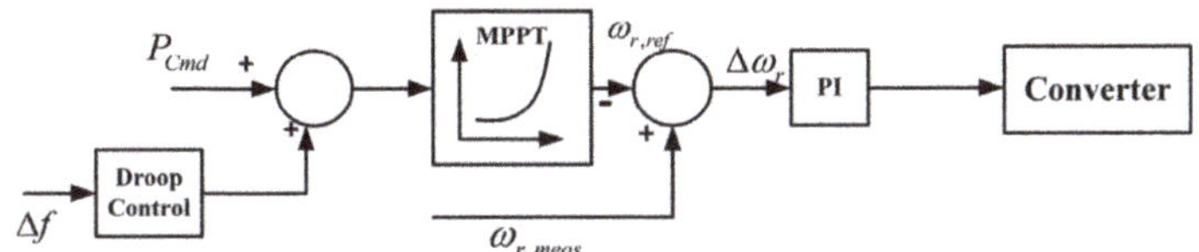

Figure 14. Speed control for wind turbine.

Therefore, the reference power given for the wind turbine can be known as [29],

$$P_{ref} = P_{del} + (P_{max} - P_{del}) * \left[\frac{\omega_{r,del} - \omega_{r,meas}}{\omega_{r,del} - \omega_{r,max}} \right] \tag{9}$$

where P_{max} is maximum power, P_{del} is de-loaded power, $\omega_{r,max}$ is rotor speed at P_{max}, $\omega_{r,del}$ is de-loaded rotor speed at P_{del}.

In the rotor speed control of wind turbines, there are two different possibilities to regulate the active power output: over-speeding and under-speeding of the turbine. In case of over-speed mode, the wind turbine delivers the KE until the operating point has reached the maximum power point (P_{MPP}). Whereas, in under-speed mode the wind turbine absorbs the KE until the operating point has reached the maximum power point [30]. The under-speeding of the rotor control results in stability problems. Hence, over-speeding of the rotor control is used.

The active power reference from Equation (9) uses a de-loaded power extraction curve shown in Figure 14. This technique regulates output of the active power from the wind turbine under a frequency event. Generally, the speed control technique is appropriate for low wind speeds.

The rotor speed control of the wind turbine is presented in [23] and regulates the active power output under a frequency event. In this paper, the author utilized the optimum power extraction curve to regulate the active power of a wind turbine. The power transferred to the grid is dependent on the slip for a doubly-fed induction generator. If the slip is increased, then the power transferred from the wind generator to the grid is increased.

(b)   Pitch Angle control

Another type of de-loading control for the wind turbines is created by changing the blade angle, known as "pitch angle control". The pitch angle control can be applied to both variable speed and fixed speed wind turbines. Figure 15 illustrates the power-speed characteristics of a wind turbine for different pitch angles.

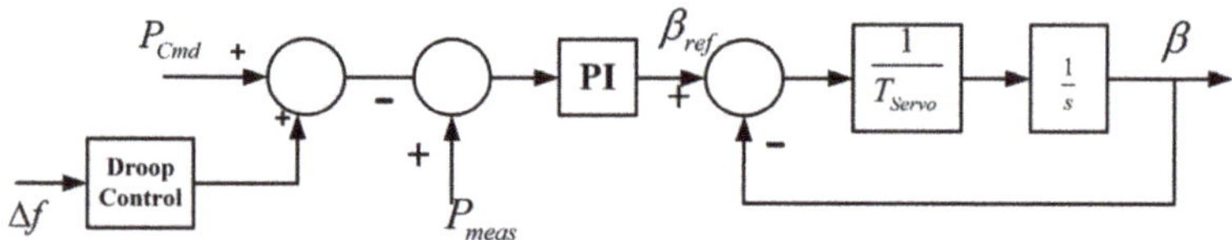

Figure 15. Pitch angle controller.

To de-load the wind turbine in pitch angle control, the pitch angle (beta) needs be increased to receive the active power reserve which is shown in Figure 16. Whenever the frequency event occurs, the pitch angle controller increases the value of beta and the operating point without changing the rotor speed. In pitch angle control, the pitch angle of the wind turbine is set at a sub-optimal point and reaches optimum value under frequency events to deliver/absorb active power. Pitch angle control is suited to high wind speed conditions [31].

Generally, the de-loading control can be selected depending upon the wind speeds. In low wind speeds, the de-loading is realized by the rotor speed control. The coordinated control of both pitch angle and rotor speed control is needed in medium wind speed conditions. The pitch angle control alone is used in high wind speed conditions.

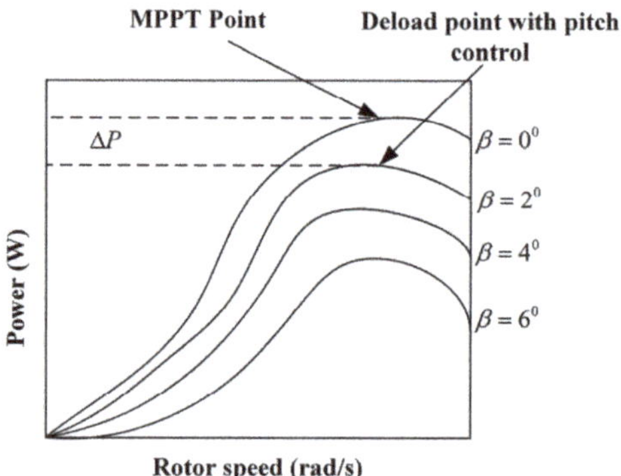

Figure 16. De-loaded curves with pitch angle control for wind turbine.

3.1.5. Solar Photovoltaic Array in Frequency Regulation

The installed photovoltaic (PV) generators do not contribute to the active reserve power for frequency regulation. All the PV systems are operated at the maximum power point using MPPT algorithms and do not alter their active power output when load variation occurs. The PV systems also need to participate in the frequency regulation to maintain the power system stability, with an increasing penetration level of PV system in the power system. The frequency regulation is employed in two ways for the PV generators:

1. De-loaded operation of PV.
2. Use of ESS.

In solar photovoltaic arrays, both the inertial and primary frequency response can be implemented by using de-loaded operation of PV and ESS with proper control algorithms.

3.1.6. De-Loaded Operation of a PV Generator [32]

In the de-loaded operation of PV, the power electronic converters need to inject the required amount of power under imbalance conditions depending on the control signal generated in Figure 17.

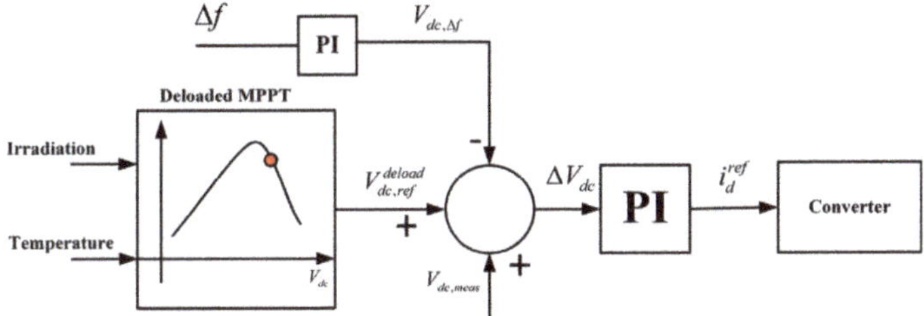

Figure 17. De-loaded operation control for solar photovoltaic (PV).

The solar power plants need to be operated under a sub-optimal power point below the maximum power point to maintain the active power reserve for the PV plants, as is shown in Figure 18. When the frequency event occurs, the operating voltage of the PV is decreased from the V_{del} to V_{MPP}, to increase the active power output. This reserve power will not be released until the frequency is deviated. When the frequency deviation occurs, the DC-link voltage is changed based on the frequency. The active power reserve available under de-loaded operation is known as,

$$P_{reserve} = P_{MPP} - P_{del} \tag{10}$$

The voltage reference calculated in the de-loaded operation is,

$$V_{dc,ref} = V_{MPP} + V_{del} - V_{dc}\Delta f \tag{11}$$

where, V_{dc} is the DC-link voltage.

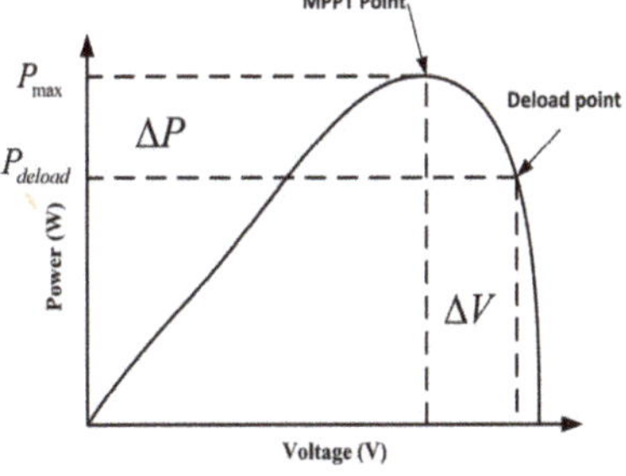

Figure 18. Frequency response in power system network.

3.2. Voltage Control Techniques

The control methods for voltage are discussed in two ways: (1) Low-voltage ride through (LVRT) and (2) High-voltage ride through (HVRT). The power disturbance in the grid is higher because of the output power from RES is irregular. Hence, LVRT should be attributed to the large capacity of grid-connected RES. The main purpose of LVRT is restricting the PCC current of the RES inverter, and power elements of RES inverter should not be damaged and tripped. The RES inverter is operated at 1.1 times the rated current in a short time, and it can supply a reactive current to support the grid voltage recovery. Thus, the control strategy of LVRT for RES is particularly dependent upon controlling the current output of the inverter during a grid fault [33]. Hence, to improve the power quality and stable operation of the grid, the RES is associated with the power system to serve the grid recovery from grid fault and to sustain the regulation of grid voltage and frequency.

Developing the penetration of distributed generation (DG) systems into the grid, one of the major problems is sudden tripping of DG from the grid during faults occurring in the system such as power outages and voltage flickers. Hence, DG units need to support the grid during fault conditions. The reactive power support diagram is shown in Figure 19. According to the theory of instantaneous reactive power, the active and reactive currents are controlled by varying the amplitude and phase of the output voltage of the inverter. In daytime situations, the active power output and the reactive power compensation (RPC) of the system are obtained concurrently [34,35]. When the PV power is not available, the RPC characteristic of the RES can be employed to enhance the utilization factor of the system.

The main factors for RPC in a system are:

(1)   Regulating the voltage,
(2)   Improving the stability of the system,
(3)   Minimizing power losses,
(4)   Effective utilization of machines associated with the system.

During the LVRT, the RES has the ability to be associated with the grid when the voltage at PCC drops to a prescribed cut-out point. When the drop is below the prescribed point, the RES can switch from the grid. Hence, a certain amount of reactive power is needed to provide effective voltage support to overcome this fault situation [36].

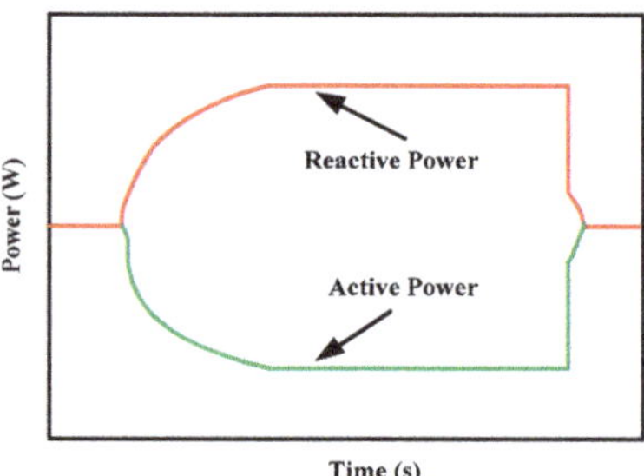

Figure 19. Reactive power support diagram.

There are various causes for the PCC voltage to be above/below rated values, such as the load has been suddenly disconnected, earth faults, and the irrational control strategies. During HVRT conditions, The RES should consume a required reactive power in order to make the PCC voltage become lower. The amount of absorbing power is determined by the capacity of the inverter and voltage at the above level of the prescribed cut-off point. Hence, the active power supplied from RES should not be changed, so the RES inverter can utilize its capacity to consume reactive power [37].

The voltage requirements during the VRT at PCC are shown in Figure 20. The HVRT handles the grid faults, which depends on voltage rise. The HVRT functionality is used for each of the three voltage phases. Hence, the functionality is actuated when the voltage rises higher than the voltage level determined by the utility grid. The control requirements for LVRT and HVRT equations are presented in Equations (12)–(14).

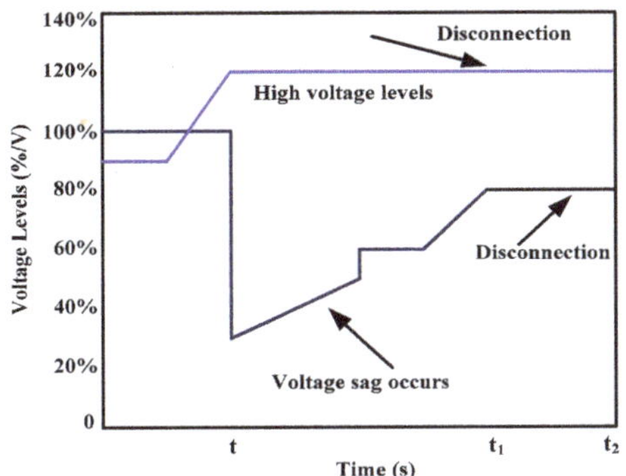

Figure 20. Voltage requisites during the low-voltage ride through (LVRT) and high-voltage ride through HVRT at PCC [38].

The equation is to satisfy the demand requirement as follows,

$$I_d^2 + I_q^2 \leq I_{max}'^2 \tag{12}$$

where I_{max} is maximum reference current from RES system, and I_d and I_q are the active and reactive currents. During a fault situation, the reactive support should meet the demand of Equation (12), and the active and reactive currents are set as follows,

$$I_d = \begin{cases} I_{d,set} \\ or \; min\left(I_{d0}, \sqrt{I_{max}'^2 - I_q^2}\right) \end{cases} \tag{13}$$

$$I_{q,refH} = \frac{\sqrt{I_{max}'^2 - \left(KI_{d,refH}\right)^2}}{K} \tag{14}$$

where I_{d0} is the pre-fault current and I_{drefH} is the reactive reference current. The active power control activities are presented in Table 2.

Table 2. Active power control activities [38].

Time	Working State	Principle
$(0, t)$	steady state	Basic control principle
(t, t_1)	steady state to fault state	I_d is from (12)
(t_1, t_2)	recovery state	recovery rate set between t_1 to t_2
$(t_2, \sim)$	steady state	Basic control principle

The control strategy of LVRT and HVRT is shown in Figure 21. In this [38], the control strategy is decided by the voltage at PCC. The I_{dref} and I_{qref} are generated various working positions, and in that way to understand the LVRT and HVRT.

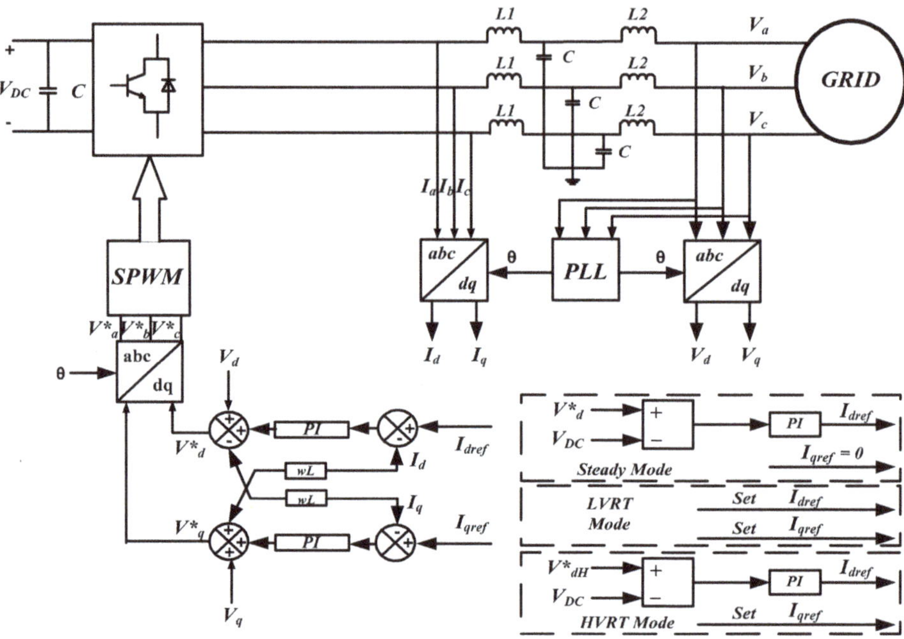

Figure 21. Control diagram of the LVRT and HVRT [38].

4. ESS Support

The above section considers the various controllers for both solar and wind and discusses the maintenance of the power system stability without ESS. RES is intermittent in nature and may lead to stability issues if we completely rely on these sources for frequency stability. Hence, ESS needs to be used along with the RES to maintain the stability of power system under high penetration of RES. The frequency of the power system can be balanced by using an energy storage device. Whenever there is an imbalance between the demand and generation, the control algorithm acts at the ESS to deliver the required amount of the active power. The ESS support for RES issues diagram is shown in Figure 22.

The performance of ESS in the smart grid is an effort towards balancing generation, consumption, reduction of variations in RES, and also delivers high quality supply and reliability of the supply. There has been an evolution of the enhanced intelligent electricity network called "smart grid", and it has a prominent contest of supporting all the sources connected with effective load control, powered from large penetration of non-dispatchable RES. These irregular energy sources are given higher levels of grid perception. Hence, ESS is the most significant smart grid advanced element to produce stable power at the grid level. ESS produces vital benefactions in defeating the complexity of irregular variation created by RES. It compensates the variation and lessens the variance among supply and demand [39]. ESS consists of pumped hydro-storage, compressed air energy storage (CAES), flywheel, superconducting magnetic energy storage (SMES), battery, and capacitors that are supposed to be widespread in RES integrated to the grid. ESS employs a power transformation system to integrate into the system network; they can add or receive both active and reactive power to balance for voltage changes in a short or medium duration. The ESS integration with the grid will increase the power quality, reliability, voltage support, backup power and decrease losses. When the irregular energy source reaches higher levels of grid penetration, ESS is the solution for providing a reliable energy supply [40,41]. Hence, proper coordination is needed for micro-grid (MG) and ESS operations. The main control methods among MG and ESS are the voltage, frequency, and active and reactive power controllers. Thus, ESS can sustain a balance within the generation and load at the instant of operation, and can also produce a firming potential of RES [42].

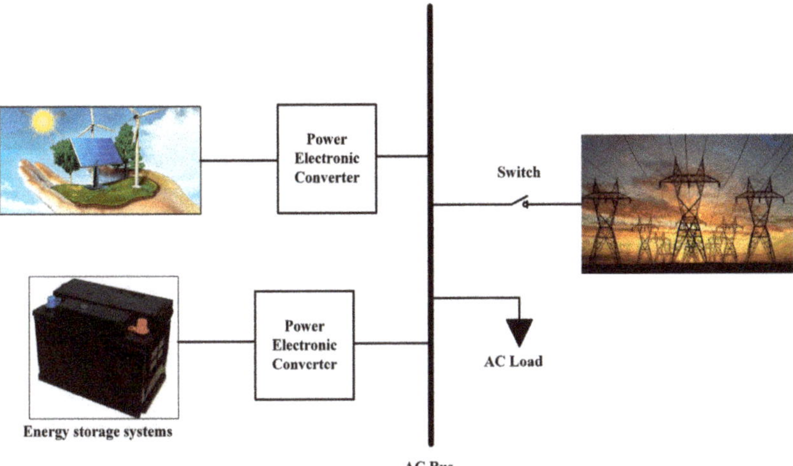

Figure 22. Energy Storage System (ESS) support for RES issues.

5. Smart Grid Features in Voltage Control with RES

Smart grid is being increasingly used in practice by electric utilities and involves the use of information, communication and control abilities to enhance the performance of the grid. The fundamental concept of a smart grid involves developing analysis, control, monitoring abilities of the traditional grid system, and reducing energy consumption. These objects are feasible over a system that can yield accurate and precise monitoring of the grid. A promising selection to minimize the challenges created by variations in RES is to add ESS into the grid. Another approach is to achieve flexible operation in energy consumption by employing demand side integration (DSI) or the usage of the micro-grid system. Moreover, the RES, ESS, and DSI are listed as one of the DERs. Hence, combining various aspects of certain sources is collectively essential in enhancing the utility of RES in the energy market [43].

6. Conclusions

The large-scale integration of RES into the power grid will have an impact on the stability of the power system. Both the frequency and voltage stability are highly affected by integrating large scale RES into the grid. This paper presented the issues related to high integration of RES in detail and reviewed their possible solutions available in the literature. To minimize frequency and voltage related issues, the several control techniques that are applied to wind and solar-based generators are summarized. Furthermore, ESS support could be used to maintain the stability and reduce the negative impacts related to grid integration as discussed.

Author Contributions: Initial idea purpose, data collection, formal analysis, original draft writing and editing: G.V.B.K. and R.K.S.; resources, supervision, review and editing: K.P., P.S., and J.B.H.-N.

Funding: No source of funding was attained for this research activity.

Acknowledgments: The authors would like to acknowledge the support and technical expertise received from the center for Bioenergy and Green Engineering, Department of Energy Technology, Aalborg University, Esbjerg, Denmark which made this publication possible.

Conflicts of Interest: The authors declare no conflict of interest.

References

1. Ulbig, A.; Borsche, T.S.; Andersson, G. Impact of low rotational inertia on power system stability and operation. *IFAC Proc.* **2014**, *19*, 7290–7297. [CrossRef]
2. Patel, M.R. *Wind and Solar Power Systems: Design, Analysis, and Operation*; CRC Press: Boca Raton, FL, USA, 2005; pp. 87–101.
3. Zhong, Q.C. Virtual Synchronous Machines: A unified interface for grid integration. *IEEE Power Electron. Mag.* **2016**, *3*, 18–27. [CrossRef]
4. Hales, D. Renewables 2018 Global Status Report, Renewable Energy Policy Network. 2018. Available online: http://www.ren21.net/status-of-renewables/global-status-report/ (accessed on 20 April 2019).
5. Shah, R.; Mithulananthan, N.; Bansal, R.C.; Ramachandaramurthy, V.K. A review of key power system stability challenges for large-scale PV integration. *Renew. Sustain. Energy Rev.* **2015**, *41*, 1423–1436. [CrossRef]
6. Eriksen, P.B.; Ackermann, T.; Smith, P.; Winter, W.; Garcia, J.M. System operation with high wind penetration. *IEEE Power Energy Mag.* **2005**, *3*, 65–74. [CrossRef]
7. Huang, Z.M. Research of the problems of renewable energy orderly combined to the grid in smart grid. In Proceedings of the Asia-Pacific Power and Energy Engineering Conference (APPEEC), Chengdu, China, 28–31 March 2010.
8. Lee, C.-T.; Hsu, C.-W.; Cheng, P.-T. A Low-Voltage Ride- Through Technique for Grid-Connected Converters of Distributed Energy Resources. *IEEE Trans. Ind. Appl.* **2011**, *47*, 1821–1832. [CrossRef]
9. Zaman, B.A. 100\% Variable Renewable Energy Grid: Survey of Possibilities. Master's Thesis, University of Michigan, Ann Arbor, MI, USA, 2018.
10. Tamrakar, U.; Shrestha, D.; Maharjan, M.; Bhattarai, B.; Hansen, T.; Tonkoski, R. Virtual Inertia: Current Trends and Future Directions. *Appl. Sci.* **2017**, *7*, 654. [CrossRef]
11. Cochran, J.; Denholm, P.; Speer, B.; Miller, M.; Cochran, J.; Denholm, P.; Speer, B.; Miller, M. *Grid Integration and the Carrying Capacity of the US Grid to Incorporate Variable Renewable Energy*; National Renewable Energy Laboratory: Golden, CO, USA, 2015.
12. Carvalho, P.M.S.; Correia, P.F.; Ferreira, L. Distributed reactive power generation control for voltage rise mitigation in distribution networks. *IEEE Trans. Power Syst.* **2008**, *23*, 766–772. [CrossRef]
13. El-Tamaly, H.H.; Wahab, M.A.A.; Kasem, A.H. Simulation of directly grid-connected wind turbines for voltage fluctuation evaluation. *Int. J. Appl. Eng. Res.* **2007**, *2*, 15–39.
14. Katiraei, F.; Agüero, J.R. Solar PV integration challenges. *IEEE Power Energy Mag.* **2011**, *9*, 62–71. [CrossRef]
15. Viawan, F.A.; Sannino, A.; Daalder, J. Voltage control with on-load tap changers in medium voltage feeders in presence of distributed generation. *Electr. Power Syst. Res.* **2007**, *77*, 1314–1322. [CrossRef]
16. Knudsen, J.N.N.H. Introduction to the modelling of wind turbines. In *Wind Power in Power Systems*; Wiley: Chichester, UK, 2005.
17. Sun, Y.; Member, S.; Zhang, Z.; Li, G.; Lin, J. Review on Frequency Control of Power Systems with Wind Power Penetration. In Proceedings of the International Conference on Power System Technology, Hangzhou, China, 24–28 October 2010; pp. 1–8. [CrossRef]
18. Nguyen, H.T.; Yang, G.; Nielsen, A.H.; Jensen, P.H. Frequency Stability Enhancement for Low Inertia Systems using Synthetic Inertia of Wind Power. In Proceedings of the 2017 IEEE Power & Energy Society General Meeting, Chicago, IL, USA, 16–20 July 2017.
19. Keung, P.; Li, P.; Banakar, H.; Ooi, B.T. Kinetic Energy of Wind-Turbine Generators for System Frequency Support. *IEEE Trans. Power Syst.* **2009**, *24*, 279–287. [CrossRef]
20. El Itani, S.; Member, S.; Annakkage, U.D.; Member, S.; Joos, G. Short-Term Frequency Support Utilizing Inertial Response of DFIG Wind Turbines. In Proceedings of the 2011 IEEE Power and Energy Society General Meeting, San Diego, CA, USA, 24–29 July 2011; pp. 1–8. [CrossRef]
21. Ye, H.; Pei, W.; Qi, Z. Analytical Modeling of Inertial and Droop Responses From a Wind Farm for Short-Term Frequency Regulation in Power Systems. *IEEE Trans. Power Syst.* **2016**, *31*, 3414–3423. [CrossRef]
22. Wu, Y.; Yang, W.; Hu, Y.; Dzung, P.Q. Frequency Regulation at a Wind Farm Using Time-Varying Inertia and Droop Controls. *IEEE Trans. Ind. Appl.* **2019**, *55*, 213–224. [CrossRef]
23. Ekanayake, J.; Holdsworth, L.; Jenkins, N. Control of DFIG wind turbines. *Power Eng.* **2003**, *117*, 28–32. [CrossRef]

24. De Almeida, R.G.; Lopes, J.A.P. Participation of doubly fed induction wind generators in system frequency regulation. *Power Syst. IEEE Trans.* **2007**, *22*, 944–950. [CrossRef]

25. El Mokadem, M.; Courtecuisse, V.; Saudemont, C.; Robyns, B.; Deuse, J. Experimental study of variable speed wind generator contribution to primary frequency control. *Renew. Energy* **2009**, *34*, 833–844. [CrossRef]

26. Pradhan, C.; Bhende, C.N. Enhancement in Primary Frequency Regulation of Wind Generator using Fuzzy-based Control. *Electr. Power Components Syst.* **2016**, *44*, 1669–1682. [CrossRef]

27. El Mokadem, M.; Courtecuisse, V.; Saudemont, C.; Robyns, B. Fuzzy Logic Supervisor-Based Primary Frequency Control Experiments of a Variable-Speed Wind Generator. *IEEE Trans. Power Syst.* **2009**, *24*, 407–417. [CrossRef]

28. Zertek, A.; Member, S.; Verbi, G.; Member, S.; Pantoš, M. A Novel Strategy for Variable-Speed Wind Turbines' Participation in Primary Frequency Control. *IEEE Trans. Sustain. Energy* **2012**, *3*, 791–799. [CrossRef]

29. Vidyanandan, K.V.; Senroy, N. Primary Frequency Regulation by Deloaded Wind Turbines Using Variable Droop. *IEEE Trans. Power Syst.* **2013**, *28*, 837–846. [CrossRef]

30. Ghosh, S.; Member, S.; Senroy, N. Electromechanical Dynamics of Controlled Variable-Speed Wind Turbines. *IEEE Syst. J.* **2015**, *9*, 639–646. [CrossRef]

31. Moutis, P.; Member, G.S.; Loukarakis, E. Primary Load-Frequency Control from Pitch- Controlled Wind Turbines. In Proceedings of the 2009 IEEE Bucharest PowerTech, Bucharest, Romania, 28 June–2 July 2009; pp. 1–7. [CrossRef]

32. Rahmann, C.; Castilo, A. Fast frequency response capability of photovoltaic power plants: The necessity of new grid requirements and definitions. *Energies* **2014**, *7*, 6306–6322. [CrossRef]

33. Bae, Y.; Vu, Y.-K.; Kim, R.-Y. Implemental Control Strategy for Grid Stabilization of Grid-Connected PV System Based on German Grid Code in Symmetrical Low-to-Medium Voltage Network. *IEEE Trans. Energy Convers.* **2013**, *28*, 619–631. [CrossRef]

34. Yu, H.; Pan, J.; Xiang, A. A multi-function grid-connected PV system with reactive power compensation for the grid. *Solar Energy* **2005**, *79*, 101–106. [CrossRef]

35. El Moursi, M.S.; Xiao, W.; Kirtley, J.L. Fault ride through capability for grid interfacing large scale PV power plants. *IET Gener. Transm. Distrib.* **2013**, *7*, 1027–1036. [CrossRef]

36. Firouzi, M.; Gharehpetian, G.B. LVRT Performance Enhancement of DFIG-Based Wind Farms by Capacitive Bridge-Type Fault Current Limiter. *IEEE Trans. Sustain. Energy* **2018**, *9*, 1118–1125. [CrossRef]

37. Xie, Z.; Zhang, X.; Zhang, X.; Yang, S.; Wang, L. Improved Ride-Through Control of DFIG during Grid Voltage Swell. *IEEE Trans. Ind. Electron.* **2015**, *62*, 3584–3594. [CrossRef]

38. Fan, S.; Chao, P.; Zhang, F. Modelling and simulation of the photovoltaic power station considering the LVRT and HVRT. *J. Eng.* **2017**, *2017*, 1206–1209. [CrossRef]

39. Li, C.; Wang, R. Building integrated energy storage opportunities in China. *Renew. Sustain. Energy Rev.* **2012**, *16*, 6191–6211. [CrossRef]

40. Basu, A.K.; Chowdhury, S.P.; Chowdhury, S.; Paul, S. Microgrids: Energy management by strategic deployment of DERs—A comprehensive survey. *Renew. Sustain. Energy Rev.* **2011**, *15*, 4348–4356. [CrossRef]

41. Kumar, G.B.; Kumar, G.A.; Eswararao, S.; Gehlot, D. Modelling and control of bess for solar integration for pv ramp rate control. In Proceedings of the 2018 International Conference on Computation of Power, Energy, Information and Communication (ICCPEIC), Chennai, India, 28–29 March 2018; pp. 368–374.

42. Michael, M.; Robert, S.; Percy, H.; Russ, N.; Robert, Y. Islands in the storm: Integrating microgrids into the larger grid. *IEEE Power Energy Mag.* **2013**, *11*, 33–39.

43. Roscoe, A.J.; Ault, G. Supporting high penetrations of renewable generation via implementation of real-time electricity pricing and demand response. *IET Renew. Power Gener.* **2010**, *4*, 369–382. [CrossRef]

Article

A Hybrid PV-Battery System for ON-Grid and OFF-Grid Applications—Controller-In-Loop Simulation Validation

Umashankar Subramaniam [1], Sridhar Vavilapalli [2], Sanjeevikumar Padmanaban [3,*], Frede Blaabjerg [4], Jens Bo Holm-Nielsen [3] and Dhafer Almakhles [1]

[1] Renewable Energy Lab, College of Engineering, Prince Sultan University, Riyadh 12435, Saudi Arabia; usubramaniam@psu.edu.sa (U.S.); dalmakhles@psu.edu.sa (D.A.)
[2] Department of Power Electronics, Bharat Heavy Electricals Limited (BHEL), Bengaluru 560026, India; sridhar.spark@gmail.com
[3] Center for Bioenergy and Green Engineering, Department of Energy Technology, Aalborg University, 6700 Esbjerg, Denmark; jhn@et.aau.dk
[4] Center of Reliable Power Electronics (CORPE), Department of Energy Technology, Aalborg University, 9220 Aalborg, Denmark; fbl@et.aau.dk
* Correspondence: san@et.aau.dk

Received: 24 December 2019; Accepted: 7 February 2020; Published: 9 February 2020

Abstract: In remote locations such as villages, islands and hilly areas, there is a possibility of frequent power failures, voltage drops or power fluctuations due to grid-side faults. Grid-connected renewable energy systems or micro-grid systems are preferable for such remote locations to meet the local critical load requirements during grid-side failures. In renewable energy systems, solar photovoltaic (PV) power systems are accessible and hybrid PV-battery systems or energy storage systems (ESS) are more capable of providing uninterruptible power to the local critical loads during grid-side faults. This energy storage system also improves the system dynamics during power fluctuations. In present work, a PV-battery hybrid system with DC-side coupling is considered, and a power balancing control (PBC) is proposed to transfer the power to grid/load and the battery. In this system, a solar power conditioning system (PCS) acts as an interface across PV source, battery and the load/central grid. With the proposed PBC technique, the system can operate in following operational modes: (a) PCS can be able to work in grid-connected mode during regular operation; (b) PCS can be able to charge the batteries and (c) PCS can be able to operate in standalone mode during grid side faults and deliver power to the local loads. The proposed controls are explained, and the system response during transient and steady-state conditions is described. With the help of controller-in-loop simulation results, the proposed power balancing controls are validated, for both off-grid and on-grid conditions.

Keywords: battery; cascaded H-Bridge; chopper; energy storage; multi-level; PV inverter

---

1. Introduction

A low voltage (LV)-rated solar PCS containing a conventional inverter with two-level topologies is preferable for PV systems rated for lower power. For higher power solar power stations, it is better to opt for the system with medium voltage (MV) rating. Multi-level inverters (MLI) are more suitable for MV applications. The cascaded H-bridge (CHB) inverter is a popular MLI configuration which requires isolated DC sources/DC links. Hence CHB configuration is highly suitable for static compensator (STATCOM) and solar applications due to the availability of isolated DC links [1]. CHB inverter needs multiple H-bridge modules, and to control the multiple H-bridge modules, the required input-output channels in the processor are more when compared to other MLI configurations, but in MV high power systems, the CHB inverter enables the independent maximum power point (MPP) controls to attain enhanced efficiency [2]. In high power applications, the number of H-bridge modules to be used in CHB is more which results in better power quality [3,4]. The transformer on the AC side can also be eliminated with this configuration.

Solar PV stations can reduce carbon emissions and provide clean energy but may not be able to supply the load requirements due to sudden changes in weather conditions and when the solar irradiation is weak. The system remains in an idle state during nighttime, which affects the utilization factor of the system drastically. Hence there is much attention to battery energy storage systems along with PV to reduce power disturbances in the system, to improve the stability, for providing continuous power to the load and for improving utilization factor of the system. In such hybrid PV-battery stations, power is transferred from PV array to battery and load during daytime and batteries transfer power to the load during night time. The battery storage system also improves the system dynamics for sudden weather changes [5,6]. Importance of hybrid PV-battery stations and operation strategy is discussed in [7]. A review of energy storage for large-scale PV systems, grid integration issues, stability concerns and the selection of batteries is available in [8]. With the battery storage system in PV applications, the utilization factor of the PCS can also be improved since the system can be made operational for all the time. To keep the grid power non-negative always, an approach called 'Solar Plus's which is the combination of energy storage, PV and load controls are presented in [9].

Cost of the battery storage systems is also decreasing over time due to the advancements in the areas of different battery technologies, battery charging and discharging methods. In [10], various battery types such as lithium-ion, lead-acid, aluminium–ion, sodium-sulphur (NaS), flow batteries, etc. suitable for large-scale PV ESS systems are explained and compared. Various discharge strategies are presented for grid-connected hybrid PV-battery applications in [11] and different battery storage technologies suitable for residential applications are also elaborated.

In a conventional grid-connected PV power conditioning system, the anti-islanding feature will be incorporated as a feature of protection. But as mentioned earlier, it may be required to operate the PCS in stand-alone mode also when there are frequent disconnections from the grid. In such scenarios, a power management system at a higher level of control architecture needs to isolate the grid from the critical loads.

Hybrid PV-battery stations can also be operated in standalone mode; hence it is possible to provide power to local critical loads during grid-side faults or during maintenance on the grid side. Such systems help in catering continuous power supply in remote locations which are not connected to the central grid or which are facing problems of regular power supply failures from the grid. Optimal design for a PV + Diesel + Battery storage hybrid system is presented in [12].

From the above, the following research gaps are identified:

√　In a conventional PV power station, the system remains in an idle state when irradiation is weak; hence the utilization factor (UF) of the system is very small. By incorporating ESS in the solar power system, the overall utilization factor of the system can be improved significantly.

√　Commercially available PV inverter configurations are based on two-level or three-level inverters with low voltage ratings. These configurations are not suitable for large scale applications, but the medium voltage systems are more suitable for high power applications. The use of conventional two-level inverter for high power applications results in poor power quality, large filter requirement and higher dv/dt across power switches.

√　A CHB-based PV inverter can nullify the disadvantages of conventional low voltage inverter configurations. An investigation on energy storage systems suitable for CHB inverters is required.

√　In previous works, controls for PV-battery hybrid system either in grid-connected mode or in standalone modes of operations are discussed but the system controls for operating in both modes are not covered.

Present work mainly focuses on medium voltage CHB-based power condition system to meet the large-scale PV system requirements. For the uninterrupted power supply and to improve the utilization factor of the system, battery energy storage is incorporated in to the PV power system. The following are the contributions of present work to address the research gaps in earlier systems:

√ A literature review is carried out to investigate various ESS configurations suitable for CHB based inverters. A power conditioning system based on a CHB MLI and battery charger based on a chopper with multiple battery banks is studied thoroughly due to the advantages of chopper-based systems.

√ A novel power balancing control (PBC) is proposed in this work which enables the system to operate both in standalone mode and in grid-connected mode, unlike earlier works where the controls proposed were for either grid-connected systems or for standalone systems only.

√ The controls proposed in this work also enable the smooth transition in changing modes of operation.

√ Controller-In-Loop simulations are carried out with the help of real-time simulator to validate the proposed power balancing controls.

This paper is organized, as explained below: In Section 2, ESS configurations suitable for CHB inverter and advantages of chopper-based ESS are explained. Operation of buck chopper and bi-directional chopper are also elaborated in this section. Section 3 covers the procedure for selection of components such as PV array, batteries, filter components and other accessories. Controls adapted for inverter and battery charger are discussed thoroughly in Sections 4 and 5, respectively. The setup for the controller-in-loop simulation validation is explained in Section 6 and results are presented in Section 7. Contributions and the conclusions of the present work are presented in Sections 8 and 9, respectively.

2. Chopper-Based ESS for CHB Inverter

An AC coupled ESS configuration suitable for CHB inverters based on a voltage regulator is presented in [13], but AC-side coupled configurations are not suitable for higher rated energy storage systems as the power quality may get affected due to the multiple interfaces connected on the grid side. DC-coupled systems are more suitable for large-scale energy storage systems and for obtaining better power quality. A dual active bridge (DAB)-based ESS configuration is presented in [14], it is it has been observed that independent power control through PCS and battery charger is possible with this configuration but the control is complex due to a greater number of power modules. Since each DAB needs eight gate pulses, the controller hardware requirements are more. To reduce the cost, controller hardware requirements and the control complexity, it is preferable to select chopper-based systems. In this work, two types of choppers namely buck-chopper and bi-directional choppers, are discussed to operate as a battery charger.

Figure 1 shows the circuit diagram for a buck chopper-based battery charger. In this system, PV voltage shall be more than battery voltage during charging mode. By regulating the duty cycle of IGBT-S1, the charging current is regulated during charging mode. The output filter is used to limit the ripple content of output current and voltage. During the nighttime, PV voltage tends to be less than the battery voltage. Then battery starts giving power to DC link through the third diode 'D3'. The changeover is instantaneous; hence the system response for transient conditions is also instantaneous. After the transition, the DC voltage is clamped to battery voltage. Since the above charger is operational in charging mode alone, the battery charger and filter inductor need to be selected for the charging current, i.e., usually less than the battery rated current. This reduces the cost of the system. Since the discharging of the battery is through D3, this diode is selected for full current to handle current during discharging mode.

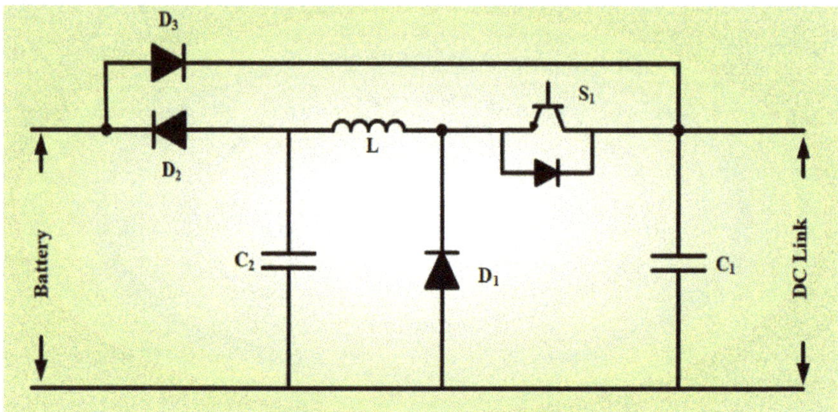

Figure 1. Battery charger based on a buck-chopper.

The main drawback with this type of charger is that the battery starts discharging only when the irradiation is weak. In grid-connected mode, the battery cannot transfer power to DC link even when the local load demand is higher than the power available at PV array since the PV array is operating at its MPP voltage which is more than the battery voltage. In this case, the power required for the local load needs to be taken from the grid. In standalone mode, the battery needs to provide power along with the PV source for the local critical load requirement due to the absence of the grid. In this case, the PV array cannot be operated at its MPP voltage but operates at battery voltage which reduces the efficiency of PV [15]. So, this charger is more suitable for grid applications. For standalone PV systems, it is required to have control during the discharging mode of operation also which is possible through bi-directional chopper based ESS.

Figure 2 shows a battery charger based on the bidirectional chopper. Unlike the buck chopper, the battery voltage can also be higher than the DC link in this configuration. Battery terminals need to be connected to output terminals of the charger in case the battery voltage is less than DC link. Similarly, battery terminals need to be connected to input terminals of the charger in a case when the battery voltage is more than DC link. This DC-DC converter needs two gate pulses, whereas buck-chopper based system requires only one gate pulse. Cost of this system is less than the DAB based system but slightly more than the buck-chopper based system since the battery charger is selected to the full capacity of the battery. With this charger, battery current in discharging mode can also be controlled hence the PV source can always be operated at MPP voltage in both on and off-grid operations.

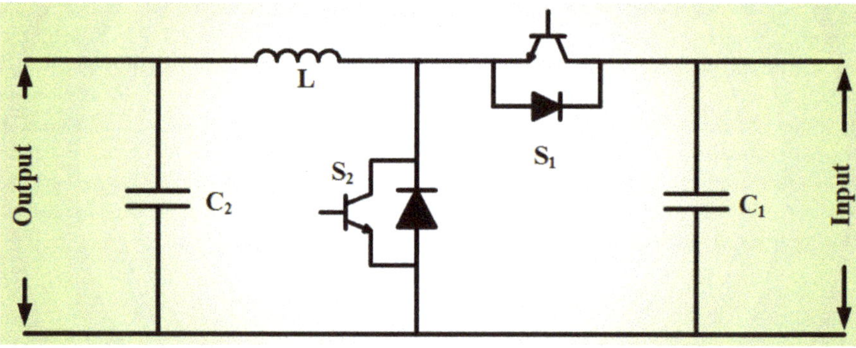

Figure 2. Bi-directional chopper-based battery charger.

3. Design Calculations for the System

In present work, a high-power PV system with a solar inverter based on a CHB MLI and a bi-directional chopper-based battery charger is studied. To discuss the controls proposed at various operating points and in different modes such as grid-connected and standalone modes, a single-phase system with ratings given in Figure 3 is chosen. In this work, a single-phase system is considered as a case study due to limitations in the controller hardware. However, the system controls can be upgraded to the three-phase system, which is suitable for high power applications.

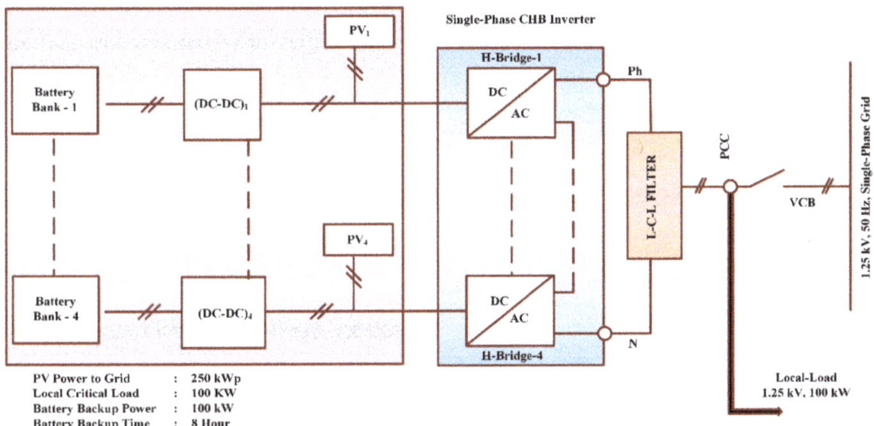

Figure 3. DC-coupled ESS for CHB-based PCS.

Design calculations for CHB inverters, selection of devices, device loss calculations etc. are discussed in [16,17]. The procedure for the selection of CHB inverter and AC side filter components is shown in Table 1.

Table 1. Design calculations for CHB Inverter and L-C-L Filter.

Electrical Parameter	Value		Remarks
CHB power rating	350	kW	PCHB = Grid + Load
Inverter voltage	1250	V	Vac
No. of H-bridges	4	No's	N_H
H-bridge power	87.5	kW	CHB Rating/N_H
H-bridge voltage	312.5	V	V_H = Vac/N_H
Output current (Iac)	280	A	P_{CHB}/Vac
Minimum DC voltage	422	V	Vdc_{min} = 1.35 × V_H
Carrier frequency	1	kHz	Fcr
Switching frequency	8	kHz	Fsw = 2 × N_H × Fcr
Corner frequency (Fc)	2	kHz	Fc selected = Fsw/4
L-C-L Filter Component Selection			
Maximum % reactive power allowed	5	%	Max reactive power through Filter Capacitor
Maximum filter voltage drops	3	%	Max voltage drop across filter Inductors
Maximum reactive power (Qc)	17.5	kW	5% of 350 kW
Current rating of the capacitor (Ic)	14	A	Qc/Grid Voltage
Maximum capacitance of filter capacitor	35.67	uF	Ic/(2 × pi × F × Grid Voltage)
Capacitance of filter capacitor selected	30	uF	
Inductance of filter inductor L1	211	uH	Fc = 1/[2 × pi × √(LC)]
% Voltage drop in L1	1.49	%	(Iac × 2 × pi × F × L1)/Grid Voltage
Maximum drop allowed in L2	1.51	%	Max % voltage drop–% drop across L1
Filter inductance L2	215	uH	

To cater to the power needed for local load and power to be transferred to the grid, the required power rating of CHB inverter is 350 kW. In this system, 4 H-bridges are selected hence the levels achieved on the output PWM voltage is nine. The voltage and the power ratings of each H-bridge are 1/4th of CHB ratings and the current rating is equal to the rated inverter output current i.e., 280 amperes. Minimum DC input required for each H-bridge is calculated based on the AC output. In this work, level-shifted PWM is adapted with a carrier frequency of 1 kHz. A filter is used at AC terminals of CHB MLI for obtaining better voltage and current THDs.

In this system, the minimum battery backup power is selected as 100 kW so that the critical load requirement can be met during the standalone mode of operation also. In this work, the battery is selected in such a way that the maximum voltage of the battery is less than the nominal voltage of PV array which is suitable for both the types of chopper based ESS configurations. The lithium-ion battery is selected with a minimum battery voltage depending on the minimum DC input required for each H-Bridge. Procedure for obtaining battery ampere-hour value is also shown in Table 2. After selecting the battery, the PV array is selected by considering that the nominal voltage of the PV array is always higher than the maximum battery voltage. Short circuit current and open-circuit voltage of the PV module is obtained from the datasheet. Minimum MPP voltage of PV array is at the maximum operating temperature range i.e., at 75 °C. Hence ten PV modules in the series are selected to meet the system requirements at 75 °C also. Procedure for selection of PV array is also explained in Table 2.

Table 2. Selection of battery and PV sources.

Electrical Parameter	Value		Remarks
Battery ratings			
Battery backup power	100	Kw	Equal to the critical load
Battery back-up time	8	Hr	T
No. of battery banks	4	No's	Nb = No. of H-bridges
The power rating of each battery bank (Pnom)	25	Kw	Total Power/Nb
Type of battery			Lithium-Ion
Minimum battery voltage (VB_min)	422	V	Minimum DC link voltage
Nominal battery voltage (VB_nom)	482	V	Vmin = 87.5% of Vnom
Maximum battery voltage (VB_max)	560	V	VB_max = 116% of Vnom
Nominal battery current (IB_nom)	52	A	Pnom/Vnom
AH Rating of each battery	416	AH	Inom × T
PV Array Ratings			
PV array power	450	Kw	Grid + load + battery
No of PV arrays	4	No's	Number of H-bridges
Rating of PV array	112.5	Kw	Total power/4
Minimum PV voltage	560	V	>VB_max
PV module M/s SunPower makes SPR-435NE-WHT-D			
PV module power	435	W	From datasheet
Number of modules	260	No's	112.5kw/435W
Isc of PV array	6.43	A	From datasheet
Voc of PV array	85.6	V	From datasheet
MPP voltage at 25 °C	72.9	V	From datasheet
MPP Voltage at 75 °C	61.2	V	From datasheet
No of series modules (Nse)	10	No's	Minimum PV voltage/MPP Voltage @ 75 °C
No of parallel modules (Np)	26	No's	No of modules/Nse
Nominal PV voltage at 25 °C	729	V	Vmpp_25 × Nse
Minimum MPP voltage in operating range	612	V	Vmpp_75 × Nse
Maximum PV current	167	A	Isc × np
The maximum power rating of each PV array	113	Kw	Nse XNp × Pmodule

Input source to the battery charger is a PV array and the battery is connected to the output terminals. In the case of a buck chopper-based system, the battery charger rating is decided based on charging current which is obtained from the charging time. Ratings of bi-directional chopper-based battery chargers are obtained by a maximum of battery charging and discharging currents. In this work, battery charging and discharging times are selected equal, to maintain equal ratings for

Buck-Chopper and bi-directional chopper-based battery chargers. The switching frequency of 5 kHz is selected for the IGBT based battery chargers. An L-C filter is connected to chopper output terminals. Design calculations for filter inductance and capacitance are also shown in Table 3.

Table 3. Design calculations for the battery charger.

Electrical Parameter	Value		Remarks
Selected Battery Ratings			
AH rating of each battery	416	AH	Inom × T
Back-up time (Tb)	8	Hr	
Charging time (Tch)	8	Hr	
Nominal battery voltage	482	V	(VB_nom)
Battery charging current (Ich)	52	A	AH Rating/Tch
Battery dis-charging current (Idisc)	52	A	AH Rating/Tb
Battery Charger Ratings			
Rating of buck chopper	25	kW	Vnom × Ich
Rating of a bidirectional chopper	25	kW	Vnom × Idisc
Input side voltage (Vin_dc)	856	V	Voc of PV Array
Minimum output voltage (Vout_dc)	422	V	(VB_min)
L-C Filter Design for Chopper			
Minimum duty cycle (D)	0.49		Vout_dc/Vin_dc
Switching frequency selected	5	kHz	Fsw_brc
Maximum off time (Toff)	101	us	Calculated from D and Fsw_brc
Ripple current allowed (di)	2.5	A	5% of rated current is selected
Inductance of filter inductor	16	mH	L = Vout_dc * Toff/di
Ripple voltage allowed (dv)	24	V	5% of rated output voltage
Capacitance of filter capacitor	270	uF	C = (dv/Toff)/output current

4. Inverter Controls in Different Modes of Operation

Figure 4 shows the building blocks of the CHB inverter which consists of PV switch, IGBT based H-Bridge, bypass module, and transducers. Each building block is fed by independent PV arrays and also connected to independent batteries through chopper-based DC-DC converters. Transducers in the H-bridge module measure PV array voltage and currents for independent MPPT controls. A bypass switch module consisting of antiparallel thyristors and a bypass contactor is used along with H-bridge. Bypass module bypasses the H-bridge during fault so that the system can continue to be operated at reduced power rating.

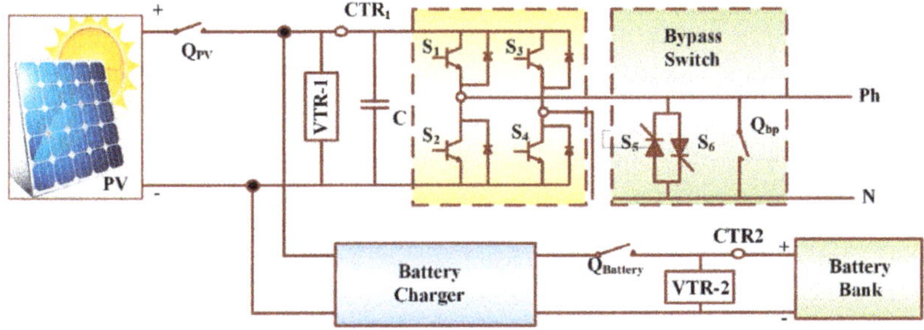

Figure 4. The basic building block of a CHB-based PCS.

In [18–20], regulation of power flow through CHB MLI based PCS are discussed. The control diagram for the power regulation through the inverter is as shown in Figure 5. In grid-connected mode, closed-loop current regulation is implemented and closed-loop voltage regulation is adapted during the standalone mode.

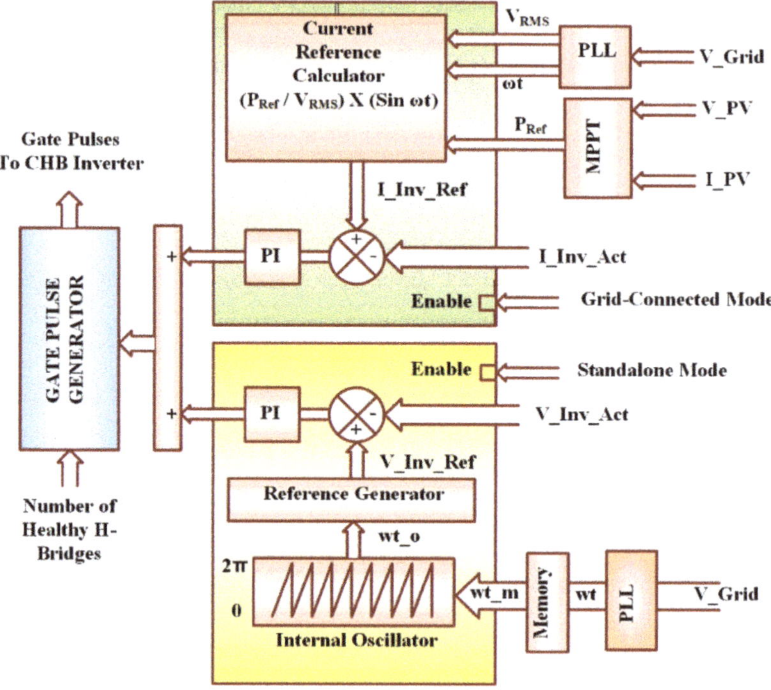

Figure 5. Control diagram for the CHB inverter in hybrid PV-ESS.

In the grid-connected mode of operation:

- Closed-loop voltage regulators are disabled and the closed-loop current regulators are enabled.
- Through a phase-locked loop (PLL), the phase angle of grid voltage 'wt' is obtained.
- PV Current (I_PV) and PV voltage (V_PV) are monitored for MPP tracking to obtain power reference (P_Ref) for the inverter.
- Reference inverter current (I_Inv_Ref) is obtained from 'P_Ref', grid voltage, and 'wt',
- By comparing actual and reference inverter currents, current regulators provide a modulating signal.
- Based on the number of healthy H-bridges, carrier waves are generated in the gate pulse generator module
- By comparing carrier and modulating signal, gate signals for CHB MLI are obtained.

In standalone mode, closed-loop voltage control is adapted to provide the rated voltage to the load. For smooth changeover in the mode of operation, it is required to match the phase angle of inverter voltage with the earlier grid voltage. Following activities are carried out in the inverter controls when the system is in standalone mode:

- The grid voltage is monitored and the phase angle 'wt' of grid voltage is obtained through PLL.
- At the moment of occurrence of grid fault, the latest value of angle 'wt_m' of grid voltage is stored and the internal oscillator is enabled.
- The internal oscillator generates a phase angle 'wt_o' which starts from the value of stored phase angle wt_m.
- On receiving the phase angle input, reference generator gives inverter reference voltage signal to the closed-loop voltage controller.

- By comparing actual and reference inverter voltages, the voltage regulators provide a modulating signal.
- Based on the number of healthy H-bridges, carrier waves are generated in the gate pulse generator module
- By comparing carrier and modulating signal, gate signals for CHB MLI are obtained.

5. Battery Charger Controls in Different Modes of Operation

With a bi-directional chopper-based battery charger, regulation of battery current in charging and discharging modes is possible which allows the PV array to operate at its MPP in all operating conditions. In grid-connected mode, the power available at PV array is used for the charging of the battery and the balance power is transferred to the grid/load as long as irradiance is available. In standalone mode, the power available at PV array power is transferred to the grid/load and the balance power if any is used for charging the battery as long as irradiance is available. In both these modes, if the battery is fully charged or if the power demand on the grid side is more, then the battery can also feed the additional power demand. In this work, battery voltage less than PV voltage is selected, hence the converter acts like a buck-chopper during charging mode and acts like a boost-chopper during discharging mode. Battery charging is always through current control whereas discharging of the battery can be carried out either through the current controller or uncontrolled. Figure 6 shows the control diagram of the bi-directional charger.

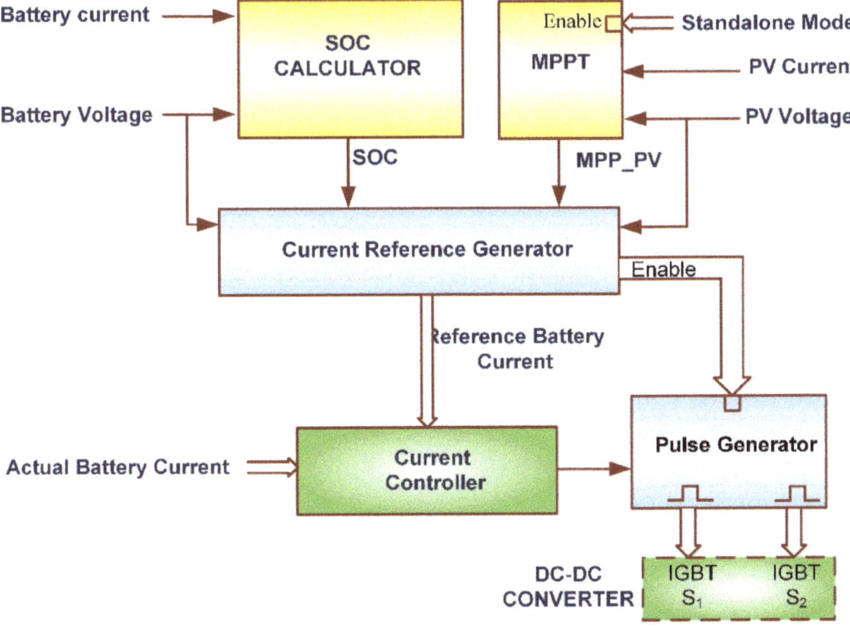

Figure 6. Control diagram for the bi-directional chopper in PV-ESS.

In grid-connected mode:

- When PV voltage is more than the minimum nominal voltage of PV array, battery charging/discharging current reference is obtained based on battery SOC.
- Convention of battery current is taken as positive during charging mode and negative during discharging mode.

- During charging mode, the current controller compares the actual battery current with reference battery current and generates gate pulse so that the chopper operates as a Buck-Chopper.
- When the reference battery current is negative and the PV voltage is more than the minimum nominal voltage of PV array, then the battery discharging current is controlled through the current controller.
- When the reference battery current is negative, and if the PV voltage is less than the minimum nominal voltage of the PV array, then the pulse generator is disabled. No gate pulse is given to IGBTs and the battery discharges through the inductor (L) and the diode across IGBT-S1. In this mode, the current through the battery charger varies based on the reference inverter power.

In the standalone mode of operation:

- If PV voltage is more than the minimum nominal voltage of PV array, battery charging current reference is obtained based on battery SOC and represented as Ibatt_SOC.
- Another battery current reference is generated through MPPT controls and represented as Ibatt_MPP.
- The minimum value in Ibatt_SOC and Ibatt_MPP is selected as a current reference for the charger.
- While charging the battery, MPPT controls are carried out in the battery charger and adjust the battery reference current to maintain the MPP of the PV array.
- Current reference may also become negative to supply power from the battery when the irradiance on PV array is poor. In this case, also, PV array operates at its MPP voltage.
- When the irradiance becomes zero, then the PV voltage becomes less than the minimum nominal voltage of the PV array and the pulse generator module is disabled. Then battery discharges through inductor L and the diode across IGBT-S1. Based on the load on the AC side, battery current varies.

In this work, a bi-directional chopper-based battery charger is considered for ease of understanding and due to limitations in the controller hardware. For high power applications interleaved buck-boost converters presented in [21] may be adapted with minor modifications in the control logic.

6. Validation of Control Algorithm for Chopper Based ESS Configurations

To validate the system, controller-in-loop simulation validation is adapted. With the controller-in-loop simulations, the control software can be tested prior to the site trials and can be tested at various operating points which are difficult to test with a real plant [22]. The following activities are carried out as part of controller-in-loop simulation validation:

- √ The power circuit also is known as the plant which consists of a grid, input breaker, local load, CHB inverter, PV array, batteries, and battery chargers are simulated as shown in Figure 7 with the help of real-time simulator library
- √ The simulated model is compiled and loaded in the high-speed processors of the real-time simulator
- √ Input-output channels of the real-time simulator are interfaced with the processor cards and other user interfaces. The real-time simulator enables the simulated plant to operate as a real plant
- √ The user interface consists of pushbuttons to provide start/stop commands and for mode selection. Potentiometers on the user interface are used for varying irradiance value on PV arrays
- √ Due to the limitations in the analogue inputs in the used processor, two processor cards of the same type are used in this work
- √ Control algorithm in processor cards is programmed through MatLab-embedded coder
- √ In controller card-1, the controls for inverter are programmed. It receives start command and mode selection from the user and receives module faulty signals from the plant as digital inputs. Analogue inputs such as grid voltage, inverter voltage, and currents are received from the simulated plant

√ From the simulated plant, controller-2 receives analogue inputs such as battery current, battery voltage, PV current, and PV. Based on the mode of operation, MPPT is carried out. Battery charger controls are programmed in this controller card

√ When the start command is given from the user interface, the controllers process the proposed control algorithm to control battery chargers and the CHB inverter

√ When the system is in grid-connected mode, based on MPPT of PV arrays, the reference power signal is communicated to controller card-1 through a serial communication interface (SCI). Transmission of data is by RS232 protocol

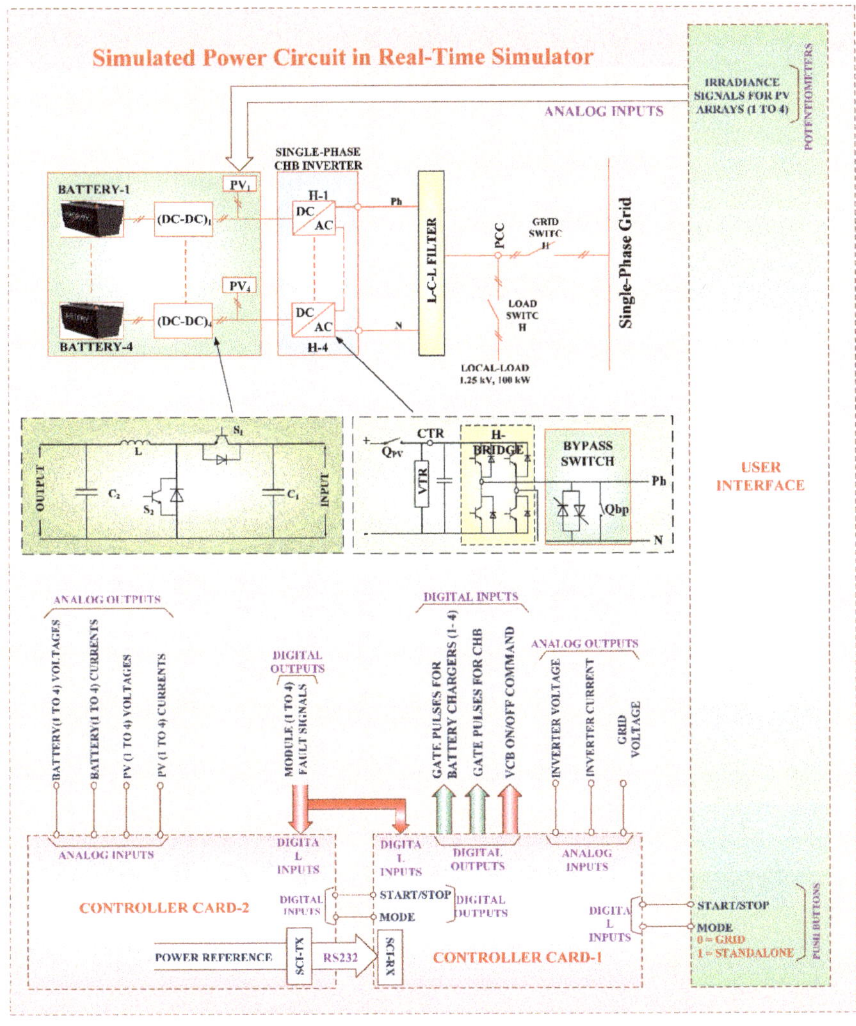

Figure 7. Block diagram of the controller-in-loop simulation setup.

Controller-in-loop simulation setup is shown in Figure 8, Table 4 lists the particulars of the real-time simulator hardware.

Figure 8. Controller-in-loop simulation setup.

Table 4. Hardware details of real-time simulator setup.

Electrical Parameter	Value
Real-time simulator	
Manufacturer name	Opal-RT
Processor	Intel Xeon Quad core, 2.5 GHz
Modelling platform	Matlab-Simulink
Analog channels	-10 to $+10$ V
Digital channels	0 V: Logic Low, +15V: Logic High
Controller Cards 1 & 2	
DSP processor	TI Make TMS320F2812
Analog input	-7.5 V to $+7.5$ V
Digital input channels	0 V: Logic Low, +15V: Logic High
Digital output channels	0 V: Logic Low, +15V: Logic High
No. of analog inputs	16 No's
No. of digital inputs	8 No's
No. of digital outputs	22 (Including PWM outputs)
User interface	
Potentiometers	To provide variable voltage to AI
Pushbuttons	To provide Digital inputs

7. Results and Discussion

In grid-connected mode, the advantage with the bi-directional chopper-based system is the possibility of discharging even when the PV array is active. The response of the inverter current is observed by operating the battery in charging mode and discharging modes intermittently. Irradiance is continuously maintained at 1000 watt per square meter hence the PV array current is constant. The battery reference current is varied from +50 Ampere to -50 Ampere and vice versa with an equal interval time of 1 s. When the battery current is negative, additional current is flowing through the inverter as shown in Figure 9. The dynamic response of the system in this condition is also found to be satisfactory.

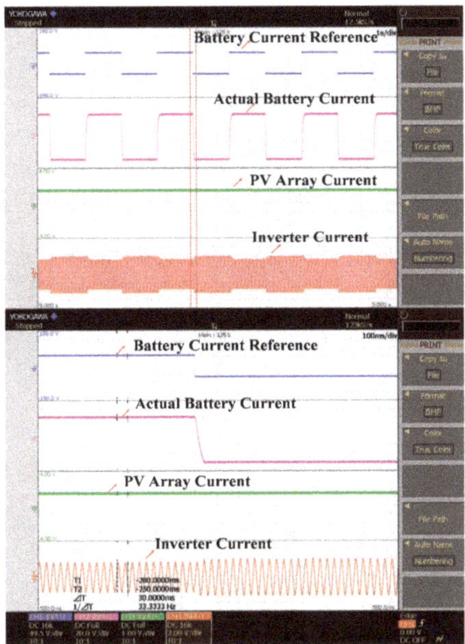

Figure 9. Change in the battery, PV array and actual inverter currents with a change in reference battery current at constant irradiance of 1000 watt per square meter in a bi-directional chopper-based ESS in grid-connected mode.

Irradiance on the PV arrays is varied from zero watts per square meter to 1000 watt per square meter in steps of 200 watts per square meter and the currents of inverter output, battery and PV array are observed.

As discussed earlier, when the irradiance is zero, the battery current is negative as it is in discharging mode and the inverter power is equal to the rated battery power and PV current is zero as displayed in Figure 10. MPP tracking is carried out in inverter controls hence the inverter current and PV currents are varied in proportion to the irradiance inputs whereas the battery current is almost constant.

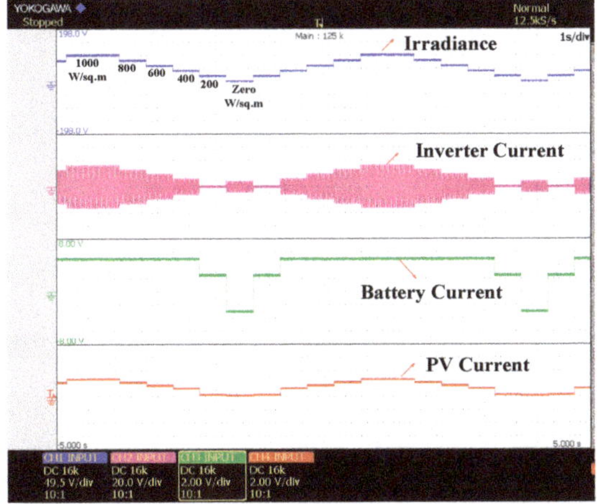

Figure 10. Inverter, battery and PV currents with varying irradiance value in grid-connected mode.

Irradiance on the PV arrays is varied from 350 watts/sq.m to 1000 watt per square meter in steps as shown in Figure 11 and inverter current is observed. When the grid side is healthy, the current required for charging the battery is provided by PV array and the remaining power is supplied to the grid/load. So, the current through inverter decreases with the reduction in irradiance. PV power is less than the power required for the battery charging when irradiance is 350 watts/sq.m; hence the inverter current is zero. As the irradiance value increases, the power transferred to the grid also increases and shows good dynamic during a sudden change in irradiance input.

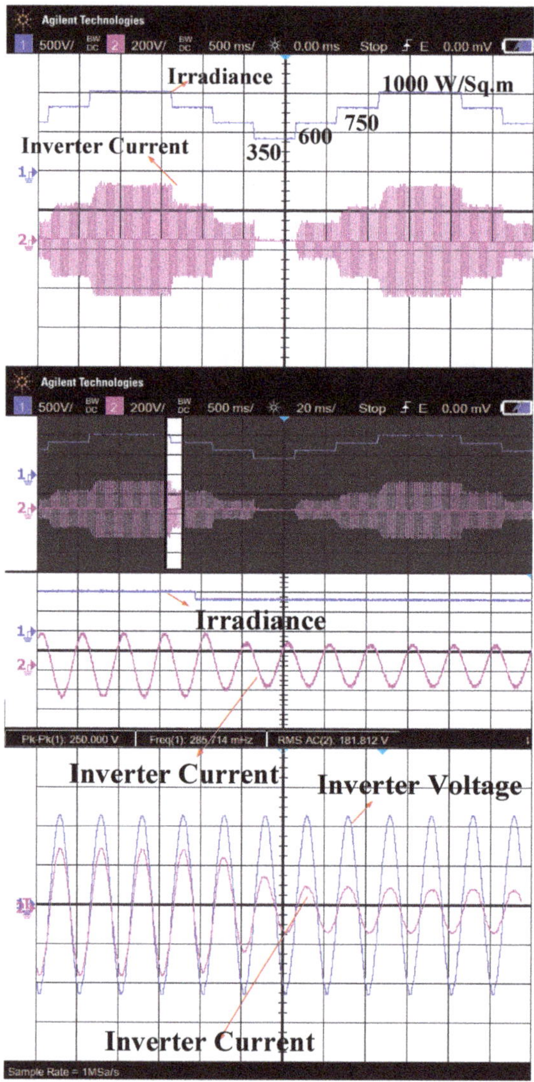

Figure 11. Change in inverter current with varying irradiance value in grid-connected Mode.

A fault on the H-Bridge module is created to verify system performance. The system continues to operate with reduced power as shown in Figure 12. The proposed algorithm enables the fault-tolerant operation instantaneously and the obtained results are in line with the simulation results presented in [1].

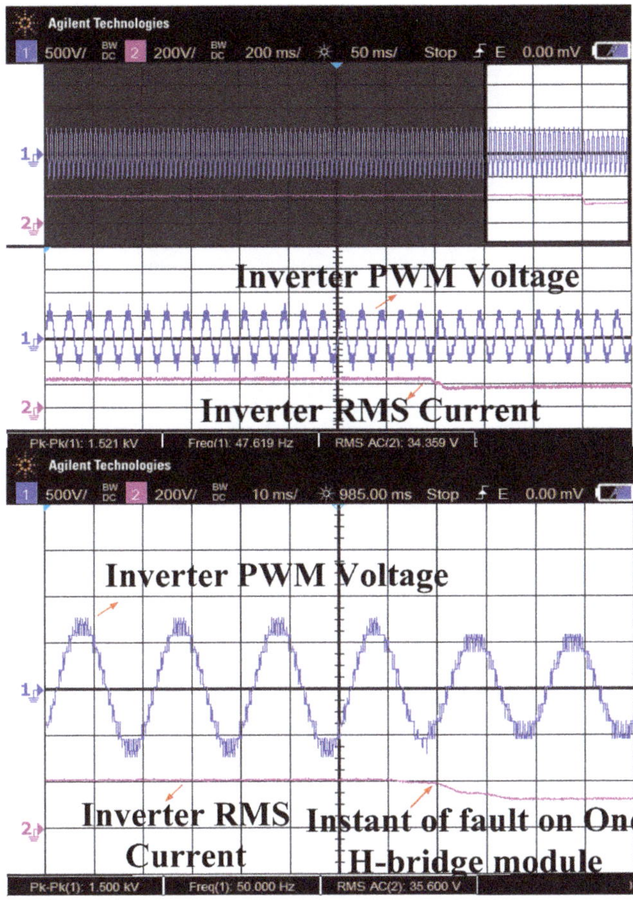

Figure 12. Inverter PWM voltage and RMS current during a fault in one H-bridge module

Mode of operation is changed from grid-connected to the standalone operation and the transients in the inverter AC voltage i.e., the input voltage to the load are observed. Smooth transition in the inverter voltage is observed during mode transfer as shown in Figure 13. The operation mode is changed from grid-connected to an off-grid mode for a bi-directional chopper-based system. Operation of the system for a fixed load is studied by varying irradiance value from zero to 1000 watt per square meter in steps of 200 Watts/sq.m. Since the load is fixed, inverter current is constant for all the values of irradiance inputs. MPP tracking is carried out in battery charger controls; hence the battery current varies with the irradiance value. At the instant when the PV power is not sufficient to meet the load demand then discharge current of the battery is regulated through current control. In this mode, DC link voltage has maintained the value of the MPP voltage of PV array. As shown in Figure 14, when irradiance value is zero, PV array voltage tends to zero hence the DC link is clamped to the battery voltage level. From the presented results it is observed that the dynamic response of the battery charger system is good as the settling time is in the range of 50 milliseconds, whereas the settling time is in the range of 200 milliseconds in the battery charger systems based on voltage regulator and dual active bridge configurations proposed in [13,14], respectively.

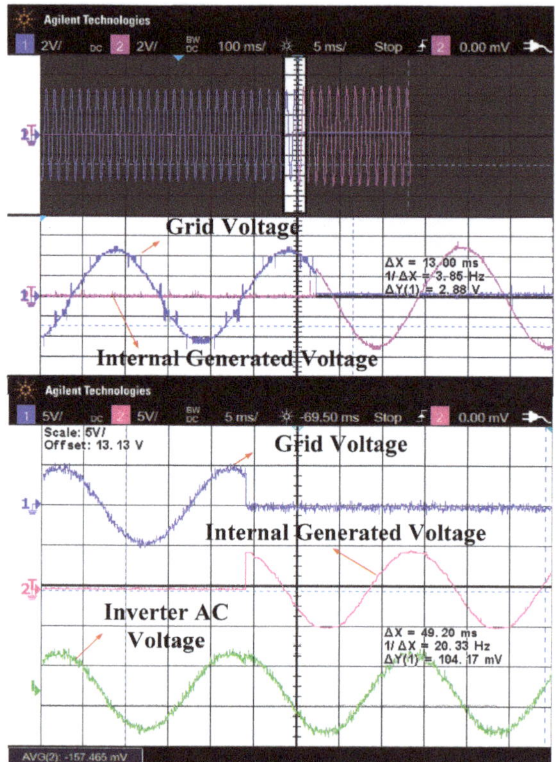

Figure 13. Grid voltage, internal oscillator voltage and inverter output voltage during a changeover in the mode of operation.

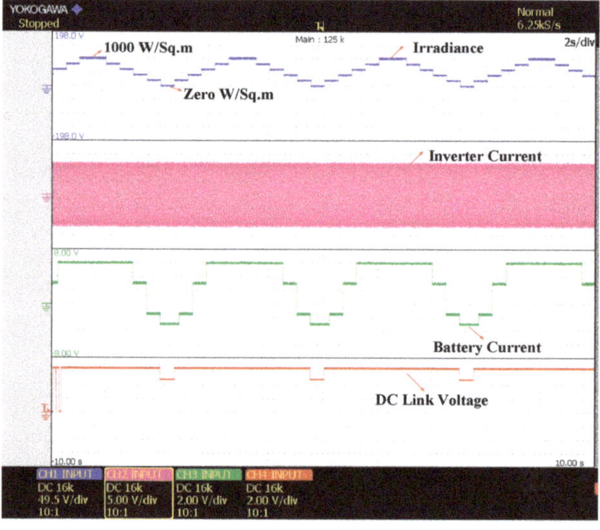

Figure 14. Change in battery current, inverter current and DC link voltages with a change in irradiance in standalone mode operation.

It is observed that the dynamic response of the solar PV inverter for sudden changes in irradiance input is found satisfactory as the settling time of inverter current is in the range of 50 milliseconds whereas the settling time is in the range of 300 milliseconds in PV inverters with improved perturb and observe method presented in [23].

8. Contributions and Future Scope

In this work, various energy storage system configurations suitable for cascaded H-bridge-based PV inverters are discussed. Design calculations for the chopper-based ESS for PV applications are presented in detail. Controls for the bi-directional chopper-based energy storage systems are studied and a control algorithm is developed and validated through real-time simulations. As future work, the proposed control algorithm can be validated on prototype models of the chopper-based systems or the AC/DC buck-boost converters proposed in [24]. The controls can be extended further for interleaved buck-boost converters to improve the power rating further. In this work, the operation of these systems is explained considering equal irradiances on each PV arrays and the state of charge of each battery bank is also considered equal. System operation for unequal irradiances can be studied as future work in line with the controls proposed in [25]. In present work, a single-phase system is considered; the controls can be extended to the three-phase system in future work. An additional feature of reactive power compensation can also be planned. The present system provides the solution for AC micro-grids for rural areas with frequent disruptions in grid supply. From earlier studies, it found that DC micro-grids are more feasible as PV sources, wind power generator and other renewable sources provide DC current and DC micro-grids are more stable compared to an AC micro-grid [26]. Feasibility of the energy storage system and its controls can be studied for DC grid systems proposed in [27].

9. Conclusions

In this paper, a chopper-based ESS suitable for CHB-based PCS is presented. With the proposed system, the power to the grid/load can be supplied without any interruption. Cost, control complexity and controller hardware requirements for the chopper-based system are less compared to other configurations such as voltage regulator-based ESS and DAB-based ESS configurations. Buck-chopper based ESS configuration identified to be more suitable for grid-connected system and the bi-directional chopper based ESS configuration is suitable for standalone operation as well. The controls proposed for the bi-directional chopper-based ESS is analyzed with the help of controller-in-loop simulations by using a real-time simulator. Good dynamic response of the system for sudden changes in irradiance is observed from the presented results. Smooth transition in the system controls is also achieved during mode change over.

Author Contributions: All authors were involved to articulate the research work for its final depiction as the full research paper. "Conceptualization, U.S, S.V, S.P.; Methodology, U.S., S.P., D.A.; Software, S.V.; Validation, S.P, F.B. and J.B.H.-N.; Formal Analysis, U.S.; Investigation, S.V.; Resources, J.H.N and F.B.; Data Curation, S.P., D.A.; Writing-Original Draft Preparation, U.S., S.P., S.V., D.A.; Writing-Review & Editing, F.B., and J.B.H.-N.; Project Administration, F.B., S.P., J.B.H.-N.; Funding Acquisition, S.P." All authors have read and agreed to the published version of the manuscript.

Funding: This research received no external funding.

Acknowledgments: The authors like to express sincere gratitude to Renewable Energy lab, Prince Sultan University, Saudi Arabia for making this research work for execution and technical implementation in real-time, and Center for Bioenergy and Green Engineering/Center of Reliable Power Electronics (CORPE), Aalborg University, Esbjerg/Aalborg, Denmark.

Conflicts of Interest: The authors declare no conflict of interest.

References

1. Sridhar, V.; Umashankar, S. A comprehensive review on CHB MLI based PV inverter and feasibility study of CHB MLI based PV-STATCOM. *Renew. Sustain. Energy Rev.* **2017**, *78*, 138–156. [CrossRef]
2. Yu, Y.; Konstantinou, G.; Hredzak, B.; Agelidis, V.G. Operation of Cascaded H-Bridge Multilevel Converters for Large-Scale Photovoltaic Power Plants under Bridge Failures. *IEEE Trans. Ind. Electron.* **2015**, *62*, 7228–7236. [CrossRef]
3. Dragonas, F.A.; Neretti, G.; Sanjeevikumar, P.; Grandi, G. High-Voltage High-Frequency Arbitrary Waveform Multilevel Generator for DBD Plasma Actuators. *IEEE Trans. Ind. Appl.* **2015**, *51*, 3334–3342. [CrossRef]
4. Sridhar, V.; Umashankar, S.; Sanjeevikumar, P.; Ramachandaramurthy, V.K.; Mihet-Popa, L.; Fedák, V. Control Architecture for Cascaded H-Bridge Inverters in Large-Scale PV Systems. *Energy Procedia* **2018**, *145*, 549–557. [CrossRef]
5. Chen, L.; Chen, H.; Li, Y.; Li, G.; Yang, J.; Liu, X.; Xu, Y.; Ren, L.; Tang, Y. SMES-Battery Energy Storage System for the Stabilization of a Photovoltaic-Based Microgrid. *IEEE Trans. Appl. Supercond.* **2018**, *28*, 1–7. [CrossRef]
6. Rallabandi, V.; Akeyo, O.M.; Jewell, N.; Ionel, D.M. Incorporating Battery Energy Storage Systems Into Multi-MW Grid Connected PV Systems. *IEEE Trans. Ind. Appl.* **2018**, *55*, 638–647. [CrossRef]
7. Yang, Y.; Ye, Q.; Tung, L.J.; Greenleaf, M.; Li, H. Integrated Size and Energy Management Design of Battery Storage to Enhance Grid Integration of Large-Scale PV Power Plants. *IEEE Trans. Ind. Electron.* **2018**, *65*, 394–402. [CrossRef]
8. Lai, C.S.; Jia, Y.; Lai, L.L.; Xu, Z.; McCulloch, M.D.; Wong, K.P. A comprehensive review on large-scale photovoltaic system with applications of electrical energy storage. *Renew. Sustain. Energy Rev.* **2017**, *78*, 439–451. [CrossRef]
9. O'Shaughnessy, E.; Cutler, D.; Ardani, K.; Margolis, R. Solar plus: Optimization of distributed solar PV through battery storage and dispatchable load in residential buildings. *Appl. Energy* **2018**, *213*, 11–21. [CrossRef]
10. Zhang, C.; Wei, Y.-L.; Cao, P.-F.; Lin, M.-C. Energy storage system: Current studies on batteries and power condition system. *Renew. Sustain. Energy Rev.* **2018**, *82*, 3091–3106. [CrossRef]
11. Olaszi, B.D.; Ladanyi, J. Comparison of different discharge strategies of grid-connected residential PV systems with energy storage in perspective of optimal battery energy storage system sizing. *Renew. Sustain. Energy Rev.* **2017**, *75*, 710–718. [CrossRef]
12. Rodríguez-Gallegos, C.D.; Gandhi, O.; Yang, D.; Alvarez-Alvarado, M.S.; Zhang, W.; Reindl, T.; Panda, S.K. A siting and sizing optimization approach for PV–battery–diesel hybrid systems. *IEEE Trans. Ind. Appl.* **2017**, *54*, 2637–2645. [CrossRef]
13. Vavilapalli, S.; Subramaniam, U.; Padmanaban, S.; Ramachandaramurthy, V.K. Design and Real-Time Simulation of an AC Voltage Regulator Based Battery Charger for Large-Scale PV-Grid Energy Storage Systems. *IEEE Access* **2017**, *5*, 25158–25170. [CrossRef]
14. Vavilapalli, S.; Padmanaban, S.; Subramaniam, U.; Mihet-Popa, L. Power Balancing Control for Grid Energy Storage System in Photovoltaic Applications—Real Time Digital Simulation Implementation. *Energies* **2017**, *10*, 928. [CrossRef]
15. Vavilapalli, S.; Umashankar, S.; Sanjeevikumar, P.; Fedák, V.; Mihet-Popa, L.; Ramachandaramurthy, V.K. A Buck-Chopper Based Energy Storage System for the Cascaded H-Bridge Inverters in PV Applications. *Energy Procedia* **2018**, *145*, 534–541. [CrossRef]
16. Sastry, J.; Bakas, P.; Kim, H.; Wang, L.; Marinopoulos, A. Evaluation of cascaded H-bridge inverter for utility-scale photovoltaic systems. *Renew. Energy* **2014**, *69*, 208–218. [CrossRef]
17. Wang, Z.; Fan, S.; Zheng, Y.; Cheng, M. Design and Analysis of a CHB Converter Based PV-Battery Hybrid System for Better Electromagnetic Compatibility. *IEEE Trans. Magn.* **2012**, *48*, 4530–4533. [CrossRef]
18. Kumar, N.; Saha, T.K.; Dey, J.; Barman, J.C. Modelling, control, and performance study of cascaded inverter based grid connected PV system. In Proceedings of the IREC2015 the Sixth International Renewable Energy Congress, Sousse, Tunisia, 24–26 March 2015; pp. 1–6.
19. Xiao, B.; Hang, L.; Mei, J.; Riley, C.; Tolbert, L.M.; Ozpineci, B. Modular Cascaded H-Bridge Multilevel PV Inverter With Distributed MPPT for Grid-Connected Applications. *IEEE Trans. Ind. Appl.* **2014**, *51*, 1722–1731. [CrossRef]

20. Iman-Eini, H.; Bacha, S.; Frey, D. Improved control algorithm for grid-connected cascaded H-bridge photovoltaic inverters under asymmetric operating conditions. *IET Power Electron.* **2017**, *11*, 407–415. [CrossRef]

21. Samavatian, V.; Radan, A. A novel low-ripple interleaved buck–boost converter with high efficiency and low oscillation for fuel-cell applications. *Int. J. Electr. Power Energy Syst.* **2014**, *63*, 446–454. [CrossRef]

22. Vavilapalli, S.; Subramaniam, U.; Padmanaban, S.; Blaabjerg, F. Design and Controller-In-Loop Simulations of a Low Cost Two-Stage PV-Simulator. *Energies* **2018**, *11*, 2774. [CrossRef]

23. Kamran, M.; Mudassar, M.; Fazal, M.R.; Asghar, M.U.; Bilal, M.; Asghar, R. Implementation of improved Perturb & Observe MPPT technique with confined search space for standalone photovoltaic system. *J. King Saud Univ. Eng. Sci.* **2018**, *30*. [CrossRef]

24. Li, X.; Wu, W.; Wang, H.; Gao, N.; Chung, H.S.-H.; Blaabjerg, F. A New Buck-Boost AC/DC Converter with Two-Terminal Output Voltage for DC Nano-Grid. *Energies* **2019**, *12*, 3808. [CrossRef]

25. Vavilapalli, S.; Umashankar, S.; Sanjeevikumar, P.; Ramachandaramurthy, V.K.; Mihet-Popa, L.; Fedák, V. Three-stage control architecture for cascaded H-Bridge inverters in large-scale PV systems—Real time simulation validation. *Appl. Energy* **2018**, *229*, 1111–1127. [CrossRef]

26. Marcon, P.; Szabo, Z.; Vesely, I.; Zezulka, F.; Sajdl, O.; Roubal, Z.; Dohnal, P. A Real Model of a Micro-Grid to Improve Network Stability. *Appl. Sci.* **2017**, *7*, 757. [CrossRef]

27. Arunkumar, G.; Elangovan, D.; Sanjeevikumar, P.; Nielsen, J.B.H.; Leonowicz, Z.; Joseph, P.K. DC Grid for Domestic Electrification. *Energies* **2019**, *12*, 2157. [CrossRef]

Article

Three-Port Converter for Integrating Energy Storage and Wireless Power Transfer Systems in Future Residential Applications

Hyeon-Seok Lee [1] and Jae-Jung Yun [2],*

[1]   Department of Electrical Engineering, POSTECH, Pohang 37673, Korea; hsasdf@postech.ac.kr
[2]   Department of Electronics and Electrical Engineering, Daegu University, Gyeongsan 38453, Korea
*   Correspondence: jjyun@daegu.ac.kr; Tel.: +82-53-850-6612

Received: 15 November 2019; Accepted: 2 January 2020; Published: 5 January 2020

Abstract: This paper presents a highly efficient three-port converter to integrate energy storage (ES) and wireless power transfer (WPT) systems. The proposed converter consists of a bidirectional DC-DC converter and an AC-DC converter with a resonant capacitor. By sharing an inductor and four switches in the bidirectional DC-DC converter, the bidirectional DC-DC converter operates as a DC-DC converter for ES systems and simultaneously as a DC-AC converter for WPT systems. Here, four switches are turned on under the zero voltage switching conditions. The AC-DC converter for WPT system achieves high voltage gain by using a resonance between the resonant capacitor and the leakage inductance of a receiving coil. A 100-W prototype was built and tested to verify the effectiveness of the converter; it had a maximum power-conversion efficiency of 95.9% for the battery load and of 93.8% for the wireless charging load.

Keywords: energy storage system; wireless power transfer system; DC-DC power conversion; photovoltaic power system

---

1. Introduction

Photovoltaics (PVs) constitute a promising alternative energy source due to diverse applications and the ubiquity of sunlight [1–4]. Residential PV systems consist of a PV module and a PV inverter, and it converts sunlight into electricity. However, energy output by PV systems is not constant because it is affected by weather conditions and the day/night cycle. Therefore, energy storage (ES) systems are used to efficiently manage the PV energy. They consist of a battery and a bidirectional DC-DC converter that connects to the PV system [5–7]. Besides, in the near future, the wireless power transfer (WPT) systems will be widely used to wirelessly charge laptops as well as cell phones in many households [8–11]. Therefore, future PV energy delivery and management infrastructure for residential applications will consist of PV systems, ES systems, and WPT systems (Figure 1a).

Three-port converters have been used to reduce the cost and size of infrastructure, such as micro-grids and smart grids [12–18]. The infrastructure consists of several systems, so its cost and size increase in proportion to the number of systems. To alleviate this problem, two systems are integrated through on a three-port converter. The three-port converters usually add a port to a typical converter that has two ports, either by using a three-winding transformer instead of a two-winding transformer [13–15] or by using the storage capacitor in the typical converter as a third port [16–18].

Many three-port converters have been introduced to integrate renewable energy sources (e.g., PV, fuel cell, wind turbine) with ES systems, but a three-port converter to integrate ES and WPT systems has not been considered. Therefore, this paper presents a three-port converter that can integrate the ES and WPT systems that will be used in future PV energy delivery and management infrastructure for residential applications (Figure 1b). The proposed three-port converter consists of a bidirectional

DC-DC converter for an ES system and an AC-DC converter with a resonant capacitor. The bidirectional DC-DC converter can also operate as a DC-AC converter of the WPT system because the inductor in the bidirectional DC-DC converter is also used as a transmitting coil for a WPT system. With these few components, the proposed converter can store energy for the ES system and simultaneously transfer the energy by using a WPT system. The proposed converter has a high power-conversion efficiency by achieving zero voltage switching (ZVS) turn-on for switches, and has high voltage gain for WPT by using a resonance between a resonant capacitor and a leakage inductance of a receiving coil.

Section 2 describes the circuit structure and operating principles of the proposed converter. Section 3 presents experimental results, and Section 4 concludes the paper.

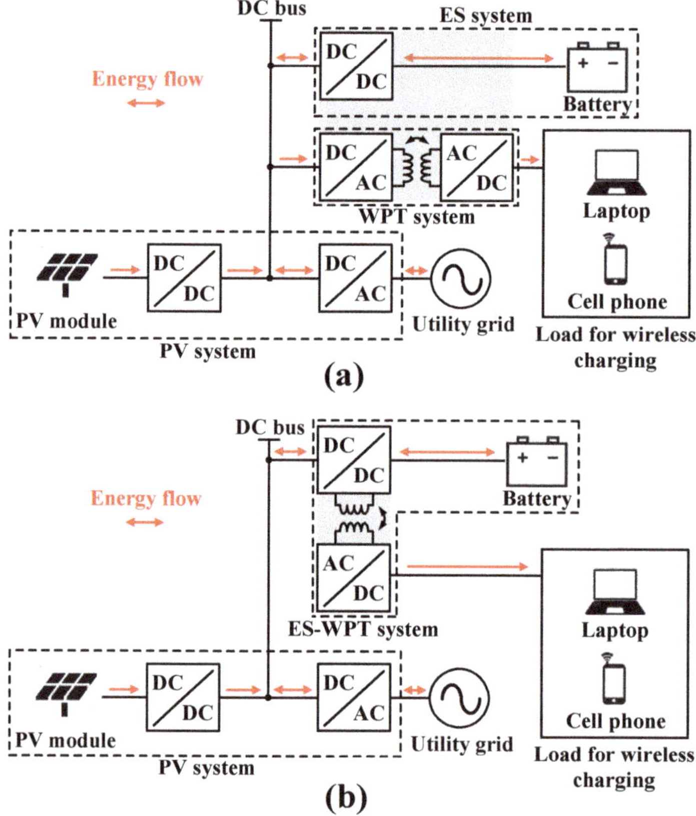

Figure 1. Future photovoltaic (PV) energy delivery and management infrastructure; (**a**) typical topology, (**b**) proposed topology.

2. Proposed Three-Port Converter

2.1. Circuit Structure

The proposed converter (Figure 2a) combines the structures of a bidirectional DC-DC converter and an AC-DC converter.

The bidirectional DC-DC converter is located between a DC bus and a battery to transfer the energy in both directions (DC bus ↔ battery). This converter consists of four switches (S_1, S_2, S_3, S_4), two filter capacitors (bus capacitor, C_{bus}, with capacitance, C_{bus}, and battery capacitor, C_{bat}, with capacitance,

C_{bat}), and an inductor (L_1 with inductance L_1). In addition, it can transfer the energy of the DC bus (or battery) to the wireless charging load because L_1 also acts as a transmitting coil for WPT.

The AC-DC converter is connected to a wireless charging load, such as a cell phone or laptop, and it has a receiving coil (L_2 with inductance L_2) for WPT, a filter capacitor (WPT capacitor, C_{wpt}, with capacitance, C_{wpt}), a resonant capacitor (C_r with capacitance, C_r), and a voltage doubler rectifier that consists of two diodes (D_1, D_2) and two doubler capacitors (C_1 with capacitance, C_1, and C_2 with capacitance, C_2).

L_1 and L_2 are parts of the two-coil structure; they are coupled magnetically with a coupling coefficient, k, to transfer the energy wirelessly. Based on [19,20], the two-coil structure can be represented as a transformer with a leakage inductor (L_{lk} with inductance, L_{lk}), an effective turn ratio (N_e), and a magnetizing inductor (L_m with inductance, L_m), where $L_{lk} = (1 - k^2)L_2$, $N_e = k\sqrt{L_2/L_1}$, and $L_m = L_1$ (Figure 2b).

Four switches achieve the ZVS turn-on by using the stored energy in L_1. L_{lk} resonates with C_r, and high voltage gain between the DC bus and wireless charging load is achieved by setting the switching frequency, f_S, to the resonant frequency, f_r, between L_{lk} and C_r. The voltage doubler rectifier converts AC voltage to DC voltage for wireless charging load, and clamps the reverse voltages of D_1 and D_2 to WPT voltage, V_{wpt}.

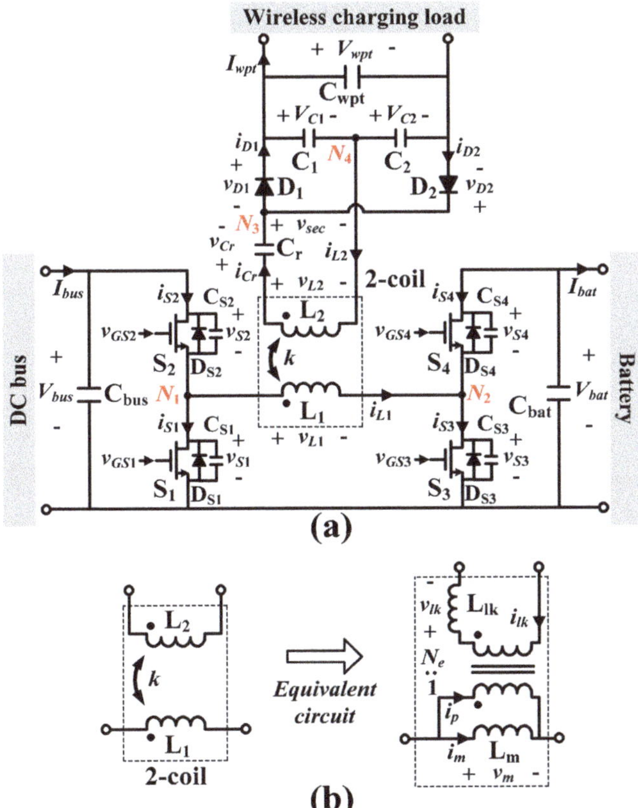

Figure 2. (a) Circuit structure of the proposed converter. (b) Equivalent circuit of the two-coil structure.

2.2. Principle of Operation

The proposed converter operates at a fixed switching frequency ($f_S = 1/T_S$), where T_S is a switching period, and it controls the voltage gain between the DC bus voltage, V_{bus}, and battery voltage, V_{bat}, by changing the duty ratios of S_1, S_2, S_3, and S_4; the duty ratios of S_2 and S_3 are defined as D, and the duty ratios of S_1 and S_4 are defined as $1-D$. In addition, the voltage gain between V_{bus} and V_{wpt} is adjusted by using N_e.

The equivalent circuits (Figure 3) and operating waveforms (Figure 4) were obtained under the following assumptions and conditions: (1) All components are lossless, (2) C_1, C_2, C_{bus}, C_{bat}, and C_{wpt} are large enough to assume that V_{C1}, V_{C2}, V_{bus}, V_{bat}, and V_{wpt} are constant voltage sources, (3) $f_r = f_S$, and (4) the converter operates in a steady state. The converter operates in four modes.

Mode 1 (Figure 3a, $t_0 \leq t \leq t_1$): This mode starts at $t = t_0$ when S_2 and S_3 are turned on. At this time, S_2 and S_3 achieve ZVS turn-on because the body diodes, D_{S2} and D_{S3}, of S_2 and S_3 are turned on before $t = t_0$. During this mode, the voltage, v_m, of L_m becomes V_{bus}, and the current, i_m, of L_m is:

$$i_m(t) = i_m(t_0) + (V_{bus}/L_m)(t - t_0), \tag{1}$$

where $i_m(t_0) = I_{bus} + I_{bat} - V_{bus}DT_S/(2L_m)$. C_r has voltage $v_{Cr} = N_e V_{bus} - V_{C1} - v_{lk}$ and current $i_{Cr} = i_{lk}$, and $i_{Cr}(t_0) = 0$. Therefore:

$$i_{Cr}(t) = \frac{N_e V_{bus} - V_{wpt}/2 - v_{Cr}(t_0)}{\sqrt{L_{lk}/C_r}} \sin[\omega_r(t - t_0)], \tag{2}$$

where $v_{Cr}(t_0) = -0.5T_S I_{wpt}/C_r$ and $\omega_r = 1/\sqrt{L_{lk}C_r}$. The current, i_{D1}, of D_1 is equal to i_{Cr} for $t_0 \leq t \leq t_1$.

Mode 2 (Figure 3b, $t_1 \leq t \leq t_2$): At $t = t_1$, S_2 and S_3 are turned off, and S_1 and S_4 remain in the off-states to prevent a shoot-through problem. This mode is known as dead time. During this mode, the output capacitance, C_{S1}, of S_1 discharges from V_{bus} to 0, and the output capacitance, C_{S2}, of S_2 charges from 0 to V_{bus}. In addition, the output capacitance, C_{S4}, of S_4 discharges from V_{bat} to 0, and the output capacitance, C_{S3}, of S_3 charges from 0 to V_{bat}. Shortly after the discharging and charging processes are finished, the body diodes, D_{S1} and D_{S4}, of S_1 and S_4 are turned on.

Mode 3 (Figure 3c, $t_2 \leq t \leq t_3$): At $t = t_2$, S_1 and S_4 are turned on under ZVS conditions because the body diodes, D_{S1} and D_{S4}, are turned on before $t = t_2$. During this mode:

$$i_m(t) = i_m(t_2) - (V_{bat}/L_m)(t - t_2), \tag{3}$$

because $v_m = -V_{bat}$. i_{Cr} is obtained using $v_{Cr} = -N_e V_{bat} + V_{C2} - v_{lk}$ and $i_{Cr}(t_2) = 0$ as:

$$i_{Cr}(t) = -\frac{N_e V_{bat} - V_{wpt}/2 + v_{Cr}(t_2)}{\sqrt{L_{lk}/C_r}} \sin[\omega_r(t - t_2)], \tag{4}$$

where $v_{Cr}(t_2) = 0.5T_S I_{wpt}/C_r$. At $t = t_2$, D_2 is turned on, and the current, i_{D2}, of D_2 is $-i_{Cr}$.

Mode 4 (Figure 3d, $t_3 \leq t \leq t_4$): S_2 and S_3 remain in the off-states because this mode is the dead time interval. During this mode, C_{S2} discharges from V_{bus} to 0, and C_{S1} charges from 0 to V_{bus}. In addition, C_{S3} discharges from V_{bat} to 0, and C_{S4} charges from 0 to V_{bat}. Shortly after the discharging and charging processes are finished, D_{S2} and D_{S3} are turned on.

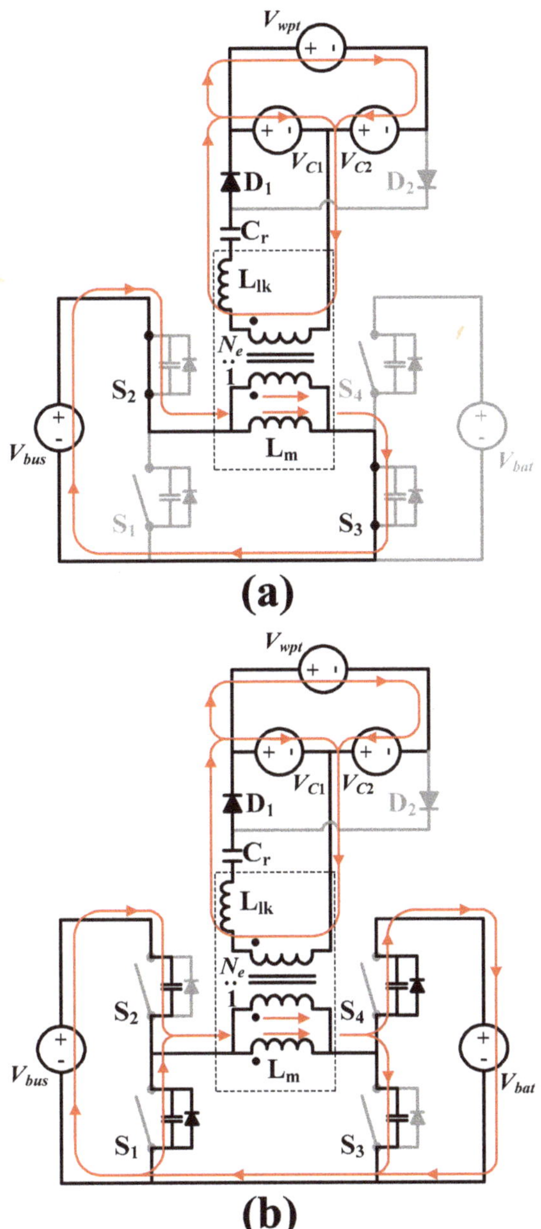

(a)

(b)

Figure 3. *Cont.*

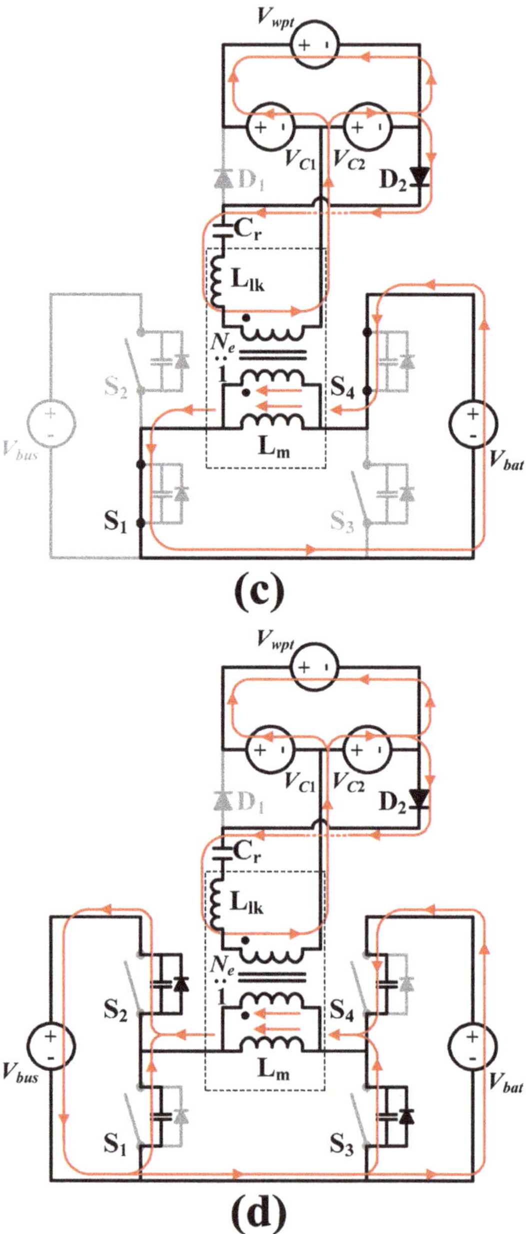

Figure 3. (**a**) Circuit diagrams for the operation in Mode 1. (**b**) Circuit diagrams for the operation in Mode 2. (**c**) Circuit diagrams for the operation in Mode 3. (**d**) Circuit diagrams for the operation in Mode 4.

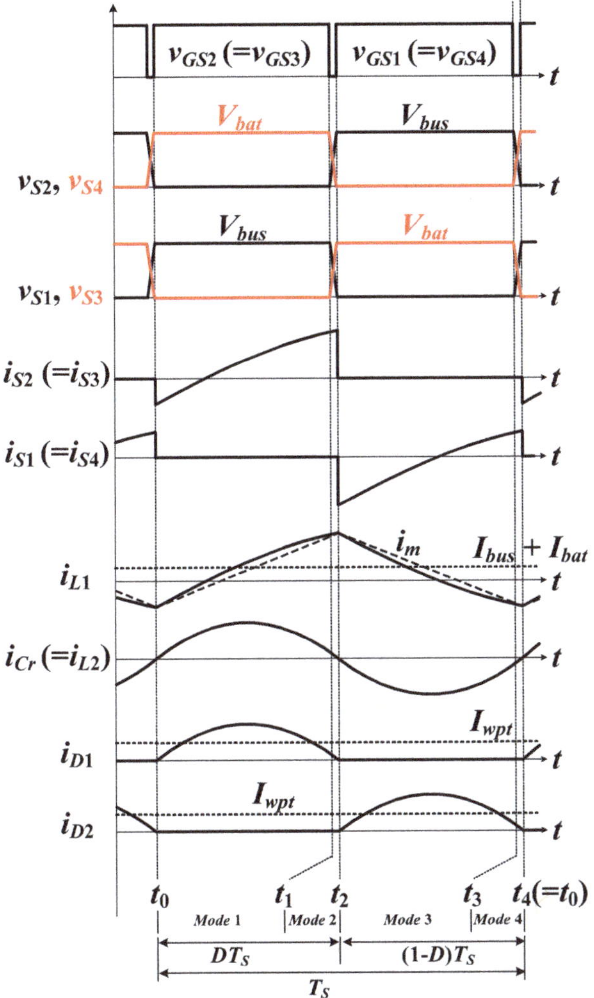

Figure 4. Operational waveforms of the proposed converter.

2.3. Voltage Gain

The proposed converter has one input port (DC bus) and two output ports (battery and wireless charging load). Therefore, it has (Equation (1)) a voltage gain, G_{bb}, between the DC bus and battery and (Equation (2)) a voltage gain, G_{wb}, between the DC bus and wireless charging load.

(1) $G_{bb} = V_{bat}/V_{bus}$

For one T_S, the average voltages of C_{bus} and C_{bat} are V_{bus} and V_{bat}, respectively. Then, the average voltage at node N_1 between S_1 and S_2 is DV_{bus}, and the average voltage at node N_2 between S_3 and S_4 is given by $(1-D)$ V_{bat}. The average voltage of L_1 is zero due to the volt-second balance law for the inductor, and the average currents of switches are given by $\langle i_{S2} \rangle = \langle i_{S3} \rangle = I_{bus}$ and $\langle i_{S1} \rangle = \langle i_{S4} \rangle = -I_{bat}$ because the average current of the capacitor is zero by the charge balance law. Therefore, the average model of the circuit can be obtained (Figure 5). In this average model, the series resistance, R_S, is the sum of R_{L1} and $2R_{on}$, where R_{L1} and R_{on} are a winding resistance of L_1 and on-resistance of a switch, respectively.

By applying Kirchhoff's voltage law (KVL) to the closed-loop that contains DV_{bus}, R_S, and (1-D) V_{bat}, the voltage gain G_{bb} ($= V_{bat}/V_{bus}$) is obtained as:

$$G_{bb} = -\frac{1}{2}\left\{\left[\frac{(1-D)R_{bat}}{R_S}+1\right]+\sqrt{\left[\frac{(1-D)R_{bat}}{R_S}+1\right]^2+\frac{4DR_{bat}}{R_S}}\right\},\tag{5}$$

where $R_{bat} = V_{bat}/I_{bat}$. In most cases, $R_{bat} \gg R_S$, so:

$$G_{bb} \approx D/(1-D).\tag{6}$$

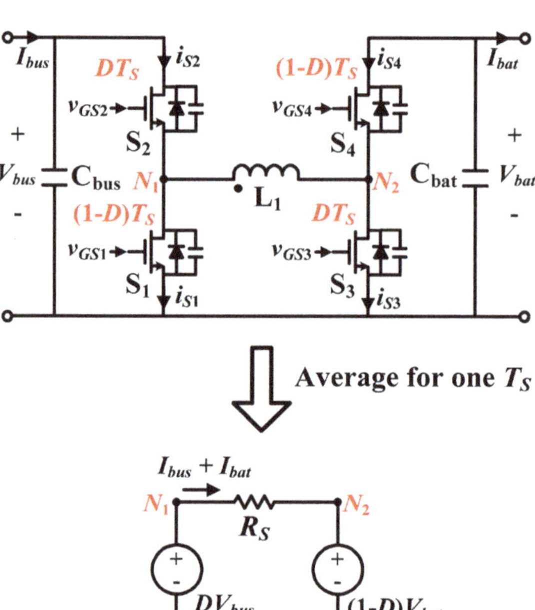

Figure 5. Average model between node N_1 and node N_2.

(2) $G_{wb} = V_{wpt}/V_{bus}$

Based on the four nodes (N_1, N_2, N_3, and N_4) in Figure 2a, the circuit of the proposed converter can be simplified (Figure 6a). Here, the square voltage, v_{ac}, between N_1 and N_2 is $V_{bus} - (I_{bus} + I_{bat})R_S$ for DT_S and $-V_{bat} - (I_{bus} + I_{bat})R_S$ for (1-D)T_S. Then, the current i_{ac} is $i_{L1} - (I_{bus} + I_{bat})$, and the square voltage, v_{ac2}, between N_3 and N_4 becomes $V_{wpt}/2$ for $0.5T_S$ and $-V_{wpt}/2$ for the other $0.5T_S$. The current i_{ac2} is given by i_{Cr}.

By applying fundamental harmonic approximation (FHA) to v_{ac}, i_{ac}, i_{ac2}, and v_{ac2}, the following root mean square (RMS) values are obtained (Figure 6b); $V_{L2,\text{rms}}$ is the RMS value of the first harmonic component in the voltage, v_{L2}, of L_2, and it is given by:

$$V_{L2,\text{rms}} = -\frac{N_e^2\pi V_{wpt}R_S}{\sqrt{2}R_{wpt}} + \frac{N_e V_{bus}}{\sqrt{2}}\left(\frac{2(1+G_{bb})}{\pi}\sin(D\pi) - \frac{T_S R_S}{\pi^2(1-D)L_m}\right),\tag{7}$$

where:

$$I_{ac2,\text{rms}} = \pi I_{wpt}/\sqrt{2},\tag{8}$$

and:

$$V_{ac2,rms} = \sqrt{2}V_{wpt}/\pi, \tag{9}$$

are RMS values of the first harmonic components in i_{ac2} and v_{ac2}, respectively. Then, the equivalent resistance, R_{ac2}, between N_3 and N_4 is obtained using Equations (8) and (9) as:

$$R_{ac2} = V_{ac2,rms}/I_{ac2,rms} = 2R_{wpt}/\pi^2, \tag{10}$$

where $R_{wpt} = V_{wpt}/I_{wpt}$. By applying KVL to the closed-loop in Figure 6b, the relation between $V_{L2,rms}$ and $V_{ac2,rms}$ is given by:

$$\frac{V_{L2,rms}}{V_{ac2,rms}} = \frac{1}{R_{ac2}}\left[j\left(\omega_S L_{lk} - \frac{1}{\omega_S C_r}\right) + R_{L2}\right] + 1, \tag{11}$$

where R_{L2} is a winding resistance of L_2. Then, substituting Equations (7) and (9) into Equation (11) yields:

$$G_{wb} = \left|\frac{V_{wpt}}{V_{bus}}\right| = \frac{N_e}{2}\left(2(1+G_{bb})\sin(D\pi) - \frac{T_S R_S}{\pi(1-D)L_m}\right)$$
$$\left/ \sqrt{\frac{1}{R_{ac2}^2}\left(\omega_S L_{lk} - \frac{1}{\omega_S C_r}\right)^2 + \left(\frac{\pi^2 R_{L2}}{2R_{wpt}} + 1 + \frac{N_e^2 \pi^2 R_S}{2R_{wpt}}\right)^2}.\right. \tag{12}$$

If $f_r = f_S$, maximum G_{wb} is obtained because $\omega_S L_{lk} = 1/(\omega_S C_r)$ at $f_r = f_S$ (Figure 7), and it is given by:

$$G_{wb}|_{f_r=f_S} = \frac{N_e}{2}\left(2(1+G_{bb})\sin(D\pi) - \frac{T_S R_S}{\pi(1-D)L_m}\right)\left/\left(\frac{\pi^2 R_{L2}}{2R_{wpt}} + 1 + \frac{N_e^2 \pi^2 R_S}{2R_{wpt}}\right).\right. \tag{13}$$

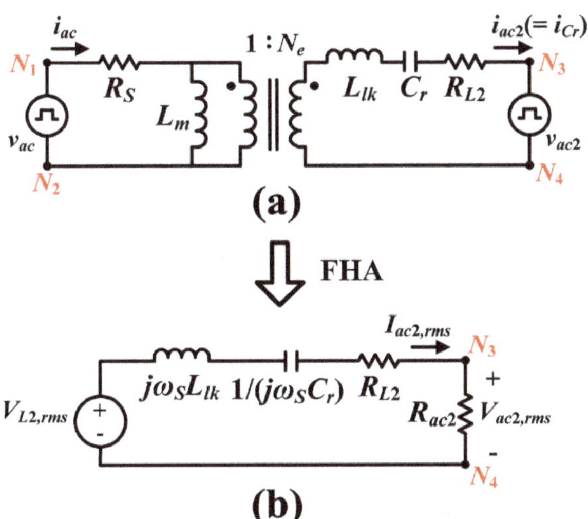

Figure 6. (a) Equivalent circuit based on four nodes (N_1, N_2, N_3, and N_4). (b) AC model in the secondary winding side after applying fundamental harmonic approximation (FHA).

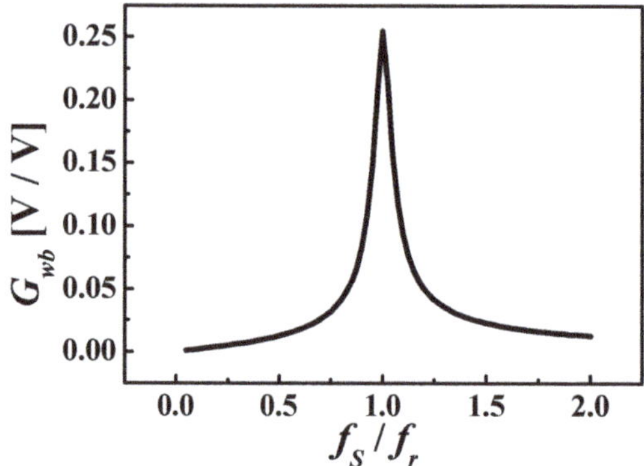

Figure 7. Voltage gain (G_{wb}) between V_{bus} and V_{wpt} according to f_S.

Therefore, C_r should be chosen to obtain the maximum voltage gain (G_{wb}) between V_{bus} and V_{wpt}. Because $\omega_r = 1/\sqrt{L_{lk}C_r}$ and $L_{lk} = (1-k^2)L_2$, C_r can be determined as:

$$C_r = \frac{1}{4\pi^2 f_S{}^2(1-k^2)L_2}.$$

2.4. Magnetic Saturation

The coil structure, which consists of a transmitting coil (L_1) and a receiving coil (L_2), can use magnetic bars to increase efficiency and reduce magnetic fields that can interfere with nearby electronics [21–23]. Based on [24,25], the magnetic flux density (B_C) of the coil structure is given by:

$$B_C = \frac{\mu_0 \mu_e N i_{m,peak}}{l_m}, \tag{14}$$

where μ_0 is a vacuum permeability, $\mu_e = (\mu_r l_m)/(2l_g \mu_r + l_m)$ is an effective relative permeability, N is the number of turns, $i_{m,peak}$ is a peak value of i_m, l_m is the mean magnetic path length, μ_r is a relative permeability, and l_g is thee air-gap length.

Because the proposed converter has a DC bias current ($=I_{bus}+I_{bat}$) of L_1, $i_{m,peak}$ is given by:

$$i_{m,peak} = I_{bus} + I_{bat} + \frac{V_{bus}DT_S}{2L_m}. \tag{15}$$

Magnetic saturation can be caused by this DC bias current because the DC bias current increases B_C by increasing $i_{m,peak}$. There are two methods to solve the problem of the DC bias current [26,27]. First, decreasing N reduces B_C. Second, increasing l_g reduces B_C by decreasing μ_e. However, l_g is determined by a distance, l_d, between L_1 and L_2. Therefore, the following condition for preventing magnetic saturation can be obtained using $B_C < B_{sat}$, (Equations (14) and (15)) as:

$$N < \frac{l_m B_{sat}}{\mu_0 \mu_e [I_{bus} + I_{bat} + V_{bus}DT_S/(2L_m)]}, \tag{16}$$

where B_{sat} is a saturation flux density of magnetic material.

3. Experimental Results

A prototype (Figure 8) of the proposed converter was fabricated using selected components and circuit parameters (Table 1), then tested to verify the operation of the proposed converter.

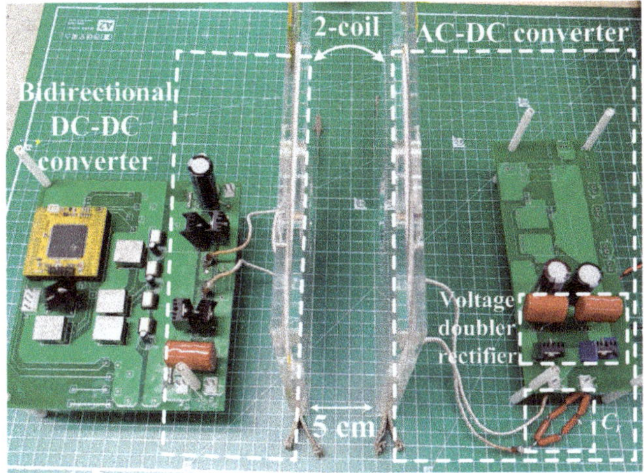

Figure 8. Photograph of the prototype.

Table 1. Values of the components used for the prototype.

Components			Values
2-coil structure	Winding type		Spiral
	Distance between two coils		5 cm
	Coupling coefficient (k)		0.227
	Equivalent turn ratio (N_e)		0.226
	Primary winding	Inductance (L_1)	117.5 µH
		Resistance (R_{L1})	0.43 Ω
		Turn number (N_1)	28 T
	Secondary winding	Inductance (L_2)	116.5 µH
		Resistance (R_{L2})	0.4 Ω
		Turn number (N_2)	28 T
	Resonant capacitor (C_r)		1.43 nF
	Doubler capacitors (C_1, C_2)		2.2 µF
	Switches ($S_1 \sim S_4$)		FCP099N60E (R_{on} = 99 mΩ)
	Diodes (D_1, D_2)		BYV29-500

The voltage and current waveforms of S_1 and S_2 were measured at V_{bus} = 400 V, V_{bat} = 400 V, f_S = 400 kHz, and P_{wpt} (or P_{bat}) = 20 and 100 W (Figure 9). At both P_{wpt} = 20 W (Figure 9a) and P_{wpt} = 100 W (Figure 9b), the voltage stresses of S_1 and S_2 were measured as 400 V, which is equal to V_{bus}, and S_1 and S_2 achieved ZVS turn-on. In addition, S_3 and S_4 achieved ZVS turn-on at these conditions because $i_{S1} = i_{S4}$ and $i_{S2} = i_{S3}$. Even at both P_{bat} = 20 W (Figure 9c) and P_{bat} = 100 W (Figure 9d), all switches were turned on under the ZVS condition, which improves the power-conversion efficiency, η_e.

The theoretical V_{wpt} obtained from Equation (13) was compared with the experimental V_{wpt} measured at V_{bus} = 400 V, V_{bat} = 400 V, and P_{wpt} = 20 W~100 W (Figure 10). The theoretical V_{wpt} was higher than experimental V_{wpt} at all P_{wpt}, but the difference was <3%. This result shows that Equation (13) predicts the experimental V_{wpt} with little error. In addition, V_{wpt} was almost constant regardless of P_{wpt}.

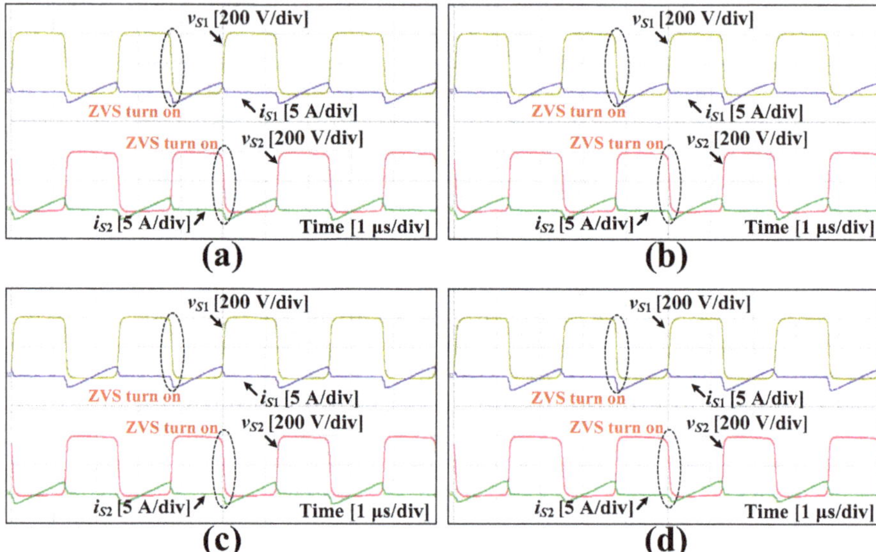

Figure 9. Voltage and current waveforms of switches (S_1, S_2) measured at P_{wpt} = (**a**) 20 W and (**b**) 100 W or at P_{bat} = (**c**) 20 and (**d**) 100 W.

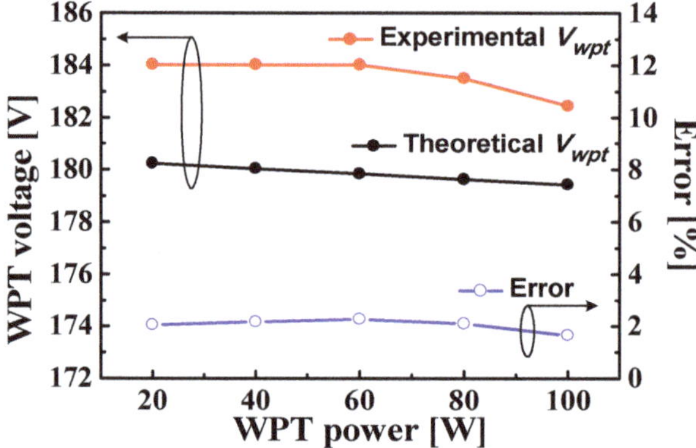

Figure 10. Comparison between experimental V_{wpt} and theoretical V_{wpt}.

The power-conversion efficiencies ($\eta_{e,\mathrm{wpt}}$ for the wireless charging load and $\eta_{e,\mathrm{bat}}$ for the battery load) were measured at V_{bus} = 400 V, V_{bat} = 400 V, f_S = 400 kHz, and P_{wpt} (or P_{bat}) = 20 W~100 W (Figure 11). The proposed converter had the highest $\eta_{e,\mathrm{wpt}}$ = 93.8% at P_{wpt} = 100 W (Figure 11a) and had the highest $\eta_{e,\mathrm{bat}}$ = 95.9% at P_{bat} = 100 W (Figure 11b). At low P_{wpt} = 20 W, the measured $\eta_{e,\mathrm{wpt}}$ was 82.5% (Figure 11a), and the measured $\eta_{e,\mathrm{bat}}$ was 83.5% at low P_{bat} = 20 W (Figure 11b). In addition, the measured $\eta_{e,\mathrm{bat}}$ was higher than the measured $\eta_{e,\mathrm{wpt}}$ because the wireless power loss during transfer between two coils is included in $\eta_{e,\mathrm{wpt}}$. These results show that the proposed converter had high $\eta_{e,\mathrm{wpt}}$ > 82% and high $\eta_{e,\mathrm{bat}}$ > 83% for all operating ranges due to the ZVS turn-on of all switches.

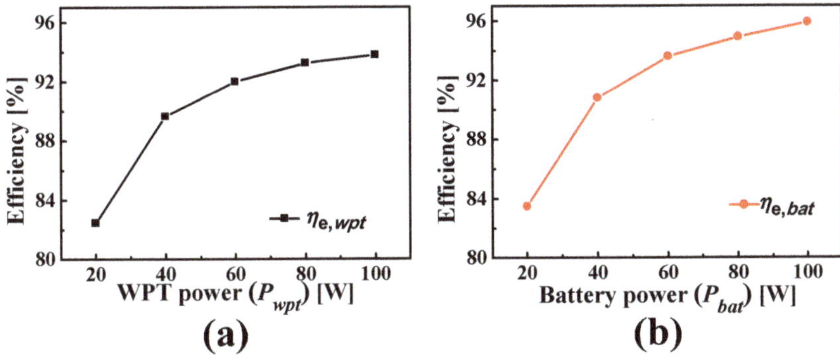

Figure 11. (**a**) Power-conversion efficiency, $\eta_{e,\mathrm{wpt}}$, measured at P_{wpt} = 20 W~100 W. (**b**) Power-conversion efficiency, $\eta_{e,\mathrm{bat}}$, measured at P_{bat} = 20 W~100 W.

The transient response of V_{wpt} was measured at V_{bus} = 400 V, V_{bat} = 400 V, and f_S = 400 kHz, while changing the wireless charging load from 20% to 100% and from 100% to 20% (Figure 12). At the load transition, the maximum voltage spike of V_{wpt} was measured as 14 $V_{\mathrm{p.p}}$, and V_{wpt} returned to the steady-state within 81 ms.

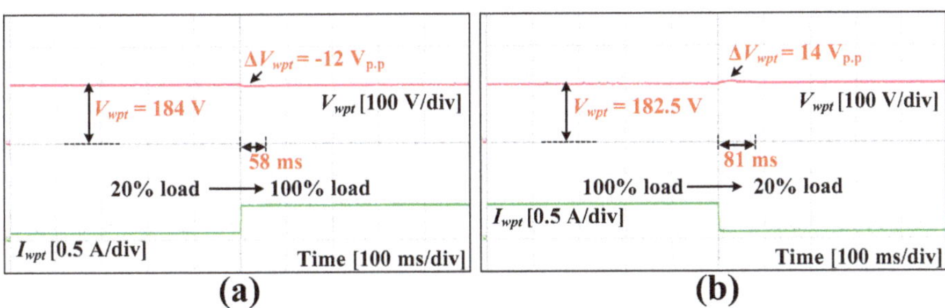

Figure 12. Transient responses of V_{wpt} while changing the wireless charging load (**a**) from 20% to 100% and (**b**) from 100% to 20%.

In addition, the transient response of V_{bat} was measured while changing the battery load from 20% to 100% and from 100% to 20% under the same conditions as in Figure 12 (Figure 13). The maximum voltage spike of V_{bat} was measured as 61 $V_{\mathrm{p.p}}$ when the load changed, and V_{bat} returned to the steady state within 125 ms. These experiment results of the transient response show that the proposed converter can operate properly despite sudden load changes.

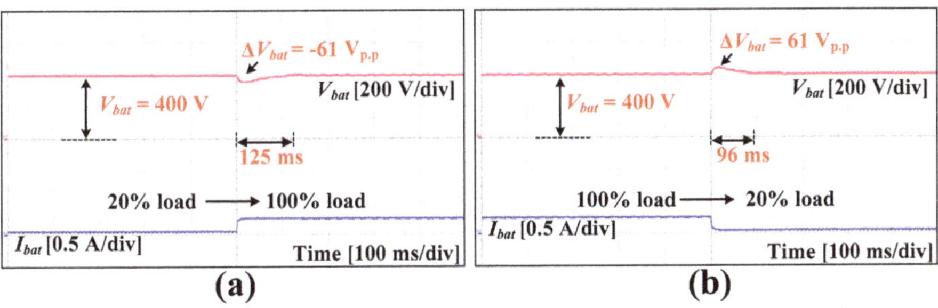

Figure 13. Transient responses of V_{bat} while changing the battery load (**a**) from 20% to 100% and (**b**) from 100% to 20%.

To show that the proposed converter can operate both when charging the battery and when charging wirelessly, V_{wpt} and V_{bat} of the proposed converter were measured under the following four conditions (Figure 14): (1) P_{wpt} = 20 W and P_{bat} = 20 W, (2) P_{wpt} = 20 W and P_{bat} = 100 W, (3) P_{wpt} = 100 W and P_{bat} = 20 W, and (4) P_{wpt} = 100 W and P_{bat} = 100 W. Figure 14 shows that V_{wpt} decreases and V_{bat} increases when the power (P_{wpt}, P_{bat}) increases. However, both V_{wpt} and V_{bat} maintained a near fixed value; V_{wpt} changed from 183.8 to 180.9 V, which is just a 1.58% change, and V_{bat} changed from 399.8 to 401 V, which is just a 0.3% change.

The measured V_{wpt} and V_{bat} are summarized in Table 2, and this experimental result shows that the proposed converter can operate in both charging the battery and charging wirelessly because both V_{wpt} and V_{bat} maintain a near fixed value regardless of P_{wpt} and P_{bat}.

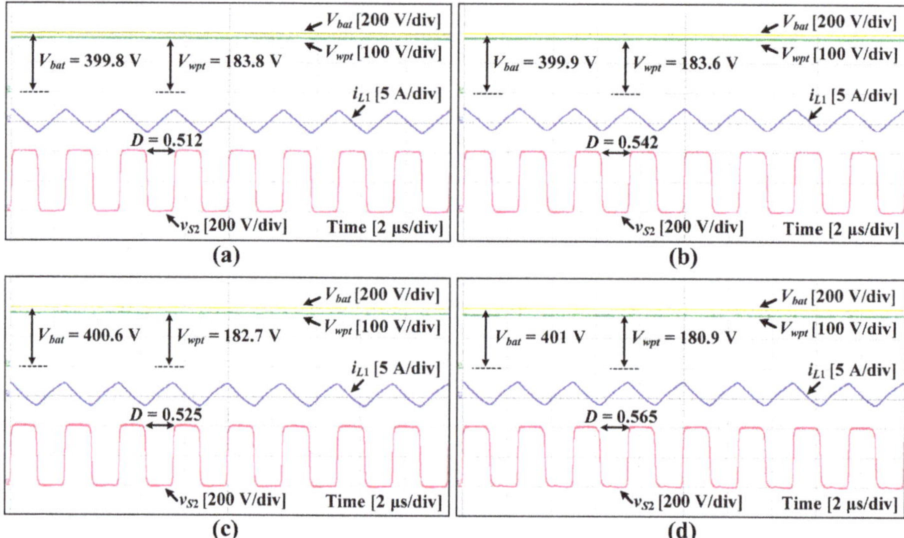

Figure 14. The WPT voltage (V_{wpt}) and the battery voltage (V_{bat}) measured at (**a**) P_{wpt} = 20 W and P_{bat} = 20 W, (**b**) P_{wpt} = 20 W and P_{bat} = 100 W, (**c**) P_{wpt} = 100 W and P_{bat} = 20 W, and (**d**) P_{wpt} = 100 W and P_{bat} = 100 W.

Table 2. WPT and battery voltages according to the WPT and battery powers.

Power		20 W	20 W	100 W	100 W
	WPT Power (P_{wpt})	20 W	20 W	100 W	100 W
	Battery Power (P_{wpt})	20 W	100 W	20 W	100 W
WPT voltage (V_{wpt})		183.8 V	183.6 V	182.7 V	180.9 V
Battery voltage (V_{bat})		399.8 V	399.9 V	400.6 V	401 V

4. Conclusions

This paper presented a three-port converter to integrate an ES system with a WPT system. If the ES and WPT systems are used separately, many converters are needed. However, these ES and WPT systems can be integrated through only one proposed converter because the proposed converter can use an inductor in the bidirectional DC-DC converter as a transmitting coil for the WPT system. Therefore, the proposed converter consists only of a bidirectional DC-DC converter and an AC-DC converter, and an ES-WPT system that uses the proposed converter can minimize the cost and circuit size. The operation of the proposed converter was verified by theoretical analysis and experimental results, and the proposed converter had high $\eta_{e,wpt}$ > 82% for 20 W $\leq P_{wpt} \leq$ 100 W, and $\eta_{e,bat}$ > 83% for 20 W $\leq P_{bat} \leq$ 100 W, due to the ZVS turn-on of all switches. These results show that the proposed converter is suitable for the ES-WPT system that is part of future PV energy delivery and management infrastructure for residential applications.

Author Contributions: Both authors contributed equally to this work. H.-S.L. presented the main idea of the three-port converter. H.-S.L. and J.-J.Y. analyzed the proposed converter and performed experiments. J.-J.Y. contributed to the overall composition and writing of the manuscript. All authors have read and agreed to the published version of the manuscript.

Funding: This work was supported by the National Research Foundation of Korea (NRF) grant funded by the government of Korea (MSIP) (NRF-2018R1A1A1A05079496).

Conflicts of Interest: The authors have no conflict of interest.

References

1. Malinowski, M.; Leon, J.I.; Abu-Rub, H. Solar photovoltaic and thermal energy systems: Current technology and future trends. *Proc. IEEE* **2017**, *105*, 2132–2146. [CrossRef]
2. Kroposki, B.; Johnson, B.; Zhang, Y.; Gevorgian, V.; Denholm, P.; Hodge, B.-M.; Hannegan, B. Achieving a 100% renewable grid: Operating electric power systems with extremely high levels of variable renewable energy. *IEEE Power Energy Mag.* **2017**, *15*, 61–73. [CrossRef]
3. Sahu, B.K. A study on global solar PV energy developments and policies with special focus on the top ten solar PV power producing countries. *Renew. Sustain. Energy Rev.* **2015**, *43*, 621–634. [CrossRef]
4. McElroy, M.B.; Chen, X. Wind and solar power in the United States: Status and prospects. *CSEE J. Power Energy Syst.* **2017**, *3*, 1–6. [CrossRef]
5. Park, L.; Jang, Y.; Cho, S.; Kim, J. Residential Demand Response for Renewable Energy Resources in Smart Grid Systems. *IEEE Trans. Ind. Inf.* **2017**, *13*, 3165–3173. [CrossRef]
6. Chaudhari, K.; Ukil, A.; Kumar, K.N.; Manandhar, U.; Kollimalla, S.K. Hybrid Optimization for Economic Deployment of ESS in PV-Integrated EV Charging Stations. *IEEE Trans. Ind. Inf.* **2018**, *14*, 106–116. [CrossRef]
7. Liu, X.; Aichhorn, A.; Liu, L.; Li, H. Coordinated Control of Distributed Energy Storage System with Tap Changer Transformers for Voltage Rise Mitigation Under High Photovoltaic Penetration. *IEEE Trans. Smart Grid* **2012**, *3*, 897–906. [CrossRef]
8. Park, J.; Kim, D.; Hwang, K.; Park, H.H.; Kwak, S.I.; Kwon, J.H.; Ahn, S. A Resonant Reactive Shielding for Planar Wireless Power Transfer System in Smartphone Application. *IEEE Trans. Electromagn. Compat.* **2017**, *59*, 695–703. [CrossRef]
9. Jeong, N.S.; Carobolante, F. Enabling Wireless Power Transfer Though a Metal Encased Handheld Device. In Proceedings of the IEEE Wireless Power Transfer Conference, Aveiro, Portugal, 5–6 May 2016.
10. Nguyen, V.T.; Kang, S.H.; Choi, J.H.; Jung, C.W. Magnetic Resonance Wireless Power Transfer using Three-Coil System with Single Planar Receiver for Laptop Applications. *IEEE Trans. Consum. Electron.* **2015**, *61*, 160–166. [CrossRef]
11. Barman, S.D.; Reza, A.W.; Kumar, N.; Karim, M.E.; Munir, A.B. Wireless powering by magnetic resonant coupling: Recent trends in wireless power transfer system and its applications. *Renew. Sustain. Energy Rev.* **2015**, *51*, 1525–1552. [CrossRef]
12. Chen, Y.M.; Huang, A.Q.; Yu, X. A High Step-Up Three-Port DC–DC Converter for Stand-Alone PV/Battery Power Systems. *IEEE Trans. Power Electron.* **2013**, *28*, 5049–5062. [CrossRef]
13. Phattanasak, M.; Gavagsaz-Ghoachani, R.; Martin, J.P.; Nahid-Mobarakeh, B.; Pierfederici, S.; Davat, B. Control of a Hybrid Energy Source Comprising a Fuel Cell and Two Storage Devices Using Isolated Three-Port Bidirectional DC–DC Converters. *IEEE Trans. Ind. Appl.* **2015**, *51*, 491–497. [CrossRef]
14. Wang, P.; Lu, X.; Wang, W.; Xu, D. Frequency Division Based Coordinated Control of Three-Port Converter Interfaced Hybrid Energy Storage Systems in Autonomous DC Microgrids. *IEEE Access* **2018**, *6*, 25389–25398. [CrossRef]
15. Duarte, J.L.; Hendrix, M.; Simoes, M.G. Three-Port Bidirectional Converter for Hybrid Fuel Cell Systems. *IEEE Trans. Power Electron.* **2007**, *22*, 480–487. [CrossRef]
16. Wang, Z.; Li, H. An Integrated Three-Port Bidirectional DC–DC Converter for PV Application on a DC Distribution System. *IEEE Trans. Power Electron.* **2013**, *28*, 4612–4624. [CrossRef]

17. Mira, M.C.; Zhang, Z.; Knott, A.; Andersen, M.A.E. Analysis, Design, Modeling, and Control of an Interleaved-Boost Full-Bridge Three-Port Converter for Hybrid Renewable Energy Systems. *IEEE Trans. Power Electron.* **2017**, *32*, 1138–1155. [CrossRef]
18. Zhang, J.; Wu, H.; Qin, X.; Xing, Y. PWM Plus Secondary-Side Phase-Shift Controlled Soft-Switching Full-Bridge Three-Port Converter for Renewable Power Systems. *IEEE Trans. Ind. Electron.* **2015**, *62*, 7061–7072. [CrossRef]
19. Hesterman, B. Analysis and Modeling of Magnetic Coupling. Denver Chapter, IEEE Power Electronics Society. 2007, pp. 35–38. Available online: http://www.verimod.com/presentations/Denver_PELS_20070410_Hesterman_Magnetic_Coupling.pdf (accessed on 16 December 2019).
20. Zhu, G.; McDonald, B.A.; Wang, K. Modeling and Analysis of Coupled Inductors in Power Converters. *IEEE Trans. Power Electron.* **2011**, *26*, 1355–1363. [CrossRef]
21. Dai, J.; Ludois, D.C. A Survey of Wireless Power Transfer and a Critical Comparison of Inductive and Capacitive Coupling for Small Gap Applications. *IEEE Trans. Power Electron.* **2015**, *30*, 6017–6029. [CrossRef]
22. Xu, H.; Wang, C.; Xia, D.; Liu, Y. Design of Magnetic Coupler for Wireless Power Transfer. *Energies* **2019**, *12*, 3000. [CrossRef]
23. Dayerizadeh, A.; Lukic, S. Saturable inductors for superior reflexive field containment in inductive power transfer systems. In Proceedings of the IEEE Applied Power Electronics Conference and Exposition (APEC), San Antonio, TX, USA, 4–8 March 2018.
24. McLyman, W.T. *Transformer and Inductor Design Handbook*, 3rd ed.; Marcel Dekker, Inc.: New York, NY, USA, 2004.
25. Mohan, N.; Undeland, T.M. *Power Electronics: Converters, Applications, and Design*; John Wiley & Sons: Hoboken, NJ, USA, 2007.
26. Orenchak, G.G. Specify Saturation Properties of Ferrite Cores to Prevent Field Failure. Available online: http://www.tscinternational.com/tech13.pdf (accessed on 16 December 2019).
27. Itoh, Y.; Hattori, F.; Kimura, S.; Imaoka, J.; Yamamoto, M. Design method considering magnetic saturation issue of coupled inductor in interleaved CCM boost PFC converter. In Proceedings of the IEEE Energy Conversion Congress and Exposition (ECCE), Montreal, QC, Canada, 20–24 September 2015.

Article

A Hybridization of Cuk and Boost Converter Using Single Switch with Higher Voltage Gain Compatibility

M. Karthikeyan [1,*], R. Elavarasu [2], P. Ramesh [3], C. Bharatiraja [4], P. Sanjeevikumar [5], Lucian Mihet-Popa [6,*] and Massimo Mitolo [7]

[1] Department of Electrical and Electronics Engineering, University College of Engineering, Pattukkottai 614701, Tamilnadu, India

[2] Department of Electrical and Electronics Engineering, Rajalakshmi Institute of Technology, Chennai 600124, Tamilnadu, India; elava3000@gmail.com

[3] Department of Electrical and Electronics Engineering, CMR Institute of Technology, Bengaluru 560037, India; ramesh8889@gmail.com

[4] Department of Electrical and Electronics Engineering, SRM Institute of Science and Technology, Chennai 603 203, India; bharatiraja@gmail.com

[5] Department of Energy Technology, Aalborg University, 6700 Esbjerg, Denmark; san@et.aau.dk

[6] Faculty of Engineering, Østfold University College, Kobberslagerstredet 5, 1671 Kråkeroy-Fredrikstad, Norway

[7] School of Integrated Design, Engineering and Automation Irvine Valley College, Irvine, CA 92618, USA; mmitolo@ivc.edu

* Correspondence: kar_thamarai82@yahoo.com (M.K.); lucian.mihet@hiof.no (L.M.-P.)

Received: 31 March 2020; Accepted: 27 April 2020; Published: 6 May 2020

Abstract: In the current era, the desire for high boost DC-to-DC converter development has increased. Notably, there has been voltage gain improvement without adding extra power switches, and a large number of passive components have advanced. Magnetically coupled isolated converters are suggested for the higher voltage gain. These converters use large size inductors, and thus the non-isolated traditional boost, Cuk and Sepic converters are modified to increase their gain by adding an extra switch, inductors and capacitors. These converters increase circuit complexity and become bulky. In this paper, we present a hybrid high voltage gain non-isolated single switch converter for photovoltaic applications. The proposed converter connects the standard conventional Cuk and boost converter in parallel for providing continuous current mode operation with the help of a single power switch, which gives less voltage stress on controlled switch and diodes. The proposed hybrid topology uses a single switch with a lower component-count and provides a higher voltage gain than non-isolated traditional converters. The converter circuit mode of operation, operating performance, mathematical derivations and steady-state exploration and circuit parameters design procedures are deliberated in detail. The proposed hybrid converter circuit components, voltage gain and performance, were compared with other topologies in the literature. The MATLAB/Simulink simulation study and microcontroller-based experimental laboratory prototype of 150 W were implemented. The simulation study and experimentation results were confirmed to be a satisfactory agreement with the theoretical analysis. This topology produced non-inverting output in continuous input current mode using a single switch with high voltage gain ($\approx$5.116 gain) with a maximum efficiency of 92.2% under full load.

Keywords: DC-to-DC converter; single switch high voltage gain converter; non-isolated DC-to-DC converter; low voltage stress; higher voltage gain

1. Introduction

Due to the increase in energy demand, large amounts of conventional energy have been consumed, which is very dangerous due to their CO2 emissions. For example, all countries are keen on replacing conventional energy sources with non-conventional sources. Researchers are currently exploring power converters and interfacing circuits to meet out-migration. Non-conventional sources such as wind energy [1,2], photovoltaic (PV) [3,4] and hydrogen-powered fuel cells (FC) [5,6] are leading sources for meeting commercial and industrial demands. A PV-powered power system consists of PV modules coupled in a series as well as parallel combinations that are fed to the required DC voltage through the DC-to-DC converter, which is then converted as a DC-to-AC source through the inverter [7]. The controller development of DC-to-DC converters using a fuzzy logic controller and sliding mode control has recently gained attention. Those converters are used in a microgrid that minimizes the instability effects in [8,9]. For optimizing the performance of the converter, an optimization algorithm was used to tune the controller's coefficients [9]. A constant power load in a shipboard DC microgrid was investigated for the finite time by adopting the finite-time disturbance observer method [10]. The estimated load power was then received by the fixed-time terminal sliding mode controller to stabilize the entire marine power grid as well as tracking the reference voltage of the DC bus in a fixed time independent of initial conditions. For a high-efficiency PV system, a dual-power stage micro-inverter (high voltage gain DC-to-DC converter + DC-to-AC inverter) was issued in the market. In many industrial applications such as those found in the automotive, telecommunication and shipping industries, systems need higher voltage gain DC-to-DC converters with large input current [11–13]. These converters typically boost the range from 24–60 V to 100–300 V. For example, automotive headlamps need 48/100/120 V range, but the vehicle battery capacity can only deliver 12/24/48 V. For these situations, high voltage converters are suggested with high voltage gain [14,15]. According to the theoretical calculations, the conventional boost converter can offer a high duty ratio with infinite voltage gain. However, in a practical case, it is limited due to the inductor saturation limits. Besides, any DC-to-DC converter, which needs to provide a high output voltage and high power conversion, draws large input current; hence, the power switches metal-oxide-semiconductor field-effect transistor (MOSFET) and diode) are needed to handle the voltage stress.

Topologies have presented numerous single switch converters in the literature to provide the high step-up voltage conversion [14–20]. These converters have some limitations for their voltage gain, which is mainly because of the inductors, power semiconductor switches and the parasitic elements of those converters. Hence, a researcher has recommended using a step-up transformer with a converter to overcome this issue (flyback converter) [21]. It may be achieved with a high current for high power uses, which is not the most efficient. Other attempts have been made by using a single switch with a forward and tapped inductor connected converters are proposed for high voltage gain [22–24]. Though these converters have a controlled degree of freedom through the transformer turns ratio adjustment, considering the transformer size, the converter is large. The high voltage gain quadratic converters are the next choice in this group [25,26]. However, a quadratic converter has high voltage stress across the first semiconductor, resulting in it being more efficient than the classical converter. The impedance (using two inductors and two capacitors) source-based converter handles the buck-boost voltage conversion with high gain; however, it needs a high voltage-handling power switch to operate shoot-through and not shoot-through operation. A high voltage gain integrated boost and flyback converter is proposed by [27] using the leakage inductor energy recirculation in the switch-off period; however, this method suffered from pulsating input currents.

The converter topology accumulated with a coupled inductor produces a high step-up conversion with high efficiency [28,29]. Here, the voltage gain is dealt with by changing the converter switch turns ratio like isolated type converters. Though this topology obtains a high gain, due to the coupled inductor-leakage inductance, the switch may suffer a high voltage spike. The passive and active clamping approaches have been established [30] by adding a coupled inductor for the high voltage conversion ratio; however, this converter is inefficient in terms of cost and size. The other choice is adapting a switched capacitor in the switching-mode converter for the high voltage gain [28,29]. Due to the pulse shape, the input current of the converter leads to a weak load and line regulation problems. The voltage-doubler concept in converters [30–34] can provide a high voltage gain with the coupled inductor. Here, the switching frequency is less than an inductor magnetizing current frequency, which is not suitable for reliable operation. Recently, using the single switch in DC-to-DC power converters, various research papers have been published to derive high voltage gain without using a higher duty ratio [35–40]. The authors in [33] proposed a 2D/1-D voltage gain single switch buck–boost converter with low input current ripple and appropriate voltage gain. Nevertheless, it has the discontinuous current in the input side. In paper [37], a transformerless high voltage gain buck–boost was proposed with a voltage gain of 3D/1-D; though this converter has a discontinuous input current. A quadratic DC–DC buck–boost converter with single-switch topology is presented in [39] for widespread voltage conversion. The high step-up single switch converter is recommended for PV-based grid applications [30]. However, in this converter, the low-level input voltage habitually roots massive input current and higher current ripples. This large amount of current affects the power switch during the higher duty ratio operation, causing a large conduction loss. Recently, Banaei et al. proposed a converter using a single switch with less switching voltage stress. Even though the converter can maintain the continuous input current for all duty ratios, the primary power switch voltage stress is strictly equal to the converter output voltage, which caused high conduction losses [38]. In [40,41], the single switch Cuk topology uses an extra inductor and capacitor to provide the extra voltage. However, in this topology, when diodes are operating with higher current and voltage, the diode reverse recovery current is predominant, which increases the switching losses. Among all converter topology, the cascade boost converter type is proficient in obtaining a higher gain with minimal duty ratio for the full range of voltage gain [19]. Nevertheless, the main switch has higher voltage stress in this topology.

Based on this discussion, and although several single switch boost converter topologies are proposed in the literature, their major approaches are concerned with the use of less magnetic elements, size, weight, conduction losses and cost-savings for the inductors. These approaches have higher voltage stresses on their switch (nearly the same as their output voltage). Therefore, these converters still have significant challenges, such as high step-up requirements for the larger duty ratio, output diode reverse recovery complication, higher switching voltage stress and satisfactory efficiency. Integrating the classical DC-to-DC converter is a better choice than modifying an existing converter. However, while cascading those classical converters with additional boosting capability, reducing the power switches and passive elements leads to an appropriate converter circuit size as well as cost reductions. With that aim, in this paper, we propose a DC–DC converter topology by integrating the conventional boost and Cuk converters with high voltage gain. Our proposed topology uses only one power switch with higher static voltage gain when compared to other conventional converters. Our proposed converter delivers voltage using two series-connected identical capacitors connected in parallel with the converter circuit. In this paper, we also deal with the maintenance of the capacitor balancing. The detailed converter mode operation, analysis, design, small-signal analysis and analytical switching losses were deliberated. The MATLAB/Simulink simulation and PIC microcontroller-based experimentation results for the integrated converter shows the advantages and practicality of the proposed converter.

The paper is organized as follows: Section 2 describes the proposed hybrid converter design and its mode of operation. Section 3 describes the converter components design and small-signal analysis. Section 4 explains the design procedure and components. The simulation, experimentation and comparison with other similar topologies are discussed in Sections 5 and 6. The conclusion is given at the end of Section 7.

2. Topology and Operation of Proposed Hybrid DC–DC Converter

The proposed integrated hybrid converter combines the conventional boost converter and classical Cuk DC-to-DC converter. The method used for the design of the proposed hybrid converter topology is illustrated in Figure 1. Figure 1a,b shows the typical conventional boost converter and classical Cuk converter, respectively. In both of these converters, the input inductor, power switches and input source are organized in the same way. Hence, there is an opportunity to merge these two converters by keeping the power switch and input inductor commonly on the input side. Except for the input side boost inductor (L_1) and the power switch T, the rest of the circuit is connected precisely in parallel with each other. Hence, the output side two capacitors (C_1 and C_3) are placed across the load. This hybrid structure increases the voltage gain by complementing the benefits of boost and Cuk converters. The converter provides continuous current mode operation with the help of a single power switch, which provides less voltage stress on the controlled switch and diodes.

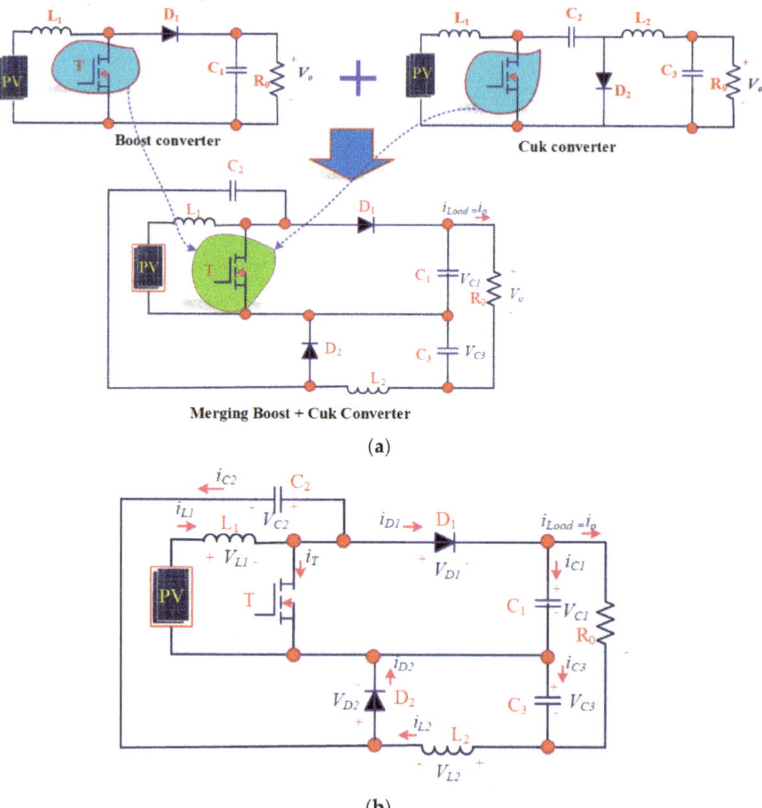

Figure 1. (**a**) Proposed converter integration; (**b**) proposed single switch hybrid DC–DC converter.

Operation of Hybrid DC–DC Converter

The proposed hybrid DC–DC converter mode operations, their capacitors (C_1, C_2 and C_3) and inductors (L_1 and L_2) charging and discharging analysis derivations were considered as follows. The two assumptions were taken for this analysis: (1) all the components are ideal; (2) the converter works under continuous conduction. Figure 3a illustrates the continuous conduction operating mode waveforms of the proposed converter. This advanced hybrid DC–DC converter mode operation has three modes of operation.

- Mode-I [t_a–t_b], presented in Figure 2a. During mode-1, when t = t_a, the power switch T is turned ON and inductors (L_1 and L_2) are charging until t_b. In the same interval, the capacitor C_2 is discharging through T, and inductor L_2 as the diodes (D_1 and D_2) are blocking concerning V_{C1} and V_{C2}.
- Mode-II [t_b–t_c], illustrated in Figure 2b. During mode-1, when t = t_b, the power switch T is in the OFF state. Now the capacitor C_1 voltage (V_{C1}) is higher than V_{C2}. Hence, after t_b interval, the C_2 is charging and inductors L_1 and L_2 are discharging. It is happening throughout t_b to t_c. In the course of this period, diode D_2 is continuously conducting since diode D_1 is still in reverse bias.
- Mode-III [t_c–t_d], presented in Figure 2c. During this mode, the power switch T remains OFF as well as the V_{C1} is equal or lesser than V_{C2}. Here, both the inductors L_1 and L_2 are discharging, and C_1 and C_2 are charging via L_1. Hence the diode D_1 and D_2 are conducting and delivering the current to load.

The proposed hybrid DC-to-DC converter operation mode waveforms are presented in Figure 3. In mode-I, the power switch T is 'ON' and turned 'OFF' in mode-II as well as mode-III. The proposed converter duty ratio versus voltage gain performance was compared to boost and Cuk converters. Figure 3b displays the voltage gain versus duty ratios for boost, Cuk and the proposed hybrid converter. Based on the plot, it can be seen that the proposed hybrid converter has a better voltage gain ratio when compared with the boost and Cuk converter, respectively. In addition to the extended voltage static gain, the proposed topology achieves lower voltage stress across the power switch and diodes, when compared to a boost converter.

The power switch voltage stress is equal to the peak voltage across the capacitor C1. Hence, voltage stress across the power switch T was lower than the total output voltage. The voltage stress of D1 and D2 is equal to the voltage across the power switch T in the OFF state. Hence, the diodes' current rating requirement was lower than power switch T. In the classical boost converters for both Cuk and boost, the power switch and diode need an equal rating. In the higher gain operations, the classical converter needs a higher duty cycle than the proposed converter. Therefore, the voltage stress for switch and diode are higher. Hence, the proposed converter efficiency in the higher gain operation is better than the conventional boost converter.

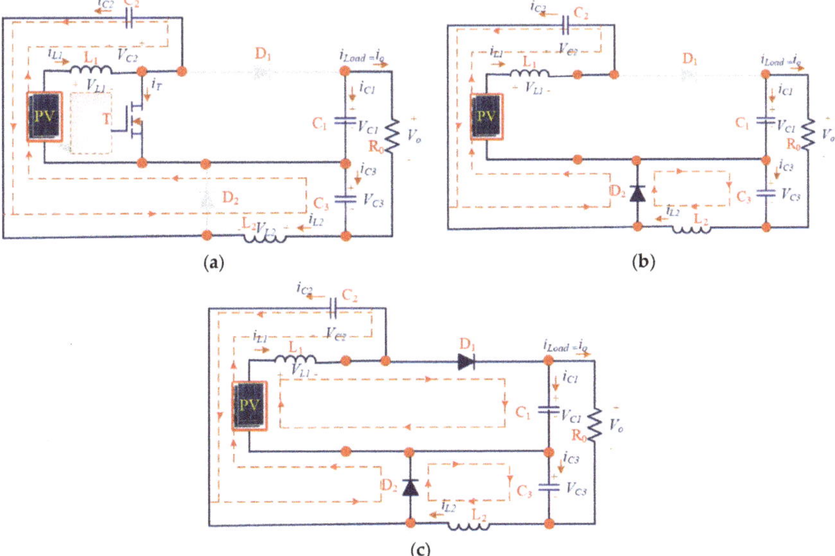

Figure 2. Modes of operation of the proposed hybrid DC-to-DC converter: (**a**) Mode-I [t_a–t_b]; (**b**) Mode-II [t_b–t_c]; (**c**) Mode-III [t_c–t_d].

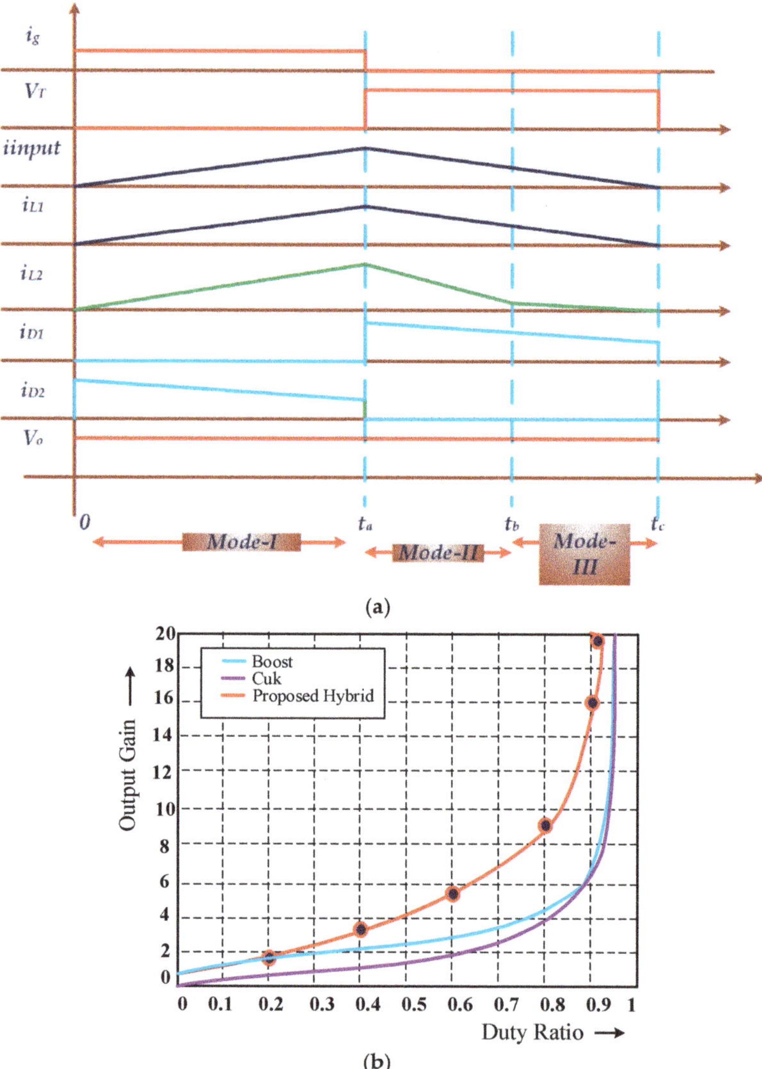

Figure 3. (a) Mode diagram of the proposed circuit; (b) voltage gain versus duty ratio for boost, Cuk and proposed hybrid converter.

In general, for any DC-to-DC converter, the input inductor selection is carried out depending on the converter conduction mode, load current requirement, and it is desirable to confirm the least output current ripple. Hence, the input inductor L_1 value was chosen with minimal current ripple Δ_{iL}. The inductor current, i_L for the proposed converter is supplied from either a PV or DC source (V_{PV} or V_{in}). When the converter receives a supply voltage from the input source, the converter power switch T is turned ON, and inductor current i_{L1} is derived as follows. Applying Kirchhoff's voltage law in mode-I [t_a–t_b], the capacitors current i_{C1} and i_{C2} are derived as

$$\begin{aligned}
-V_{PV} + v_{L1} = 0 &\Rightarrow v_{L1} = V_{PV} \\
-v_{C2} + v_{L2} + v_{C3} = 0 &\Rightarrow v_{L2} = -v_{C3} + v_{C2}
\end{aligned} \tag{1}$$

$$-i_T + i_{L1} - i_{C2} = 0$$
$$i_O - i_{L2} - i_{C3} = 0 \tag{2}$$

$$i_{C2} = -i_{L2}$$
$$i_{C1} = -i_O \tag{3}$$

During mode-II [t_b–t_c], when the power switch T is in OFF, the coil transfers energy to the capacitors C_1 and C_3. As a result, from loop 1 and loop 2, it can verify that

$$-V_{PV} + v_{L1} + v_{C2} = 0 \Rightarrow v_{L1} = -v_{C2} + V_{PV}$$
$$v_{L2} + v_{C3} = 0 \Rightarrow v_{L2} = -v_{C3} \tag{4}$$
$$v_{C2} = v_{C1}$$

From loop 1 and loop 2 (mode-II Figure 2b), the capacitor current i_{C1} and i_{C2} are derived as follows,

$$i_{C1} = i_{D1} - i_O$$
$$i_{C2} = i_{L1} - i_{D1} = i_{D2} - i_{L2} \tag{5}$$

$$I_C = \frac{1}{T} \int_0^T i_C dt = \frac{C}{T} \int_0^T dv_C(t) = \frac{C}{T}(v_C(T) - v_C(0)) \tag{6}$$

during steady-state condition $V_{C(T)} = V_{C(0)}$. Hence, the average value of the current capacitors is null. Also, the inductor coils average voltage is zero, since $v_L(t) = L\frac{di_L(t)}{dt}$. The currents in the inductors and voltage in the capacitors tend to be approximately constant. The power switch is in the ON state for a percentage of the period (δT) and OFF during the next state (δT-T). Here, T is the total switching time. Therefore, the average value inductor voltage V_{L1} and capacitor's current i_{C1} are illustrated in Figure 4 and Figure 7.

$$V_{L1} = \frac{1}{T}\left[\int_O^{\delta T} V_{PV} dt + \int_{\delta T}^T (V_{PV} - V_{C1}) dt \right] = 0 \tag{7}$$

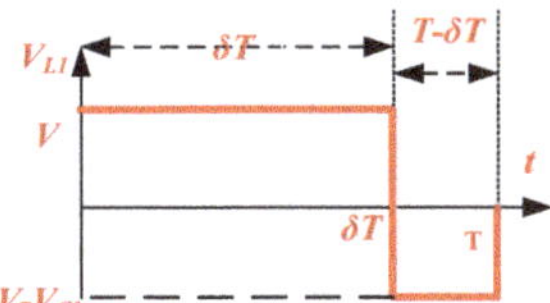

Figure 4. Inductor voltage V_{L1}.

Replacing the voltage on the calculation, we obtain the capacitor voltage V_{C1}

$$V_{C1} = \frac{1}{1-\delta} V_{PV} \tag{8}$$

The inductor L_2 voltage is illustrated in Figure 5, and inductor average voltage value is

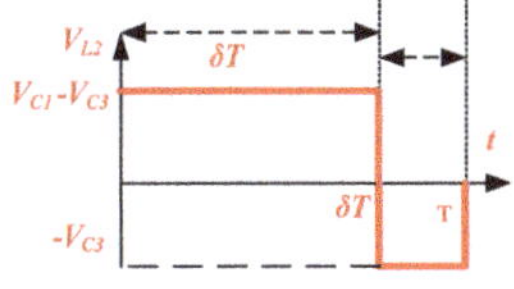

Figure 5. Inductor voltage V_{L2}.

$$V_{L2} = \frac{1}{T}\left[\int_0^{\delta T}(V_{C1} - V_{C3})dt + \int_{\delta T}^{T} -V_{C3}dt\right] = 0 \tag{9}$$

The voltage across the capacitor C_3 can be written as

$$V_{C3} = \frac{\delta}{1-\delta}V_{PV} \tag{10}$$

Hence, the converter output voltage can be calculated as

$$V_O = \frac{1+\delta}{1-\delta}V_{PV} \tag{11}$$

By equating the converter input power and output power, the input inductor current is derived as

$$V_{PV}I_{L1} = V_O I_O \Rightarrow I_{L1} = \frac{1+\delta}{1-\delta}I_O \tag{12}$$

When the Kirchhoff's current law is applied in the loop

$$i_{C3} + i_O - i_{L2} = 0 \Rightarrow i_{L2} = i_{C3} + i_O \tag{13}$$

The capacitor C_3 charge and discharge current is shown in Figure 6. Here the capacitor current average value observed is zero, and the average current in the inductor is similar to the average output current, as the inductor tends to retain that average value.

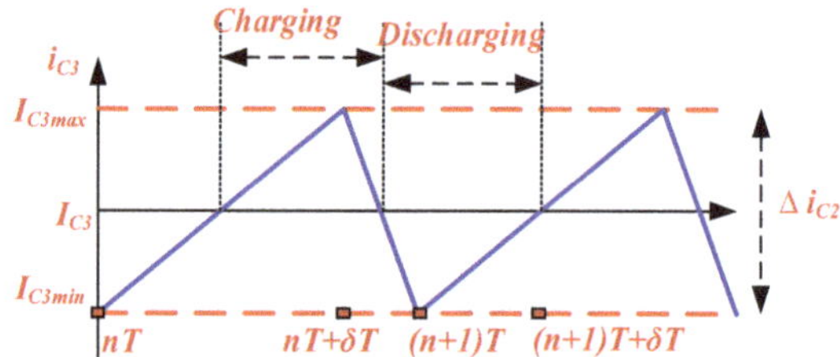

Figure 6. Capacitor current i_{C3}.

Assuming that $i_0 = i_{l2}$

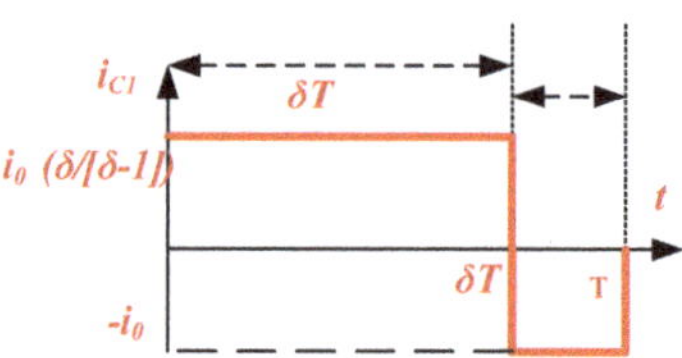

Figure 7. Capacitor current C_1.

The mean value of the capacitor C_1 current (i_{C1}) is displayed in Figure 7. From the capacitor current interval zero to δT and δ to T time, the i_{C1} is calculated as

$$I_{C1} = \frac{1}{T}\left[\int_0^{\delta T}(-i_O)dt + \int_{\delta T}^{T}(i_{D1} - i_O)dt\right] = 0 \tag{14}$$

$$i_{D1} - i_O = i_O\frac{\delta}{1-\delta} \tag{15}$$

When the power switch T is turned ON, the diode D_1 current is calculated as

$$i_{D1} = \frac{I_O}{1-\delta} \tag{16}$$

Similarly, when the power switch T is turned OFF, the diode D_2 current is calculated as

$$i_{L1} - i_{D1} = i_{D2} - i_{L2} = I_O \frac{\delta}{1-\delta} \tag{17}$$

$$i_{D2} = \frac{I_O}{1-\delta} \tag{18}$$

The current of the power switch T can be written as

$$i_T = i_{L1} - i_{C2} = \frac{2I_O}{1-\delta} \tag{19}$$

3. Scaling Converter Components Design

The proposed hybrid converter reactive components, inductors coil (L_1 and L_2) and capacitors (C_1, C_2 and C_3), are calculated for maximum values as the power switch T should support both the converter voltage and current.

3.1. Design of Inductors

The inductor coil (L_1 and L_2) values calculation and current limitation analysis are observed by precise variation concerning the average value shown in Figure 8. The differential equation of inductor voltage V_L is shown as

$$v_L(t) = L \frac{di_L(t)}{dt} \tag{20}$$

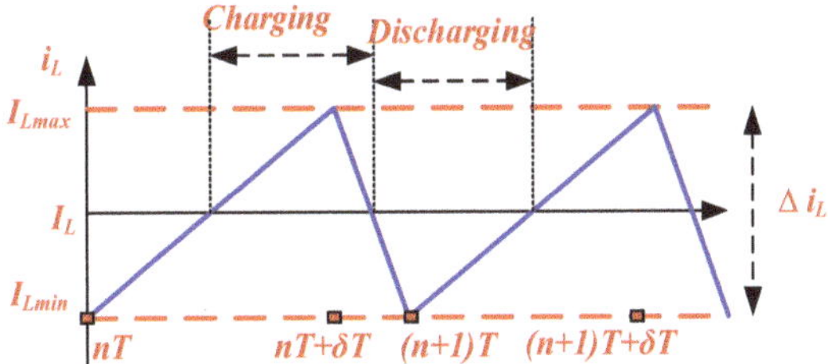

Figure 8. Evolution of the inductor current.

By assuming inductor voltage V_L nearly constant, the current equation inductor is calculated as follows

$$i_L(t) = \frac{v_L}{L} \Delta t_L + i_L(t0) \tag{21}$$

Thus, $v_L(t) = L \frac{di_L(t)}{dt}$ it becomes

$$\frac{\Delta i_L}{\Delta t_L} = \frac{v_L}{L} \tag{22}$$

Using the instantaneous inductor voltage equation, the inductor voltage will be

$$v_{L1} = \begin{cases} V_{PV}, & nT < t < nT + \delta T \\ V_{PV} - V_{C1}, & nT + \delta T < t < (n+1)T \end{cases} \tag{23}$$

During 'ON' state of the power switch, $\Delta t_L = \delta T$

$$L_1 = \frac{V_{PV}\delta T}{\Delta i_{L1}} \tag{24}$$

$$v_{L2} = \begin{cases} V_{C1} - V_{C3}, & nT < t < nT + \delta T \\ -V_{C3}, & nT + \delta T < t < (n+1)T \end{cases} \tag{25}$$

For the 'OFF' state of the power switch, it has $\Delta t_L = (1-\delta)T$

$$L_2 = \frac{v_{C3}(1-\delta)T}{\Delta i_{L2}} \tag{26}$$

3.2. Design of Capacitors

The calculation of capacitor values C_1, C_2 and C_3 are given below. The changing and discharging variation around the average value is shown in Figure 9.

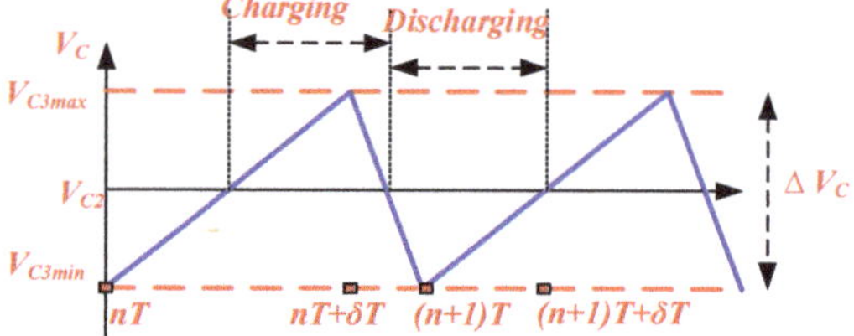

Figure 9. Evolution of the current in the coil.

The differential equation of the capacitor

$$i_c(t) = C\frac{dv_c(t)}{dt} \tag{27}$$

Similar to the calculation of inductors, the capacitor variation is calculated using Equation (27) after linearization of $i_c(t) = C\frac{dv_c(t)}{dt}$ becomes,

$$\frac{\Delta v_c}{\Delta t_c} = \frac{i_C}{C} \tag{28}$$

With the instantaneous value of the capacitor current over a certain period, the value of the capacitor can be calculated. For capacitor C_1, the current is given by:

$$i_{C1} = \begin{cases} -I_0, & nT < t < nT + \delta T \\ I_0\frac{\delta}{1-\delta}, & nT + \delta T < t < (n+1)T \end{cases} \tag{29}$$

During the first-time interval, for $\Delta t_c = \delta T$, C_1 is

$$C_1 = \frac{I_0\delta T}{\Delta v_{C1}} \tag{30}$$

The capacitor current C_2 is given by

$$i_{C2} = \begin{cases} -I_0, & nT < t < nT + \delta T \\ I_0\frac{\delta}{1-\delta}, & nT + \delta T < t < (n+1)T \end{cases} \tag{31}$$

During the time interval, when $\Delta t_c = \delta T$

$$C_2 = \frac{I_0 \delta T}{\Delta v_{C2}} \tag{32}$$

For capacitor C_3, the current does not show instantaneous values and is nearly constant during the switching state. The behavior of capacitor C_3 is opposite capacitors C_1 and C_2. When controlling the power semiconductor switch for the driving load variation, capacitor charge ΔQ is related to the inductor current $\Delta i_{L2}/2$ and time is taken T/2.

$$Q = Cv_c \Rightarrow C = \frac{Q}{v_c}, \; C_3 = \frac{\Delta Q}{\Delta v_{c3}} \tag{33}$$

$$\text{At load variation, } \Delta Q = \frac{\frac{T}{2}\frac{\Delta i_2}{2}}{2} = \frac{T \Delta i_{L2}}{8} \tag{34}$$

$$\text{The capacitor } C_3 \text{value is calculated as } C_3 = \frac{T \Delta i_{L2}}{8 \Delta v_{c3}} \tag{35}$$

In the dynamic condition, capacitors C_1 and C_3 values are deliberate in this section. The calculation is computed by including the sudden change in drive load resistance. Output voltage in the dynamic operating region is determined using the equivalent circuit (Figures 10 and 11), assuming that the current passes zero to its steady-state value, Δt_1, the settling time of the current in the inductor L_1.

$$i_{C1} = C_1 \frac{dv_{C1}}{dt} = -i_0 \quad \Rightarrow v_{C1}(t) = \frac{1}{C_1} \int_0^t (-i_0)dt + v_{C1}(0) \tag{36}$$

$$v_{C1}(\Delta t_1) = \frac{1}{C_1} \int_0^t (-i_0)dt + v_{C1}(0) = -\frac{1}{C_1}\frac{P_0}{V_0}\Delta t_1 + v_{C1}(0) \tag{37}$$

Thus, capacitor voltage VC_1 is calculated using Equation (37)

$$v_{C1}(\Delta t_1) = \frac{1}{C_1} \int_0^t (-i_0)dt + v_{C1}(0) = \frac{1}{C_1}\frac{P_0}{V_0}\Delta t_1 \Rightarrow C_1 = \frac{1}{\Delta v_{C1}}\frac{P_0}{V_0} \tag{38}$$

In a dynamic case, while changing the converter duty ratio, the inductor current increases coil rapidly. Hence, the Δt_1 value needs to calculate, in detail, from the response of $i_{LI}(s)$, which displays the input current of the response when rapid changes occur in output current $i_0(s)$. At time Δ_{t2}, the current flow through the capacitor C_3 is calculated as

$$i_{C3} = C_2 \frac{dv_{C3}}{dt} = -i_0 \tag{39}$$

$$C_3 = \frac{1}{\Delta v_{C3}}\frac{P_0}{V_0}\Delta t_2 \tag{40}$$

For calculation straightforwardness, it is assumed that the current has equal settling times.

In terms of the energetic properties, the proposal converter is similar to the boost and Cuk converters. The proposed converter input inductor, power switches and input source are organized the same way as the classical boost and Cuk converters. Except for the input side boost inductor (L_1) and the power switch T, the rest of the proposed converter circuit elements are connected precisely in parallel with each other and on the output side, two capacitors (C_1 and C_3) are placed across the load. Therefore, the proposed converter increases the voltage gain by combining the benefits of boost and Cuk converters.

3.3. Small Signal Analysis of Hybrid DC–DC Converter

The average equivalent circuit model of the proposed hybrid DC–DC converter was derived and is presented in Figure 10. The circuit analysis was derived for the switch in both the ON and OFF period. The initial conditions were an approach to obtain the average value of the coil; the current remained the same.

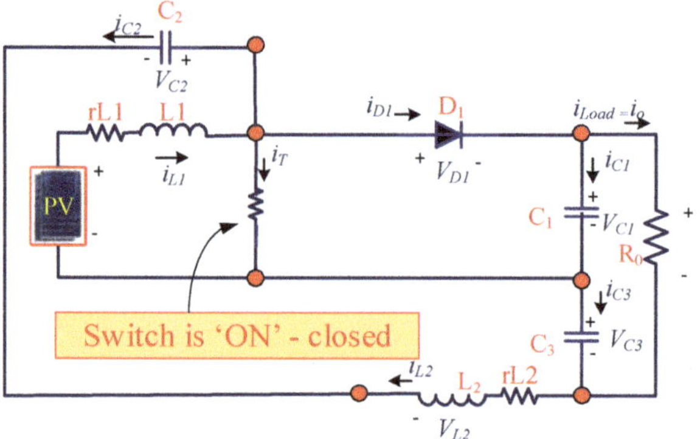

Figure 10. Equivalent circuit of the converter with losses in the mode-1 operation.

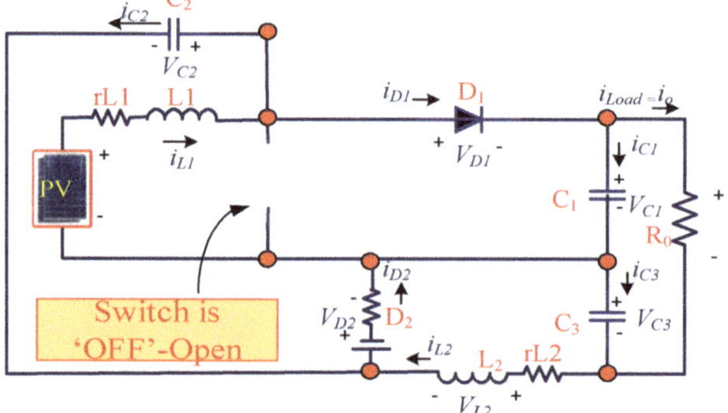

Figure 11. Equivalent circuit of the converter with losses in mode-2 and mode-3 operation.

By the application of mesh law to the circuit in Figure 10, we can verify that

$$\begin{cases} -V_{PV} + v_{L1} + r_{L1}I_{L1} + R_{DSon}i_T = 0 \\ v_{C2} - v_{L2} - R_{DSon}i_T - v_{C3} - r_{L2}I_{L2} = 0 \end{cases} \Rightarrow \begin{cases} v_{L1} = -V_{PV} - r_{L1}I_{L1} - R_{DSon}i_T \\ v_{L2} = v_{C2} - R_{DSon}i_T - v_{C3} - r_{L2}I_{L2} \end{cases} \tag{41}$$

$$\begin{cases} -V_{PV} + v_{L1} + r_{L1}I_{L1} + v_{C2} + V_{D2} + R_{D2}i_{D2} = 0 \\ v_{L2} + r_{L2}I_{L2} + v_{C3} + V_{D2} + R_{D2}i_{D2} = 0 \end{cases} \Rightarrow \begin{cases} v_{L1} = -V_{PV} - r_{L1}I_{L1} - v_{C2} - V_{D2} - R_{DSon}i_T \\ v_{L2} = -r_{L2}I_{L2} - v_{C3} - V_{D2} - R_{D2}i_{D2} \end{cases} \tag{42}$$

$$-V_{PV} + v_{L1} + r_{L1}I_{L1} + v_{C1} + v_{D1} + R_{D1}i_{D1} = 0 \tag{43}$$

$$v_{D1} = v_{D2}, \; R_{D1}i_{D1} = R_{D2}i_{D2}, \; v_{C2} = v_{C1}$$

From the waveform v_{L1} and v_{L2},

$$V_{L1A} = V_{PV} - R_{Dson}I_T - r_{L1}I_{L1} \tag{44}$$

$$V_{L1B} = V_{PV} - r_{L1}I_{L1} - V_{C1} - V_{D2} - R_{D2}I_{D2} \tag{45}$$

$$V_{L2A} = V_{C1} - R_{DSON}I_{SC} - V_{C3} - r_{L2}I_{L2} \tag{46}$$

$$V_{L2B} = -r_{L2}I_{L2} - V_{C3} - V_{D2} - R_{D2}I_{D2} \tag{47}$$

The analysis of converter input to output relation is calculated with losses and approximated with ideal semiconductor devices. The output voltage expressions of the converter were derived and are given below.

$$V_{L1} = 0 \Rightarrow V_{C1} = \frac{V_{PV} - r_{L1}I_{L1}}{1 - \delta} \tag{48}$$

$$V_{L2} = 0 \Rightarrow V_{C3} = \frac{\delta}{1 - \delta}(V_{PV} - r_{L1}I_{L1}) - r_{L2}I_{L2} \tag{49}$$

From the expressions V_{C1} and V_{C2}, the voltage gain converter is calculated as

$$\frac{V_0}{V_{PV}} = \frac{1 + \delta}{1 - \delta + \frac{r_{L1}}{R_0}\frac{(1+\delta)^2}{1-\delta} + \frac{r_{L2}}{R_0}(1 - \delta)} \tag{50}$$

where $r_{L1}/R_0 = r_{L2}/R_0$.

Figure 12 illustrates the voltage output gain versus the duty cycle for the proposed hybrid converter. The Figure indicates the ideal condition ($r_{L1}/R_0 = 0$), where losses need to be introduced in the circuit (the gain for unit value goes to zero, as expected) and other operating conditions $r_{L1}/R_0 = r_{L2}/R_0 = 0.0001$ to 0.76, where near 0.76 duty cycle, the converter gain approaches six times boosting ($V_0 = 6V_{in}$).

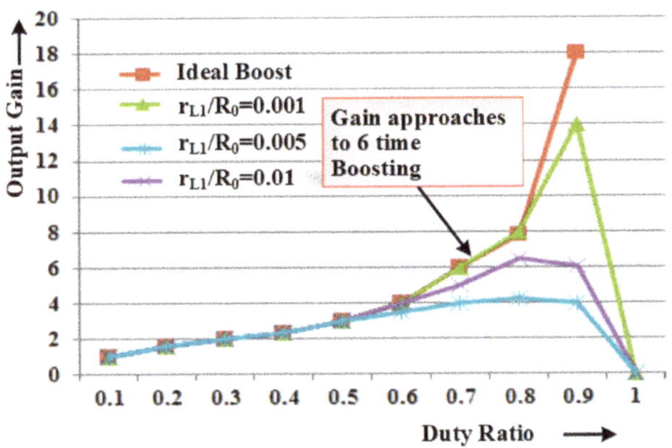

Figure 12. Voltage output gain versus duty cycle.

3.4. Analysis of Losses

The losses of each indictor L_1 and L_2 are denoted from internal resistors r_{L1} and r_{L2}, respectively. Thus, the losses in r_{L1}

$$P_{\rightleftarrows r_{L1}} = p_1 \rightleftarrows P_i \rightleftarrows = r_{L1}I_{L1rms^2} \Leftrightarrow r_{L1} = \frac{p_1 P_i}{I_{L1rms}^2} \tag{51}$$

The sufficient value is given as

$$I_{L1rms} = \sqrt{I_{L1}^2 + \left(\frac{\Delta i_{L1}}{2\sqrt{3}}\right)^2} = \sqrt{I_{L1}^2 + \left(\frac{0.1 I_{L1}}{2\sqrt{3}}\right)^2} = I_{L1}\sqrt{1 + \left(\frac{0.1}{2\sqrt{3}}\right)^2}$$

$$\Delta i_{L2} = 0.1 I_{L2} \tag{52}$$

Since both the inductors are identical, $\Delta i_{L1} = 0.1 I_{L1}$ \qquad (53)

Two kinds of losses in the semiconductor are driving and switching.

3.5. Conduction Losses in the Diodes

The losses in the diode are given as

$$P_D = \frac{1}{T}\int_0^T v_D(t)i_D(t)dt = V_D I_D + R_D I_{Drms}^2 \tag{54}$$

$$v_D(t) = V_D + R_D i_D(t) \tag{55}$$

Specific to the case of the diode, for mode-2 operation:

$$i_{D1}(\delta T < t < T) = \frac{I_0}{(1 - \delta)} \tag{56}$$

$$I_{D1} = \frac{1}{T}\int_0^T i_{ak1}(t)dt = I_0 \tag{57}$$

$$I_{D1rms} = \sqrt{\frac{1}{T}\int_0^T i_{D1}^2(t)dt} = \frac{I_0}{\sqrt{(1 - \delta)}} \tag{58}$$

Thus, resistance is calculated as

$$P_{D1} = \frac{p_3}{2}P_i = V_{D1}I_0 + R_{D1}\left(\frac{I_0}{\sqrt{(1 - \delta)}}\right)^2 \Leftrightarrow R_{D1} = \left(\frac{p_3}{2}P_i - V_{D1}I_0\right)\frac{(1 - \delta)}{I_0^2} \tag{59}$$

The same can be applied to D_2, resulting in the same results for this diode

$$R_{D2} = \left(\frac{p_3}{2}P_i - V_{D2}I_0\right)\frac{(1 - \delta)}{I_0^2} \tag{60}$$

4. Design Procedure

The component selections and other design parameters of the proposed converter at the power range of 150 W were calculated and are presented. The input power was considered a DC-fixed source. The general diagram of the converter design flow chart is shown in Figure 13.

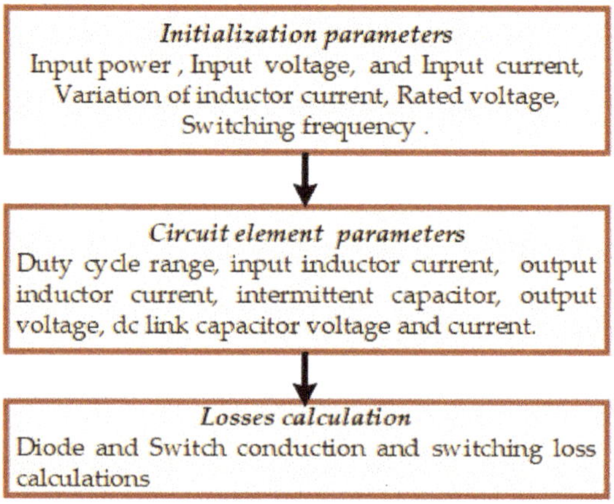

Figure 13. Design setup of converter parameter selections.

To illustrate the numerical values of converters, capacitor and inductors, the below parameters were fixed for the converter design.

i.   The input power P_{in} = 150 W, for V_{in} = 24 V, I_{in} = 6.2 A;
ii.   Power of the converter, P_i = 150 W;
iii.   Input voltage converter V_{in} = 24 V;
iv.   Duty cycle is fixed as δ = 0.8;
v.   The converter output voltage, V_o = 104 V;
vi.   The output current and inductor current were expected to be I_0 = 1.11 A and I_{L1} = 4.25 A, respectively;
vii.   The capacitor C_1 and C_3 voltages were calculated as V_{C1} = 110 V and V_{C3} = 104 V;
viii.   The general typical value sizing of capacitor C_1, C_3 and L_1, L_2 was calculated as Δi_{L1} = 10% I_{L1}, hence for L_1 = 1 mH, the change in this was Δi_{L1} = 0.425 A. The same changes can be seen for Δi_{L2} = 10% I_{L2}, L_2 = 1 mH and Δi_{L2} = 0.14 A;
ix.   When the change in the capacitor ΔV_{C1} was 1% V_{C1}, the capacitor C_1 value was 100 μF and ΔV_{C1} = 1.10 V. Similarly, for ΔV_{C2} = 1% V_{C2}, C_2 = 100 μF, ΔV_{C2} = 1.04V and ΔV_{C3} = 1% V_{C3}, C_3 = 2 μF, ΔV_{C3} = 0.08 V;
x.   For the power semiconductor switch, the maximum open-circuit voltage was Vs_{max} = 100 V, Vs_{max} = 95 V;
xi.   Diodes D_1 and D_2, Vs_{max} = 100 V, Vs_{max} = 95 V.

The D_1 and D_2 can ensure a voltage of 100V, which ensures the safety factor of the converter. The converter can support a maximum current of six amps. When using six amps, the current safety factor is reduced to 60%–65%. Hence, the semiconductor must be selected to withstand the converter to provide maximum currents and voltages with a safety factor around 50%. The n-type reinforcing MOSFET is better chosen for providing the safety factor, and the proposed converter design uses the same [35]. The diodes (D_1 and D_2), and MOSFET switching losses and conduction losses were calculated and given in Equations (61)–(73).

Conduction losses of the diode D_1 and D_2:

$$P_D = \int V_{ak}\,(t)\,i_{ak}\,(t)dt = V_D i_{ak} + R_D I_{akrms}^2 \tag{61}$$

$$R_{D1} = \left(\frac{p3}{2} P_i - V_{D1} I_o \right) \frac{(1-\delta)}{I_o^2} \tag{62}$$

$$R_{D1} = \left(\frac{p3}{2} P_i - V_{D2} I_o \right) \frac{(1-\delta)}{I_o^2} \tag{63}$$

$$P_{D1} = V_{D1} i_{ak1} + R_{D1} I_{akrms1}^2 \tag{64}$$

$$P_{D2} = V_{D2} i_{ak2} + R_{D2} I_{akrms2}^2 \tag{65}$$

Switching losses of the Diode D_1 and D_2:

$$P_{SD1} = \frac{t_{rr} - t_s}{T} V_{ak1} i_{ak1} \tag{66}$$

$$P_{SD2} = \frac{t_{rr} - t_s}{T} V_{ak2} i_{ak2} \tag{67}$$

Conduction losses of the MOSFET:

$$P_{MOSFET\ Conduction\ Loss} = R_{DS\_ON} i_{MOSFET\ rms} \tag{68}$$

$$i_{\text{MOSFET}} = \frac{2I_0}{(1 - \delta)} \tag{69}$$

$$P_{P_{\text{MOSFET switcing loss}}} = \frac{t_{\text{ON}} + t_{\text{OFF}}}{T}(V_{\text{MOS}}i_{\text{MOS}}) + \frac{1}{2T}(C_{\text{MOS}}V_{\text{MOS}})^2 \tag{70}$$

Total switching losses for the proposed converter is:

$$P_{\text{total switcing losses}} = \frac{t_{\text{ON}} + t_{\text{OFF}}}{T}(V_{\text{MOS}}i_{\text{MOS}}) + \frac{1}{2T}(C_{\text{MOS}}V_{\text{MOS}})^2 + 2\frac{t_{\text{rr}-}t_s}{T}V_{\text{ak}}i_{\text{ak}} \tag{71}$$

$$t_{\text{ON}} = t_r(i_{\text{MOS}}) + t_f(V_{\text{MOS}}) \tag{72}$$

$$t_{\text{OFF}} = t_f(i_{\text{MOS}}) + t_r(V_{\text{MOS}}) \tag{73}$$

The efficiency of the proposed converter is found using

$$\eta = \left(\frac{P_o}{P_i}\right) = \frac{P_i - \Sigma P_T}{P_i} \tag{74}$$

where P_i = input power and ΣP_T = Total losses ($P_{\text{Diode2 Con.Losses}}$ + $P_{\text{Diode2 Con.Losses}}$ + $P_{\text{MOSFET switching loss}}$ + $P_{\text{MOSFET Conduction Loss}}$).

5. Simulation Results

The hybrid DC–DC converter operation and performance estimation were modeled in the MATLAB/Simulink simulation platform and waveforms were presented. The simulation specification and parameter were as follows: The input power = 150 W, input voltage of the converter (V_{in}) = 24 V, maximum duty cycle δ = 0.8 and switching frequency f_s = 10 kHz. The converter input and output inductors were L_1 = 1 mH and L_2 = 1 mH receptivity. The capacitors were C_1 = 100 μF, C_2 = 100 μF and C_3 = 2 μF. Figures 14–17 show the proposed converter simulation results for 10 kHz switching frequency and 80% duty cycle, and the results confirm the theoretical values. The converter duty cycle was fixed to be equal to or less than 0.8 to minimize the conduction losses. From Figure 14, when the converter duty cycle was fixed at 0.8 with 24V input voltage, the converter delivered an output voltage of 124 V (5.166 times higher than the input voltage). During the continuous conduction mode, the inductance L1 current was limited within the saturation limit in the range of 3 to 4.5 A and maintained the converter input current. Figure 15 displays the input current, as well as voltage across the power switches, and Figure 17 shows D_1 and D_2 voltages, V_{D1} and V_{D2}, respectively, during the switching period. From the results, it could be seen that during the time of switching, the switches (MOSFET and diode) were maintained with their maximum allowable voltage as 100 V. It was verified that the voltage across the switches was less than that of the converter. From the i_{L2} and V_{D2} simulation results, it can be seen that the proposed converter maintains a continuous current capability. Figure 16 shows the simulation waveforms for the inductor current i_{L1} and inductor current i_{L2}. From this waveform, it is seen that the inductors were charging uniformly and delivering the current in the continuous conduction. Figure 17 illustrates the voltage across the power diode, V_{D1} and V_{D2}. When the duty cycle was reduced to 0.6, the converter performance, switching reliability and continuous current capability were linear. Hence, the proposed converter has a wide range of controllability with a controlled degree of freedom to avail wider voltage outputs. The simulation was also performed in transient conditions (changing load and sudden change in the duty cycle). During this transient period, the output voltage and current through i_{L1} changed with a small transient period and after reaching the continuous conduction and maintaining the constant output voltage.

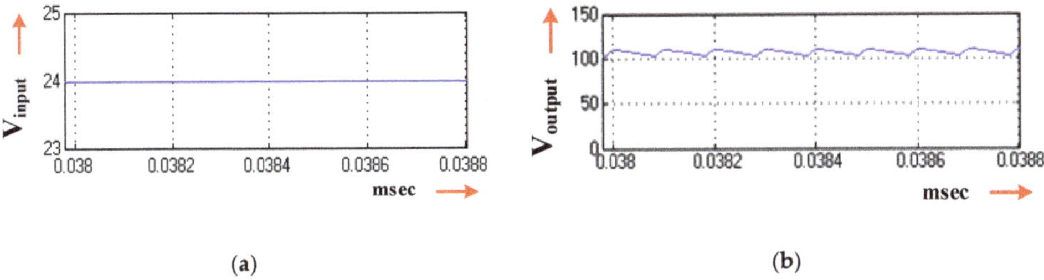

(a)   (b)

Figure 14. Simulation waveforms for input voltage 24V DC and 0.8 duty cycle; (**a**) input voltage waveform (voltage scale: 1 V/div and t: 20 μs/div) and (**b**) input voltage waveform (voltage scale 50 V/div and t: 20 μs/div).

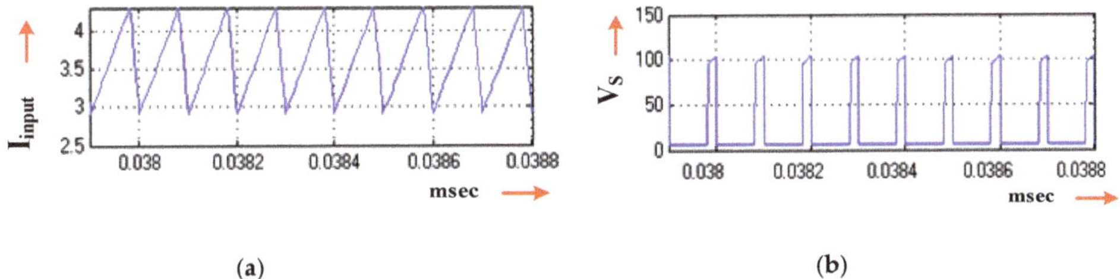

(a)   (b)

Figure 15. Simulation waveforms for input voltage 24V DC and 0.8 duty cycle; (**a**) input current waveform (current scale: 0.5 A/div and t: 20 μs/div) and (**b**) voltage across the power switch waveform (voltage scale 50 V/div and t: 20 μs/div).

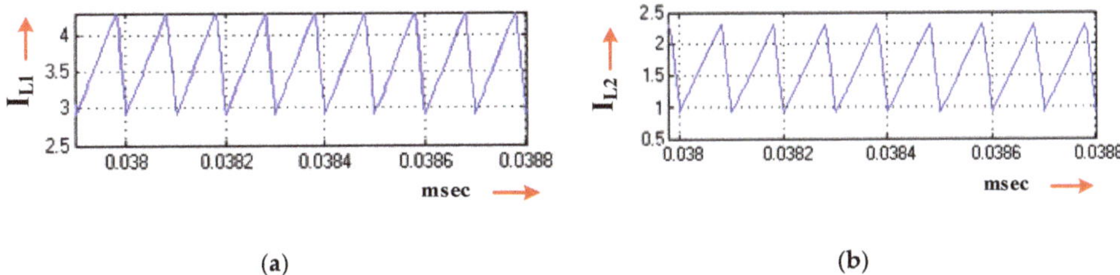

(a)   (b)

Figure 16. Simulation waveforms for input voltage 24 V DC and 0.8 duty cycle; (**a**) inductor current i_{L1} waveform (current scale: 0.5 A/div and t: 20 μs/div) and (**b**) inductor current i_{L2} waveform (current scale: 0.5 A V/div and t: 20 μs/div).

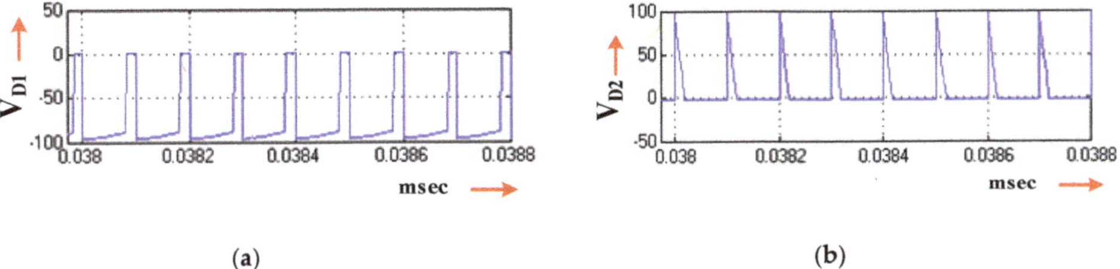

(a)   (b)

Figure 17. Simulation waveforms for input voltage 24V DC and 0.8 duty cycle; (**a**) voltage across the power diode, V_{D1} waveform (voltage scale: 50 V/div and t: 20 μs/div) and (**b**) voltage across the power diode, V_{D2} waveform (voltage scale: 50 V/div and t: 20 μs/div).

The proposed single switch hybrid DC–DC converter is compared with conventional converters (boost and Cuk) for different duty cycles from the range zero to one. The switching frequency and other circuit components for this evaluation were taken as being the same as the proposed converter values are given in the design (see Table 1). As expected, the proposed converter voltage gain was more than the boost and Cuk converter duty cycle. Figure 18 shows the voltage gain comparison of boost and Cuk with the proposed converter.

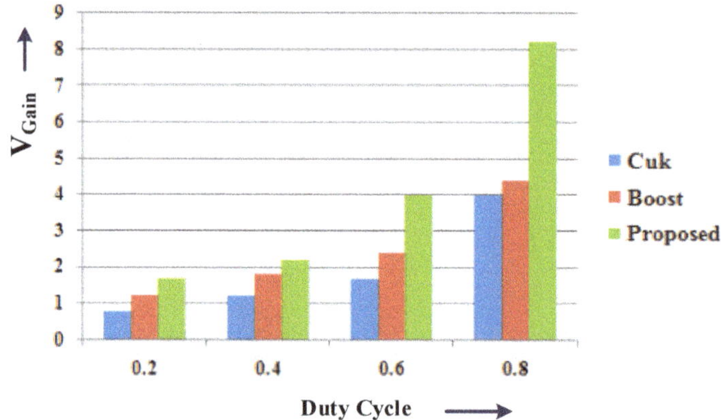

Figure 18. Voltage gain comparison of boost and Cuk with the proposed converter.

6. Experimental Results

To confirm the experimental performance of the proposed hybrid DC–DC converter, the experimental laboratory setup was developed in collaboration with a Peripheral Interface Controller (PIC) microcontroller, as shown in Figure 19. To verify the theoretical and simulation results, the experimentations were conducted with similar values considered in the simulation studies. The converter was a 150 W circuit with parameters as listed in Table 1.

Similar to the simulation verification, the converter duty cycle was fixed as equal to or less than 0.8 to minimize the conduction losses. While testing the converter, the input DC source was fixed to generate constant input voltage and power as 24 V and 150 W range. As seen in Figure 20, while the converter duty cycle was fixed as 0.8, the corresponding output voltage was observed to be 122 V (closer to the simulation results). Figure 21 shows the converter input current and output voltage.

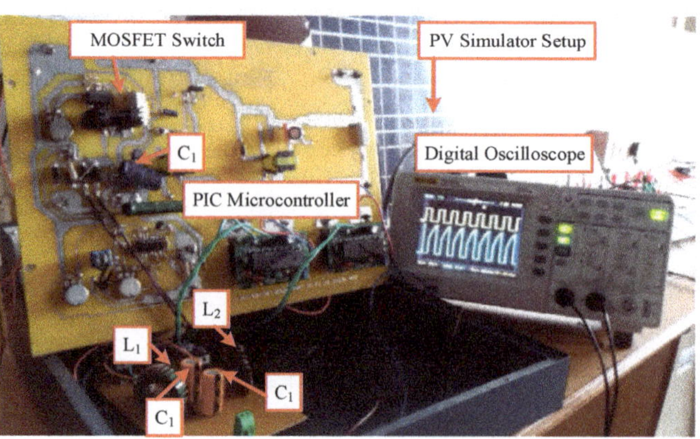

Figure 19. Prototype setup of the proposed converter.

Table 1. Parameter of components of the proposed converter.

Components	Parameter
Input power P_{input}	150 W
Input voltage V_{in}	24 V
Output power P_0	112 W
Switching frequency f_s	10 KHz
Power MOSFET	SiHB30N60E
Diode D_1 and D_2	VS-15EWX06FN-M3
Inductance L_1 and L_2	1 mH
Capacitor C_1, C_2 and C_3	100 μF, 100 μF and 2 μF
The output of Diode V_{D1} and V_{D2}	100 V and 95 V
Output Capacitor V_{C1}, V_{C2} and V_{C3}	104 V, 110 V and 8 V

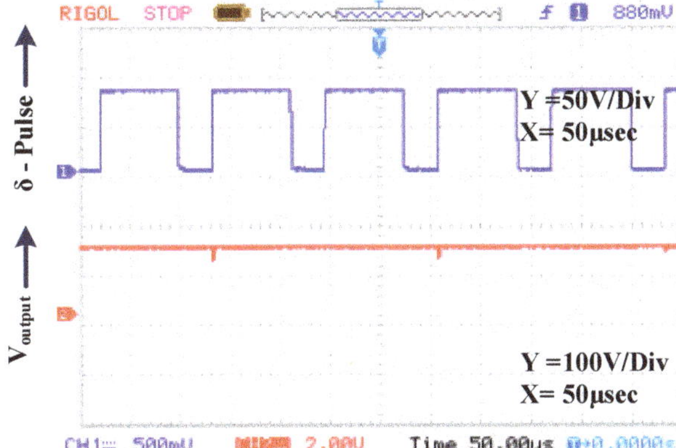

Figure 20. The experimental waveform of the duty cycle and input voltage.

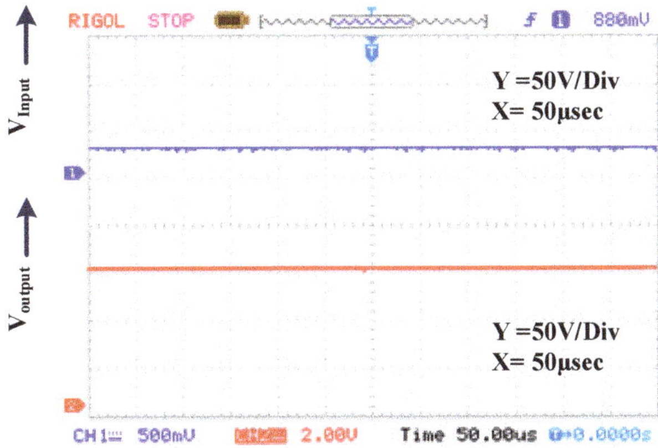

Figure 21. The experimental waveform of input voltage and output voltage for input voltage 24 V and 0.8 duty cycle.

During the DC-to-DC conversion period, the converter maintained the continuous conduction with the inductance L_1 current saturation limit range of 3 to 4.5 A, as depicted in Figure 22. Hence, the power switch was secured against the high rising current by maintaining the converter input current inductance L_1 current saturation limit, which ensures the converter reliability against the input source. Similarly, from Figures 23 and 24, during the time of switching, the MOSFET and diode were maintained with their maximum allowable voltage as 100 V, which was smaller than the converter output voltage (102 V). Here, during the switching period, the voltage across the main power switch was 100 V, and diode D_1 and D_2 were equal to $V_{D1} = 100$ V and $V_{D2} = 95$ V, respectively. During the entire mode of operation, the inductor current i_{L1} and i_{L2} maintained the identical current profile, which maintains the voltage balance between C_1 and C_2. Figure 25 shows the experimental waveform of the input inductor current, i_{L1} and voltage across i_{D2} the power switch for input voltage 24 V and 0.8 duty cycle.

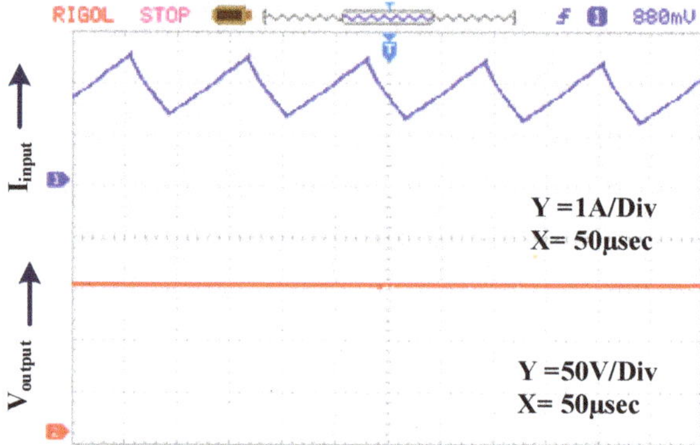

Figure 22. The experimental waveform of input current and output voltage for input voltage 24 V and 0.8 duty cycle.

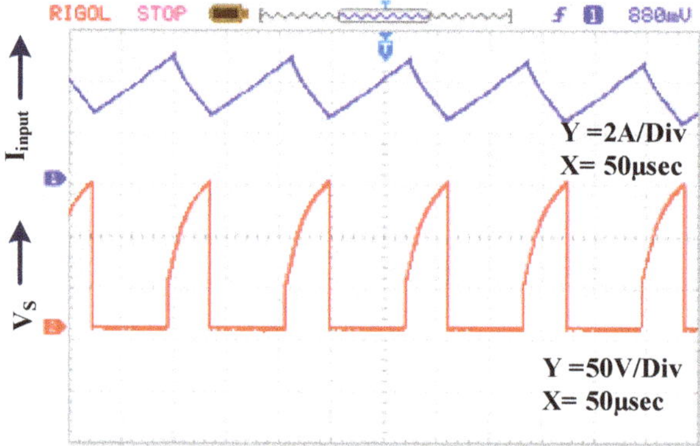

Figure 23. The experimental waveform of input current and voltage across the power switch for input voltage 24 V and 0.8 duty cycle.

Next, the converter was operated by changing the duty ratio to observe transient operation behaviour. For the period of transient operation, the converter load was kept constant as the previous value. During the trial, the switching duty-cycle varied from 0.8 to 0.5. During this period, likely the output voltage decreased and stabilized after a few milliseconds. A similar response happened in the inductor current i_{L1} and preserved in the converter in continuous conduction.

To validate the comparison of the theoretical and experimental results, the converter continuous conduction mode(CCM) state voltage gain was plotted concerning the variation duty ratio from 0.2 to 0.8 (see Figure 26a). From the results, it can be seen that the experimental values are very close to the theoretical calculations. Finally, the efficiency of the proposed converter was found under full load. The calculated experimental maximum efficiency of the proposed converter at 0.8 duty cycle is 92.2%. The calculated experimental maximum efficiency of the proposed converter is 92.2%. Figure 26b shows experimental power loss distribution operating at the rated condition. During duty ratio $\delta = 0.8$, the semiconductors (D_1, D_2 and MOSFET) switching losses were calculated as 0.4 W, 0.5 W and 1.2 W using equations (61)–(74). Hence, the total switching losses for the converter was 2.1 W. Similarly, the conduction of the power switches and other circuit parameters losses were observed. In the overall power distribution losses, the MOSFET switching loss and conduction loss alone are about 52%. As presented in Figure 26b, the I^2R losses in the MOSFET, diode and the snubber circuitry losses were accounted for as significant losses. Nevertheless, the proposed converter voltage stress reduction helps to choose the lower voltage-rating switch, and hence conduction losses are expected to reduce.

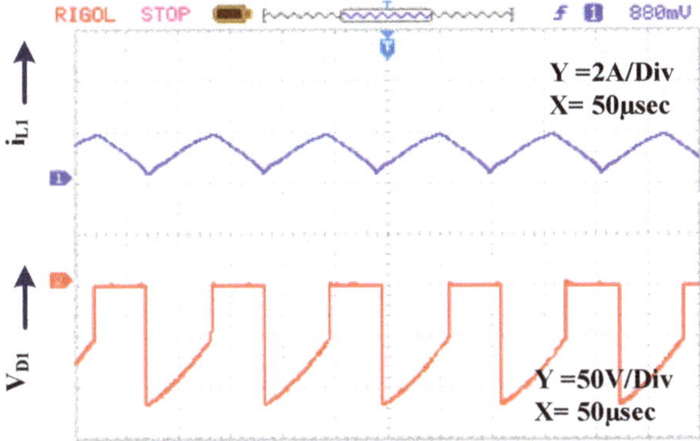

Figure 24. The experimental waveform of the input inductor current, i_{L1} and voltage across i_{D1} the power switch for input voltage 24 V and 0.8 duty cycle.

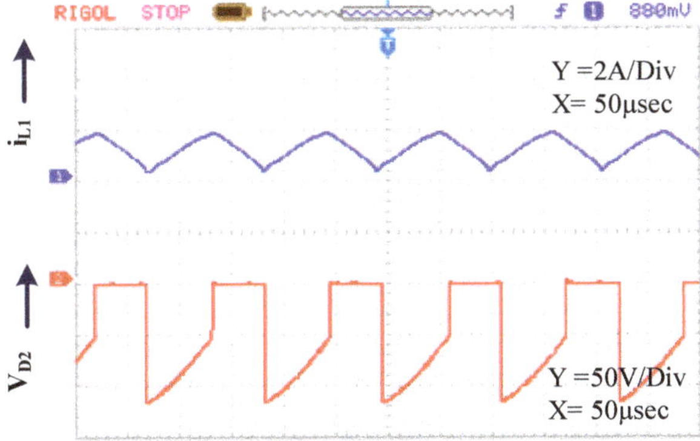

Figure 25. The experimental waveform of the input inductor current, i_{L1} and voltage across i_{D2} the power switch for input voltage 24 V and 0.8 duty cycle.

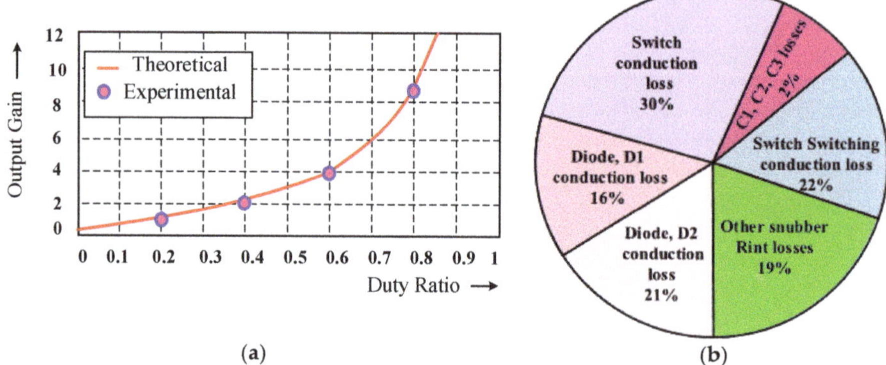

Figure 26. (**a**) Theoretical and experimental results comparison. (**b**) Experimental power loss distribution operating at rated condition (duty ratio from 0.8).

Key Performance Comparison

For validating the proposed converter performance, Table 2 shows the comparison with other similar converters. According to the table, the proposed converter provides a better voltage gain with a single active switch, and normalized voltages stress of semiconductor devices is less when compared to other converters. Based on the presented analysis and discussions, results and comparisons confirm the functionality and advantages of the proposed converter.

Table 2. Performance comparison of similar converter topology.

Similar Converter Topology	Converter [39]	Converter [9]	Converter [30]	Converter [40]	Converter [36]	Proposed Converter
Switches used	1	2	1	1	1	1
Diodes used	5	2	2	3	1	2
No. of Inductors used	3	2	2	3	2	2
No. of capacitors used	3	2	3	3	3	3
Continuous input current	Yes	No	Yes	Yes	No	Yes
Voltage gain, V_O	$\frac{(\delta)^2}{(1-\delta)^2}$ Vin	$\frac{2(1+\delta)}{(1-\delta)}$ Vin	$\frac{1}{1-\delta}$ Vin	$\frac{\delta}{(1-\delta)^2}$ Vin	$\frac{2\delta}{(1-\delta)}$ Vin	$V_O = \frac{1+\delta}{1-\delta} V_{PV}$
Efficiency	91%	90%	91%	90%	92%	92.2%
The voltage stress on the active switch	Moderate	Less	High	Less	High	Moderate

7. Conclusions

The high voltage gains and highly efficient single switch hybrid non-isolated DC–DC converter is shown in this paper. The proposed topology was derived by integrating conventional boost and Cuk converters. This topology produced a non-inverting output in continuous input current mode with a single switch having high voltage gain ($\approx$5.116 gain). When compared with the classical boost and Cuk converters, the proposed topology facilitates a substantial voltage gain with comparable lower switching stress. The steady-state analysis under the CCM condition and design calculation for the proposed hybrid was discussed in detail.

Finally, the validation test done with the proposed converter privileges, the voltage gain, power switch voltage stress and elements used in the circuit simulation studies were presented. Characterize the proposed topology for its obtained performances, PIC microcontroller based real-time experimental setup was realized under the power rating of 150 W with an efficiency of 92.2%. Experimental results confirmed the practicability in real-time applications needs.

Author Contributions: All authors are involved in developing the concept, simulation and experimental validation and in proof-reading the article. All authors have read and agreed to the published version of the manuscript.

Conflicts of Interest: The authors declare no conflict of interest.

References

1. Mansouri, N.; Lashab, A.; Sera, D.; Guerrero, J.M.; Cherif, A. Large Photovoltaic Power Plants Integration: A Review of Challenges and Solutions. *Energies* **2019**, *12*, 3798. [CrossRef]
2. Deng, F.; Chen, Z. Control of improved full-bridge three-level DC/DC converter for wind turbines in a DC grid. *IEEE Trans. Power Electron.* **2013**, *28*, 314–324. [CrossRef]
3. Mangu, B.; Akshatha, S.; Suryanarayana, D.; Fernandes, B.G. Grid-Connected PV-Wind-Battery-Based Multi-Input Transformer-Coupled Bidirectional DC-DC Converter for Household Applications. *IEEE J. Emerg. Sel. Top. Power Electron.* **2016**, *4*, 1086–1095. [CrossRef]
4. Chiu, H.J.; Yao, C.J.; Lo, Y.K. A DC-DC converter topology for renewable energy systems. *Int. J. Circuit Theory Appl.* **2009**, *37*, 485–495. [CrossRef]
5. Eccher, M.; Salemi, A.; Turrini, S.; Brusa, R.S. Measurements of power transfer efficiency in CPV cell-array models using individual DC-DC converters. *Appl. Energy* **2015**, *142*, 396–406. [CrossRef]
6. Hu, X.; Gong, C. A High Gain Input-Parallel Output-Series DC/DC Converter With Dual Coupled Inductors. *IEEE Trans. Power Electron.* **2015**, *30*, 1306–1317. [CrossRef]
7. Kouro, S.; Leon, J.I.; Vinnikov, D.; Franquelo, L.G. Grid-Connected Photovoltaic Systems: An Overview of Recent Research and Emerging PV Converter Technology. *IEEE Ind. Electron. Mag.* **2015**, *9*, 47–61. [CrossRef]
8. Khooban, M.H.; Gheisarnejad, M.; Farsizadeh, H.; Masoudian, A.; Boudjadar, J. A New Intelligent Hybrid Control Approach for DC–DC Converters in Zero-Emission Ferry Ships. *IEEE Trans. Power Electron.* **2020**, *35*, 5832–5841. [CrossRef]
9. Farsizadeh, H.; Gheisarnejad, M.; Mosayebi, M.; Rafiei, M.; Khooban, M.H. An Intelligent and Fast Controller for DC/DC Converter Feeding CPL in a DC Microgrid. *IEEE Trans. Circuits Syst. II Express Br.* **2020**. [CrossRef]
10. Sarrafan, N.; Zarei, J.; Razavi-Far, R.; Saif, M.; Khooban, M.H. A Novel On-Board DC/DC Converter Controller Feeding Uncertain Constant Power Loads. *IEEE J. Emerg. Sel. Top. Power Electron.* **2020**. [CrossRef]
11. Chen, Y.; Zhao, S.; Li, Z.; Wei, X.; Kang, Y. Modelling and Control of the Isolated DC–DC Modular Multilevel Converter for Electric Ship Medium Voltage Direct Current Power System. *IEEE J. Emerg. Sel. Top. Power Electron.* **2017**, *5*, 124–139. [CrossRef]
12. Amjadi, Z.; Williamson, S.S. Power-electronics-based solutions for plug-in hybrid electric vehicle energy storage and management systems. *IEEE Trans. Ind. Electron.* **2010**, *57*, 608–616. [CrossRef]
13. Boico, F.; Lehman, B.; Shujaee, K. Solar battery chargers for NiMH batteries. *IEEE Trans. Power Electron.* **2007**, *22*, 1600–1609. [CrossRef]
14. Chen, S.J.; Yang, S.P.; Huang, C.M.; Chou, H.M.; Shen, M.J. Interleaved High Step-Up DC-DC Converter Based on Voltage Multiplier Cell and Voltage-Stacking Techniques for Renewable Energy Applications. *Energies* **2018**, *11*, 1632. [CrossRef]
15. Hsieh, Y.P.; Chen, J.F.; Liang, T.J.; Yang, L.S. Novel high step-up DC-DC converter for distributed generation system. *IEEE Trans. Ind. Electron.* **2013**, *60*, 1473–1482. [CrossRef]
16. Todorovic, M.H.; Palma, L.; Enjeti, P.N. Design of a wide input range dc–dc converter with a robust power control scheme suitable for fuel cell power conversion. *IEEE Trans. Ind. Electron.* **2008**, *55*, 1247–1255. [CrossRef]
17. Wai, R.J.; Lin, C.Y.; . Lin, C.Y.; Duan, R.Y.; Chang, Y.R. High efficiency power conversion system for kilowatt-level stand-alone generation unit with low input voltage. *IEEE Trans. Ind. Electron.* **2008**, *55*, 3702–3714.
18. Changchien, S.K.; Liang, T.J.; Chen, J.F.; Yang, L.S. Novel high stepup dc-dc converter for fuel cell energy conversion system. *IEEE Trans. Ind. Electron.* **2010**, *57*, 2007–2017. [CrossRef]
19. Chen, S.; Liang, T.; Yang, L.; Chen, J. A cascaded high step-up dc–dc converter with single switch for micro source applications. *IEEE Trans. Power Electron.* **2011**, *26*, 1146–1153. [CrossRef]
20. Vighetti, S.; Ferrieux, J.-P.; Lembeye, Y. Optimization and design of a cascaded DC/DC converter devoted to grid-connected photovoltaic systems. *IEEE Trans. Power Electron.* **2012**, *27*, 2018–2027. [CrossRef]

21. Pan, C.T.; Lai, C.M. A high-efficiency high step-up converter with low switch voltage stress for fuel-cell system applications. *IEEE Trans. Ind. Electron.* **2010**, *57*, 1998–2006.
22. Grant, D.A.; Darroman, Y.; Suter, J. Synthesis of tapped-inductor switched-mode converters. *IEEE Trans. Power Electron.* **2007**, *22*, 1964–1969. [CrossRef]
23. Wai, R.J.; Lin, C.Y.; Duan, R.Y.; Chang, Y.R. High-efficiency DC–DC converter with high voltage gain and reduced switch stress. *IEEE Trans. Ind. Electron.* **2007**, *54*, 354–364. [CrossRef]
24. Wu, T.F.; Lai, Y.S.; Hung, J.C.; Chen, Y.M. Boost converter with coupled inductors and buck-boost type of active clamp. *IEEE Trans. Ind. Electron.* **2008**, *55*, 154–162. [CrossRef]
25. Wijeratne, D.S.; Moschopoulos, G. Quadratic power conversion for power electronics: Principles and circuits. *IEEE Trans. Circ. Syst.* **2012**, *59*, 426–438. [CrossRef]
26. Kadri, R.; Gaubert, J.P.; Champenois, G.; Mostefaï, M. Performance analysis of transformer less single switch quadratic Boost converter for grid connected photovoltaic systems. In Proceedings of the International Conference on Electrical Machines, Rome, Italy, 6–8 September 2010; pp. 1–7.
27. Liang, T.J.; Tseng, K.C. Analysis of integrated boost-flyback step up converter. *IEEE Proc. Electr. Power Appl.* **2005**, *152*, 217–225. [CrossRef]
28. Fretias, A.; Tofoli, F.L.; Jbbunior, E.M.S.; Daher, S.; Antunes, F.L.M. High-voltage gain DC-DC boost converter with coupled inductors for photovoltaic systems. *IET Power Electron.* **2015**, *8*, 1885–1892. [CrossRef]
29. Khalilzadeh, M.; Abbaszadeh, K. Non-isolated high step-up DC-DC converter based on coupled inductor with reduced voltage stress. *IET Power Electron.* **2015**, *8*, 2184–2194. [CrossRef]
30. Saravanan, S.; Ramesh Babu, N. Analysis and implementation of high step-up DC-DC converter for PV based grid application. *Appl. Energy* **2017**, *190*, 64–72. [CrossRef]
31. Wu, G.; Ruan, X.; Ye, Z. Non isolated high step-up DC-DC converters adopting switched-capacitor cell. *IEEE Trans. Ind. Electron.* **2015**, *62*, 383–393. [CrossRef]
32. Poorali, B.; Torkan, A.; Adib, E. High step-up Z-source DC-DC converter with coupled inductors and switched capacitor cell. *IET Power Electron.* **2015**, *8*, 1394–1402. [CrossRef]
33. Zhao, Y.; Li, W.; He, X. Single-phase improved active clamp coupled-inductor based converter with extended voltage doubler cell. *IEEE Power Electron.* **2012**, *27*, 2869–2878. [CrossRef]
34. Yang, L.S.; Liang, T.J.; Lee, H.C.; Chen, J.F. Novel high step-up DC-DC converter with coupled-inductor and voltage-doubler circuits. *IEEE Trans. Ind. Electron.* **2011**, *58*, 4196–4206. [CrossRef]
35. Bharatiraja, C.; Sanjeevikumar, P.; Mahesh Swathimala, A.S.; Raghu, S. Analysis, design and investigation on a new single-phase switched quasi Z-source inverter for photovoltaic application. *Int. J. Power Electron. Drive Syst.* **2017**, *8*, 853–860. [CrossRef]
36. Banaei, M.R.; Ardi, H.; Farakhor, A. Analysis and implementation of a new single-switch buck–boost DC/DC converter. *IET Power Electron.* **2014**, *7*, 1906–1914. [CrossRef]
37. Banaei, M.R.; Bonab, H.A.F. A Novel Structure for Single-Switch Non isolated Transformerless Buck–Boost DC–DC Converter. *IEEE Trans. Ind. Electron.* **2017**, *64*, 198–205. [CrossRef]
38. Banaei, M.R.; Sani, S.G. Analysis and Implementation of a New SEPIC-Based Single-Switch Buck–Boost DC–DC Converter With Continuous Input Current. *IEEE Trans. Power Electron.* **2018**, *33*, 10317–10325. [CrossRef]
39. Zhang, N.; Zhang, G.; See, K.W.; Zhang, B. A Single-Switch Quadratic Buck–Boost Converter With Continuous Input Port Current and Continuous Output Port Current. *IEEE Trans. Power Electron.* **2018**, *33*, 4157–4166. [CrossRef]
40. Maroti, P.K.; Padmanaban, S.; Wheeler, P.; Blaabjerg, F.; Rivera, M. Modified high voltage conversion inverting cuk DC-DC converter for renewable energy application. In Proceedings of the IEEE Southern Power Electronics Conference (SPEC) 2017, Puerto Varas, Chile, 4–7 December 2017; pp. 1–5.
41. Oluwafemi, A.W.; Ozsoy, E.; Padmanaban, S.; Bhaskar, M.S.; Ramachandaramurthy, V.K.; Fedák, V. A modified high output-gain cuk converter circuit configuration for renewable applications—A comprehensive investigation. In Proceedings of the IEEE Conference on Energy Conversion (CENCON) 2017, Kuala Lumpur, Malaysia, 30–31 October 2017; pp. 117–122.

Article

Double Stage Double Output DC–DC Converters for High Voltage Loads in Fuel Cell Vehicles

Mahajan Sagar Bhaskar, Sanjeevikumar Padmanaban * and Jens Bo Holm-Nielsen

Center for Bioenergy and Green Engineering, Department of Energy Technology, Aalborg University, 6700 Esbjerg, Denmark; sagar24.mahajan@gmail.com (M.S.B.); jhn@et.aau.dk (J.B.H.-N.)
* Correspondence: san@et.aau.dk

Received: 1 July 2019; Accepted: 19 September 2019; Published: 26 September 2019

Abstract: This article aims to enhance the output voltage magnitude of fuel cells (FCs), since the actual generation is low. The traditional technique is too complicated and has a cascaded or parallel connection solution to achieve high voltage for multiple loads in vehicles. In this case, electronic power converters are a viable solution with compact size and cost. Hence, double or multiple output DC–DC converters with high voltage step up are required to feed multiple high voltage loads at the same time. In this article, novel double stage double output (DSDO) DC–DC converters are formulated to feed multiple high voltage loads of FC vehicular system. Four DSDO DC–DC converters called DSDO L–L, DSDO L-2L, DSDO L-2LC, and DSDO L-2LC are developed in this research work and all the converters are derived based on the arrangement of different reactive networks. The primary power circuitry, conceptual operation, and output voltage gain derivation are given in detail with valid proof. The proposed converters are compared with possible parallel combinations of conventional converters and recently available configuration. Comprehensive numerical simulation and experimental prototype results show that our theoretical predictions are valid and that the configuration is applicable for real time application in FC technologies for 'more-electric vehicles'.

Keywords: DC–DC; double stage; double output; fuel cell; step-up; vehicular system; X–Y converter family

1. Introduction

In electric grid, hybrid 'more-electric vehicles', automobile high-intensity discharge headlamps, uninterruptible power supply, and luxury loads application multiport and multilevel power converters popular solutions [1–4]. Some advanced predictive control based on the multilevel converter is also proposed for energy storage [5,6]. Fuel cells have many features, including compatibility, size, and modularity [7,8]. However, the amalgamation of series and parallel FCs is not a suitable solution for generating high DC voltage/current. Trade-off loss increases the cost of the system and requires a large space. Furthermore, the major obstacles facing FC technology are durability, low generated voltage, and to fulfill the voltage demand of high voltage loads in FC vehicles. In such cases, power electronic DC–DC converters with high voltage conversion ratios and high efficiency play a pivotal control role [9–12]. Theoretically, a moderate or high output voltage is obtained from a traditional boost converter by operating in an extreme duty cycle. Adverse effects at extreme duty ratio lead to reduced controllability, increased switching losses, high conduction losses, large current ripple, high current and voltage ratings, and reverse diode recovery problems [13–16]. Consequently, traditional DC–DC converters are not suitable candidates for FC electric vehicle applications.

Previous research has achieved high voltage conversion ratio by using a cascaded traditional boost converter configuration. However, cascaded converter configurations have low efficiency and high cost due to the increased number of high voltage/current rating semiconductor devices and reactive elements [17,18]. Further, for implementation, they need complex control logic and increased driver modules to protect and control the semiconductor devices. Quadratic boost converters achieve a high voltage conversion ratio, but high current/voltage rating components/devices and the internal resistance of the inductors limits the output voltage [19]. Multistage diode/capacitor-based DC–DC converters have been proposed to achieve high voltage gain [20–24]. However, multiple discharging/charging loops of the capacitors lead to increased conduction loss, cost, and size, and reduced efficiency due to their parasitic nature. Converters have been proposed to get multiple outputs from a single input source by using push-pull, half-bridge, full-bridge, and fly-back converter topologies [25–27]. In all cases, high voltage is obtained with a high transformer rating on the primary side. Therefore, these converters cannot provide a proper solution for low weight/cost applications.

The parallel configurations of traditional converters such as boost, buck-boost, Cuk, single ended primary inductance converter (SEPIC), and ZETA can be possible solutions to achieve multiple outputs. The power circuitry of possible configurations without common front-end structure are shown in Figure 1a–e. These configurations provide two outputs using two different control switches. However, the voltage gain is not significantly improved, even when using a large number of components and devices. Furthermore, in order to reduce the component or device counts, common front-end structure can be a solution, as shown in Figure 1f–j. These configurations provide dual output using common front-end structures. However, only a few input side components are used, the device count is reduced, and the voltage gain is limited. Moreover, the current rating of the components is increased due to the common structure. In order to reduce the component count, hybrid converters are another possible solution. Moreover, hybrid Cuk, SEPIC, and ZETA converter configurations can achieve multiple outputs. A combination of Cuk and SEPIC converter structure was employed in [28]. Figure 2a shows the SEPIC-Cuk converter circuitry with common front-end design and dual output. A combination of ZETA and buck-boost converter structure was employed in [29].

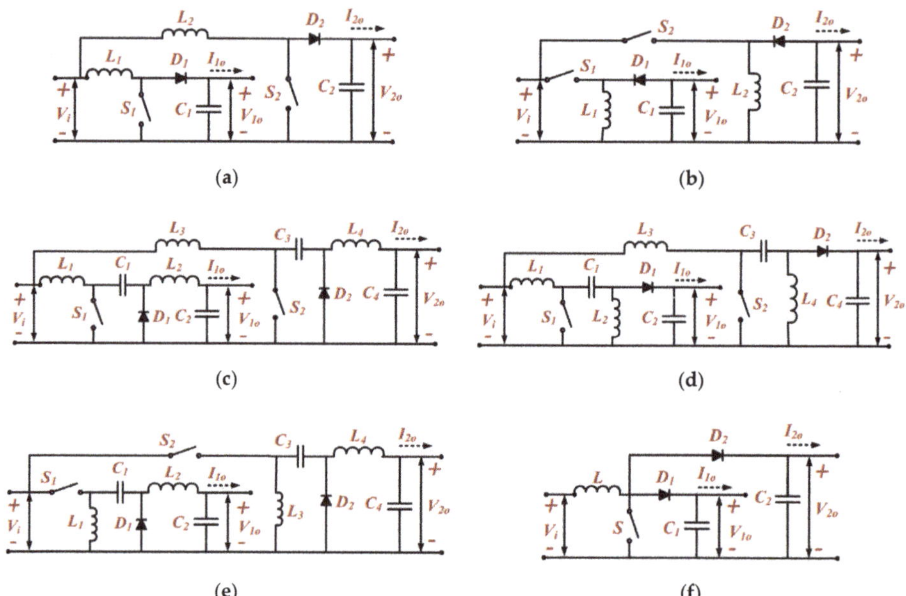

Figure 1. *Cont.*

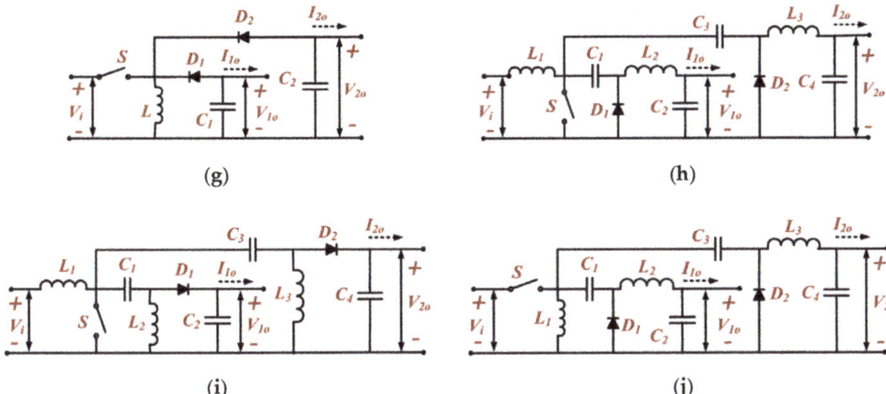

Figure 1. Power circuit of conventional parallel converters: (**a**) boost-boost converter; (**b**) buck-boost-buck-boost converter; (**c**) Cuk-Cuk converter; (**d**) single ended primary inductance converter (SEPIC)-SEPIC converter; (**e**) ZETA-ZETA converter; (**f**) boost-boost converter with common front-end structure; (**g**) buck-boost-buck-boost converter with common front end structure; (**h**) Cuk-Cuk converter with common front-end structure; (**i**) SEPIC-SEPIC converter with common front-end structure; (**j**) ZETA-ZETA converter with common front-end structure.

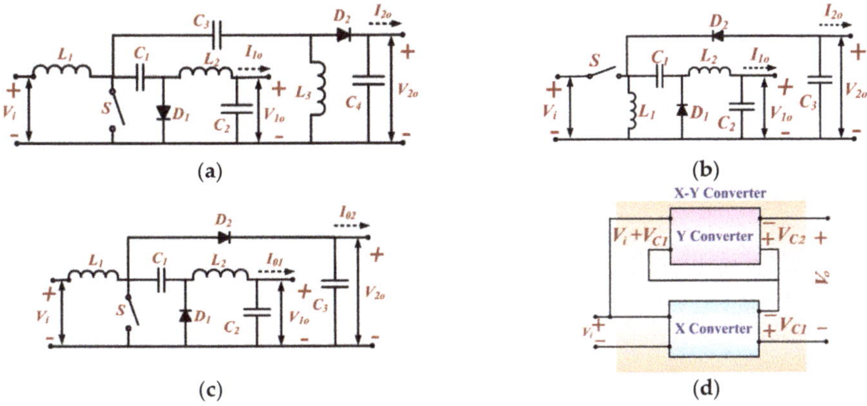

Figure 2. Power circuit of (**a**) SEPIC-Cuk converter, (**b**) ZETA–Buck-boost converter, (**c**) boost-Cuk converter; (**d**) Basic block diagram of the X–Y converter family.

Figure 2b shows the circuitry of a ZETA–Buck-boost converter with common front-end structure and two outputs. The combination of Cuk and boost converters is employed in [30]. The circuitry of a boost-Cuk converter with common front-end and two outputs is shown in Figure 2c. The voltage conversion ratio of these topologies is limited. Furthermore, efforts have been made to reduce the switching and to obtain multiple outputs [31–33]. However, these converters have low voltage conversion ratio and are more suitable for low-power applications. The "X–Y converter family" has been proposed for high voltage output and has single switching and a capacitor stack at the output side [34–38]. The block diagram of the X–Y converter family is depicted in Figure 2d. Notable, in XY converters, the X converter is directly connected to the input supply, and the Y converter is fed from the output voltage of the X converter. The output voltage of an X–Y converter is the sum of the output voltages of the X and Y converters. In order to achieve multiple outputs and high voltage conversion ratio, this article contributes the following:

- The L–Y converters, i.e., an expanded member of the XY converter family that feeds power to two different high voltage loads. At same time, the proposed converter provides high voltage conversion ratio. A diagram of a typical fuel-cell vehicle (FCV) with a DSDO converter is shown Figure 3, where low FC voltage is fed to two high voltage loads.
- Four new original converter configurations (DSDO L–L, DSDO L-2L, DSDO, L-2LC, and DSDO, L-2LC$_m$) are derived from a single switch.
- The modes of operation, characteristics waveform, and voltage gain analysis for each proposed configuration are discussed in detail.
- The performance of the proposed converters is validated through numerical simulation and experimental prototype results.

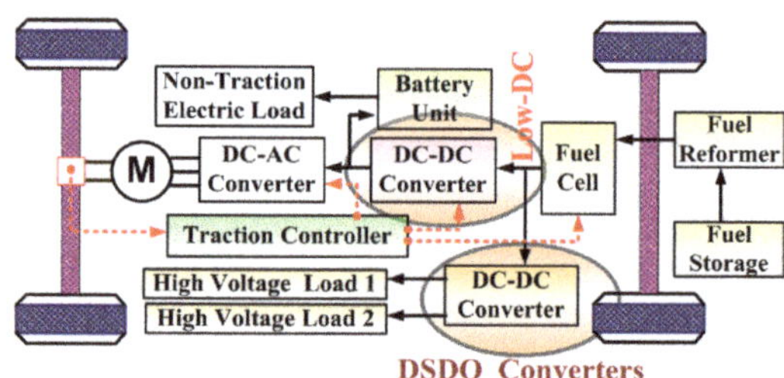

Figure 3. Typical structure of fuel cell (FC) vehicle with double stage double output (DSDO) converters.

2. Double Stage Double Output Converters

2.1. DSDO L–L Converter

Figure 4 depicts the power circuit of a DSDO L–L converter. In a DSDO L–L converter, a single switch S and input voltage V_i are arranged in two stages. Two L–L converters are employed to obtain dual output voltage. The capacitors C_1, C_2, inductors L_1, L_2, and diodes D_1, D_2, D_3 are elements of L–L converter-1. The capacitors C'_1, C'_2, inductors L'_1, L'_2, and diodes D'_1, D'_2, D'_3 are elements of L–L converter-2. The stage-1 and stage-1' voltages are obtained across capacitors C_1 and C'_1, respectively. The stage-2 and stage-2' voltages are obtained across capacitors C_2 and C'_2, respectively. The load R_{1o} is connected across capacitors C_1, C2, and load R_{2o} is connected across capacitors C'_1 and C'_2 to achieve double output voltages, i.e., V_{1o} and V_{2o}, from a single source input (V_i) as shown in Figure 4.

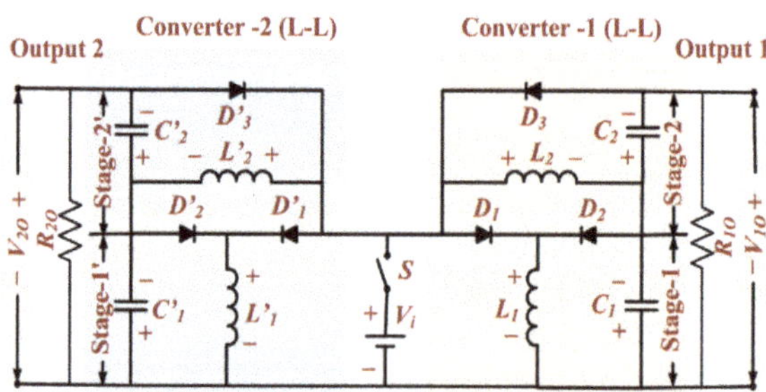

Figure 4. Power circuit of a DSDO L–L converter.

The DSDO L–L converter operates in two modes: switch S turn-ON and another when switch S turn-OFF. Figure 5 shows the characteristics of inductor voltage and current obtained for one switching cycle. Time zone A–B describes the ON time of the switch and time zone B–C describes the OFF time. The equivalent circuit for turn-ON mode is shown in Figure 6a. In this mode, inductor L_1 is magnetized through switch S and diode D_1 from the input power of voltage V_i. At the same time, input voltage V_i and the voltage across capacitor C_1 magnetizes the inductor L_2. Total energy stored in capacitors C_1 and C_2 provide load R_{1o}. Inductor L'_1 is magnetized through switch S and diode D'_1 by the input voltage V_i. At the same time, input voltage V_i and voltage across capacitor C'_1 magnetizes inductor L'_2. The energy is delivered to load R_{2o} by capacitors C'_1 and C'_2. During turn-ON, capacitors C_1, C'_1, C_2, and C'_2 are discharged, and inductors L_1, L'_1, L_2, and L'_2 are magnetized. Throughout this mode, diodes D_1, D'_1 are forward biased and diodes D_2, D_3, D'_2, D'_3 are reverse biased.

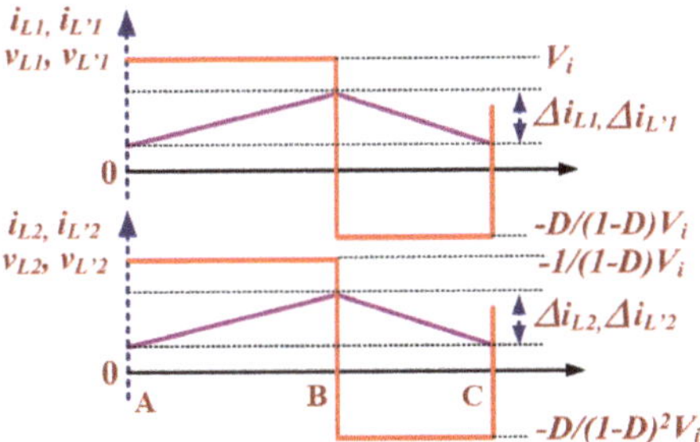

Figure 5. Waveforms of inductor voltages and currents for a DSDO L–L converter.

The voltages across inductors can be obtained as follows,

$$V_{L1} = V_i;\ V_{L2} = V_{C1} + V_i;\ V_{L'1} = V_i;\ V_{L'2} = V_{C'1} + V_i \tag{1}$$

The equivalent circuit for turn-OFF mode is shown in Figure 6b. In this mode, the capacitor C_1 is charged by stored energy in the inductor L_1 through diode D_2. At the same time, capacitor C_2 is charged by stored energy in the inductor L_2 through diode D_3. Energy is provided to load R_{1o} by the connection of inductors L_1 and L_2. Capacitor C'_1 is charged from stored energy in inductor L'_1 through diode D'_2. At the same time, capacitor C'_2 is charged from stored energy in inductor L'_2 through diode D'_3. Energy is provided to load R_{2o} by connection of inductors L'_1 and L'_2. In turn-OFF mode, capacitors C_1, C'_1, C_2, and C'_2 are charged and inductors L_1, L'_1, L_2, and L'_2 are demagnetized. In this mode, diodes D_1, D'_1 are reverse biased and diodes D_2, D_3, D'_2, and D'_3 are forward biased.

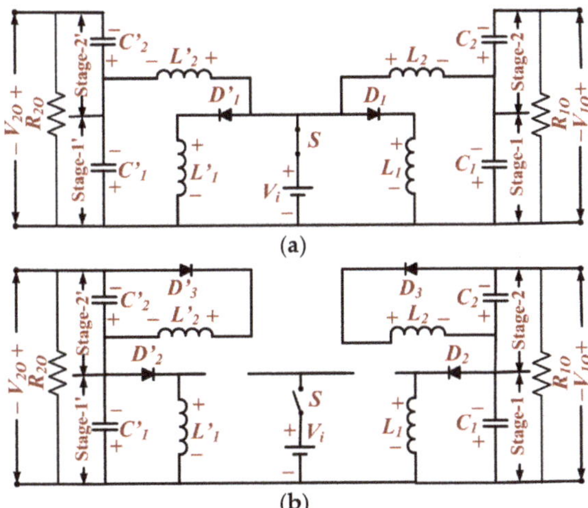

Figure 6. Equivalent circuitry of a DSDO L–L converter: (**a**) ON mode; (**b**) OFF mode.

The voltages across inductors can be obtained as follows,

$$V_{L1} = -V_{C1}; V_{L2} = -V_{C2}; V_{L\prime 1} = -V_{C\prime 1}; V_{L\prime 2} = -V_{C\prime 2} \qquad (2)$$

The output voltages V_{1o} and V_{2o} are obtained as follows,

$$\left. \begin{array}{l} \frac{V_{C1}}{V_i} = \frac{V_{C\prime 1}}{V_i} = D(1-D)^{-1}, \frac{V_{C2}}{V_i} = \frac{V_{C\prime 2}}{V_i} = D(1-D)^{-2} \\ V_{1o} = V_{2o} = -V_i \times \left(D(1-D)^{-1} + D(1-D)^{-2} \right) \end{array} \right\} \qquad (3)$$

2.2. DSDO L–2L Converter

Figure 7 illustrates the power circuit of a DSDO L–2L converter. A single switch S and input voltage V_i are employed with two-stage L–2L converters to obtain dual output voltage. The capacitors C_1, C_2 inductors L_1, L_2, L_3, and diodes D_1, D_2, ... , and D_6 are elements of L–2L converter-1. The capacitors $C\prime_1$, $C\prime_2$, inductors $L\prime_1$, $L\prime_2$, $L\prime_3$, and diodes $D\prime_1$, $D\prime_2$, ... , and $D\prime_6$ are elements of L–2L converter-2. The stage-1 and stage-2 voltages are taken across capacitors C_1 and C_2, respectively. The stage-1′ and stage-2′ voltages are taken across capacitors $C\prime_1$ and $C\prime_2$, respectively. The two output voltages are taken across capacitors $C_1 + C_2$ and $C\prime_1 + C\prime_2$, respectively. The load R_{1o} is connected across capacitors C_1 and C_2, and load R_{2o} is connected across capacitors $C\prime_1$ and $C\prime_2$.

The operation of a DSDO L–2L converter is sectioned into two modes: one when switch S turn-ON and another when switch S turn-OFF. Figure 8 shows the characteristic waveforms of voltage and current of the inductor for one switching cycle. In the characteristic waveform, time zone A–B describes the turn-ON time and time zone B–C describes the turn-OFF time.

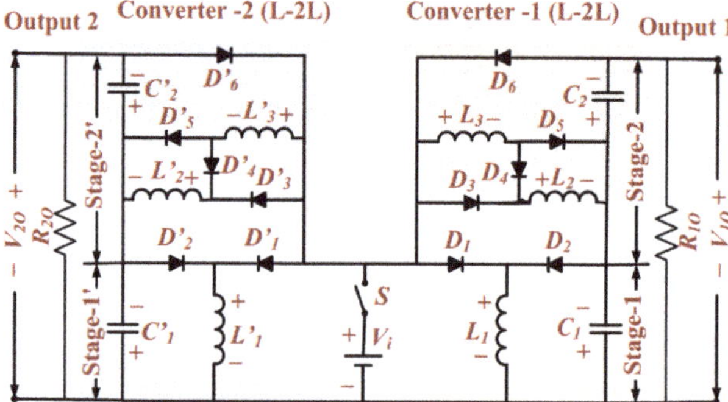

Figure 7. Power circuit of a DSDO L–2L converter.

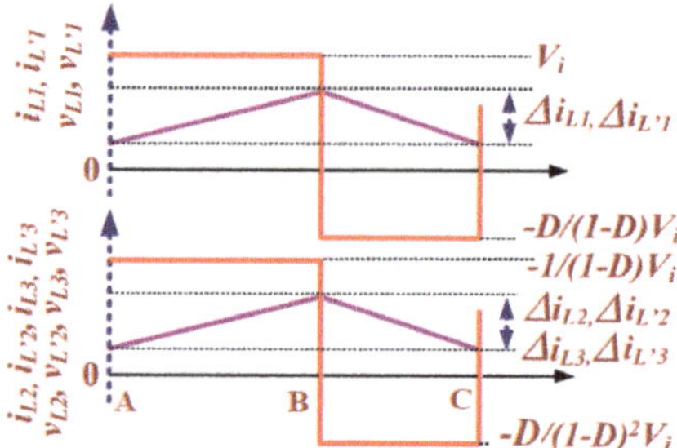

Figure 8. Waveforms of inductor voltage and current for a DSDO L–2L converter.

The equivalent circuit for turn-ON mode is shown in Figure 9a. Inductor L_1 is magnetized by the input voltage V_i. In the same interval, the input voltage V_i and voltage across capacitor C_1 magnetizes inductors L_2 and L_3. Energy is provided to load R_{1o} by capacitors C_1 and C_2. Inductor L'_1 is magnetized by the input voltage V_i. In the same interval, the input voltage V_i and voltage across capacitor C'_1 magnetize inductors L'_2 and L'_3. Energy is provided to load R_{2o} by capacitors C'_1 and C'_2. During turn-ON mode, capacitors C_1, C'_1, C_2, and C'_2 are discharged, and inductors L_1, L'_1, L_2, L'_2, L_3, and L'_3 are magnetized. In this mode, diodes D_1, D'_1, D_3, D'_3, D_5, and D'_5 are forward biased and diodes D_2, D'_2, D_4, D'_4, D_6, and D'_6 are reverse biased.

The voltage across inductors can be obtained as follows,

$$V_{L1} = V_{L'1} = V_i; \; V_{L2} = V_{L'2} = V_{L3} = V_{L'3} = V_{C1} + V_i \tag{4}$$

The equivalent circuit of the turn-OFF mode of a DSDO L–2L converter is shown in Figure 9b. In this mode, capacitor C_1 is charged by the stored energy of inductor L_1. At same time, capacitor C_2 is charged by the series connection of inductors L_2 and L_3. Energy is provided to load R_{1o} by inductors L_1, L_2, and L_3. Capacitor C'_1 is charged by the stored energy of inductor L'_1, whereas the capacitor C'_2 is charged by the series connection of inductors L'_2 and L'_3. Energy is provided to load R_{2o} by inductors L'_1, L'_2, and L'_3. In turn-OFF mode, capacitors C_1, C'_1, C_2, and C'_2 are charged and inductors L_1, L'_1, L_2, L'_2, L_3, and L'_3 are demagnetized. In this mode, diodes D_1, D'_1, D_3, D'_3, D_5, and D'_5 are reverse biased and D_2, D'_2, D_4, D'_4, D_6, and D'_6 are forward biased.

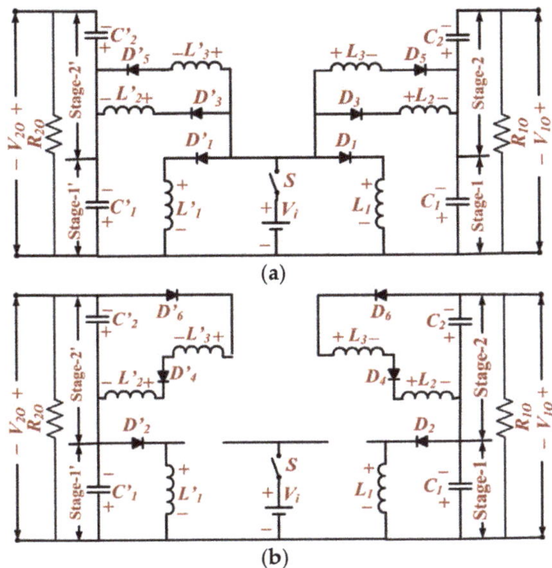

Figure 9. Equivalent circuitry of a DSDO L–2L converter: (a) ON mode; (b) OFF mode.

The voltage across inductors can be obtained as follows,

$$V_{L1} = V_{L'1} = -V_{C1}; \; V_{L2} = V_{L'2} = V_{L3} = V_{L'3} = -V_{C2}/2 \tag{5}$$

The output voltages V_{1o} and V_{2o} are obtained as follows,

$$\left. \begin{array}{l} \frac{V_{C1}}{V_i} = \frac{V_{C'1}}{V_i} = D(1-D)^{-1}, \frac{V_{C2}}{V_i} = \frac{V_{C'2}}{V_i} = 2D(1-D)^{-2} \\ V_{1o} = V_{2o} = -V_i \times \left(D(1-D)^{-1} + 2D(1-D)^{-2} \right) \end{array} \right\} \tag{6}$$

2.3. DSDO L–2LC Converter

Figure 10 illustrates the power circuit of a DSDO L–2LC converter. A single switch S and input voltage V_i are employed with two-stage L–2LC converters to obtain dual output voltages. The capacitors C, C_1, and C_2, inductors L_1, L_2, and L_3, diodes D_1, D_2, ... , and D_7 are elements of L–2LC converter-1. The capacitors C′, C'_1, and C'_2, inductors L'_1, L'_2, and L'_3, diodes D'_1, D'_2, ... , and D'_7 are elements of L–2LC converter-2. The stage-1 and stage-2 voltages are taken across C_1 and C_2, respectively. The stage-1′ and stage-2′ voltages are taken across capacitors C'_1 and C'_2, respectively. The load R_{1o} is connected across capacitors C_1 and C_2, and load R_{2o} is connected across capacitors C'_1 and C'_2.

The operation of a DSDO L–2LC converter is sectioned into two modes: one when switch S turn-ON and another when switch S turn-OFF. Figure 11 shows the characteristic waveforms of voltage and current of the inductor for one switching cycle. Time zone A–B describes the turn-ON time and time zone B–C describes the turn-OFF time of the switch. The equivalent circuit for turn-ON mode is shown in Figure 12a. In this interval, inductor L_1 is magnetized by the input voltage V_i. The input voltage V_i and voltage across capacitor C_1 magnetize inductors L_2 and L_3, and charge capacitor C in parallel. Energy is provided to load R_{1o} by capacitors C_1 and C_2. The inductor L'_1 magnetized by the input voltage V_i. The input voltage V_i and voltage across capacitor C'_1 magnetize inductors L'_2 and L'_3 and charge the capacitor C′ in parallel. Energy is provided to load R_{2o} by capacitors C'_1 and C'_2. In turn-ON mode, capacitors C_1, C'_1, C_2, and C'_2 are discharged, capacitors C and C′ are charged, and inductors L_1, L'_1, L_2, L'_2, L_3, and L'_3 are magnetized. In this mode, diodes D_1, D'_1, D_3, D'_3, D_5, D'_5, D_6, and D'_6 are forward biased and diodes D_2, D'_2, D_4, D'_4, D_7, and D'_7 are reverse biased.

The voltages across the inductors are obtained as follows,

$$V_{L1} = V_{L'1} = V_i; \; V_{L2} = V_{L3} = V_{L'2} = V_{L'3} = V_{C1} + V_i \tag{7}$$

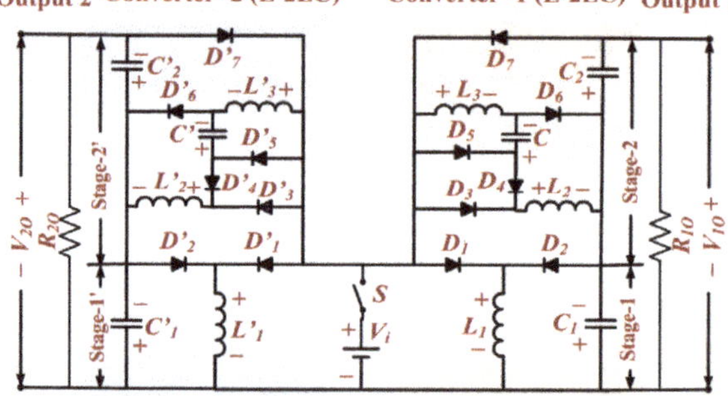

Figure 10. Power circuit of a DSDO L–2LC converter.

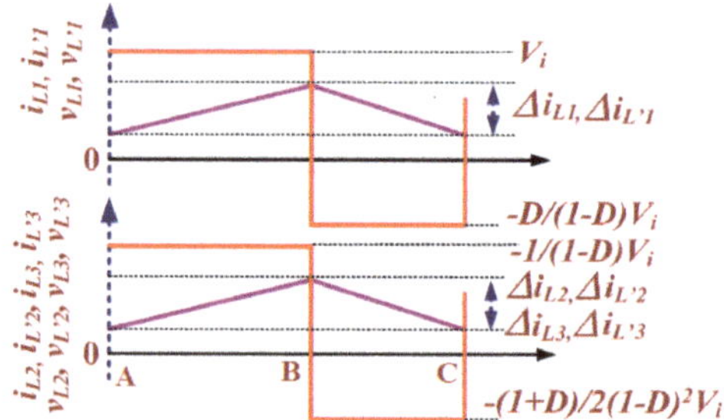

Figure 11. Waveforms of inductor voltage and current for a DSDO L–2LC converter.

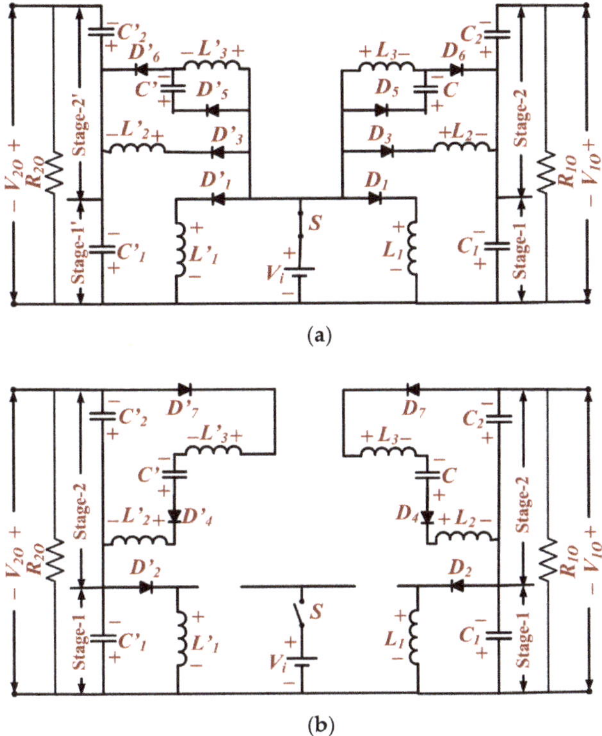

Figure 12. Equivalent circuitry of DSDO L–2LC converter: (**a**) ON mode; (**b**) OFF mode.

The equivalent circuit for turn-OFF mode of a DSDO L–2LC converter is shown in Figure 12b. The capacitor C_1 is charged by energy stored in inductor L_1. The capacitor C_2 is charged by the series connection of inductors L_2, L_3 and capacitor C. Energy is provided to load R_{1o} by inductors L_1, L_2, L_3, and capacitor C. The capacitor C'_1 is charged by stored energy of inductor L'_1. The capacitor C'_2 is charged by the series connection of inductors L'_2, L'_3, and capacitor C'. Energy is provided to load R_{2o} by inductors L'_1, L'_2, L'_3, and capacitor C'. Hence, the capacitors C_1, C'_1, C_2, and C'_2 are charged, capacitors C and C' are discharged, and inductors L_1, L'_1, L_2, L'_2, L_3, and L'_3 are demagnetized. In this mode, diodes D_1, D'_1, D_3, D'_3, D_5, D'_5, D_6, and D'_6 are reverse biased and diodes D_2, D'_2, D_4, D'_4, D_7, and D'_7 are forward biased.

The voltages across the inductors can be obtained as follows,

$$V_{L1} = V_{L'1} = -V_{C1}; \ (V_{L2} = V_{L'2} = V_{L3} = V_{L'3}) = \frac{-(V_{C2} - V_C)}{2} = \frac{-(V_{C2} - V_{C'})}{2} \tag{8}$$

The output voltages V_{1o} and V_{2o} are obtained as follows,

$$\left. \begin{aligned} \frac{V_{C1}}{V_i} = \frac{V_{C'1}}{V_i} = D(1-D)^{-1}, \frac{V_{C2}}{V_i} = \frac{V_{C'2}}{V_i} = \frac{(1+D)}{(1-D)^2} \\ V_{1o} = V_{2o} = -V_i \times \left(D(1-D)^{-1} + (1+D)(1-D)^{-2}\right) \end{aligned} \right\} \tag{9}$$

2.4. DSDO L–2LC$_m$ Converter

The DSDO L–2LC$_m$ converter is a modified version of the DSDO L–2LC converter, obtained by eliminating two diodes. Figure 13 shows the power circuit of the DSDO L–2LC$_m$ converter in which a single switch S and input voltage V_i are employed with two-stage L–2LC$_m$ converters to obtain dual output voltages. The capacitors C, C_1, and C_2, inductors L_1, L_2, and L_3, diodes D_1, D_2, ... , and D_5 are elements of L–2LC$_m$ converter-1. The capacitors C', C'_1, and C'_2, inductors L'_1, L'_2, and L'_3, and diodes D'_1, D'_2, ... , and D'_5 are elements of L–2LC$_m$ converter-2. The stage-1 and stage-2 voltages are taken across capacitors C_1 and C_2, respectively. The stage-1' and stage-2' voltages are taken across capacitors C'_1 and C'_2, respectively. Load R_{1o} is connected across capacitors C_1 and C_2 and load R_{2o} is connected across capacitors C'_1 and C'_2.

The operation of a DSDO L–2LC$_m$ converter is sectioned into two modes: one when switch S turn-ON and another when switch S turn-OFF. Figure 14 shows the characteristic waveforms of voltage and current for inductors for one switching cycle. The time zone A–B describes the turn-ON time and B–C describes the turn-OFF time. The equivalent circuit for turn-ON mode is shown in Figure 15a. The inductor L_1 is magnetized by the input voltage V_i. In the same interval, the inductor V_i and voltage across capacitor C_1 magnetize inductors L_2 and L_3 and charge capacitor C in parallel. The energy is provided to load R_{1o} by capacitors C_1 and C_2. The inductor L'_1 magnetized by the input voltage V_i. In the same interval, input voltage V_i and voltage across capacitor C'_1 magnetize inductors L'_2 and L'_3 and charge capacitor C' in parallel. The energy is provided to load R_{2o} by capacitors C'_1 and C'_2. Capacitors C_1, C'_1, C_2, and C'_2 are discharged, capacitors C and C' charged, and inductors L_1, L'_1, L_2, L'_2, L_3, and L'_3 are magnetized. In this mode, diodes D_1, D'_1, D_3, D'_3, D_4, and D'_4 are forward biased and diodes D_2, D'_2, D_5, and D'_5 are reverse biased.

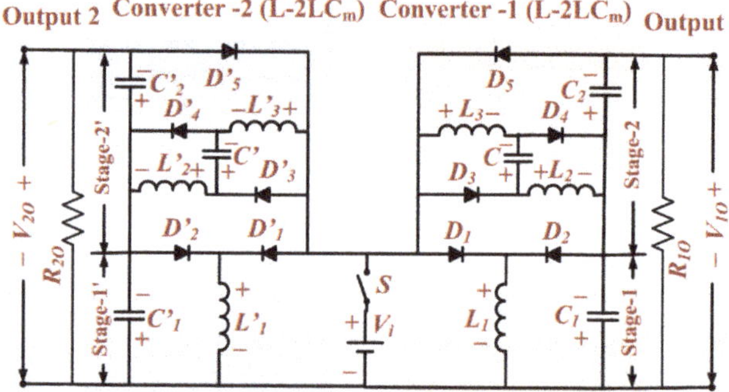

Figure 13. Power circuit of the DSDO L–2LC$_m$ converter.

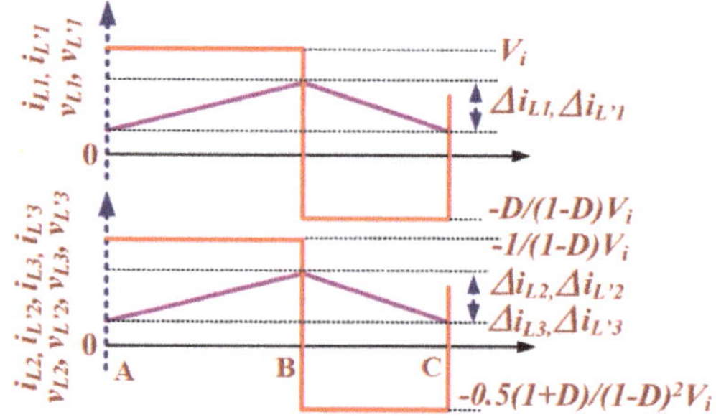

Figure 14. Waveforms of inductor voltage and current for a DSDO L–2LC$_m$ converter.

The voltages across inductors can be obtained as follows,

$$V_{L1} = V_{L'1} = V_i;\ V_{L2} = V_{L3} = V_{L'2} = V_{L'3} = V_{C1} + V_i \tag{10}$$

The equivalent circuit for turn-OFF mode is shown in Figure 15b. Capacitor C_1 is charged by the energy of inductor L_1, and capacitor C_2 is charged by the series connection of inductors L_2, L_3 and capacitor C. Energy is provided to load R_{1o} by inductors L_1, L_2, L_3, and capacitor C. Capacitor C'_1 is charged by energy of inductor L'_1. Capacitor C'_2 is charged by the series connection of inductors L'_2, L'_3 and capacitor C'. Energy is provided to load R_{2o} by inductors L'_1, L'_2, L'_3, and capacitor C'. The capacitors C_1, C'_1, C_2, and C'_2 are charged, capacitors C and C' are discharged, and inductors L_1, L'_1, L_2, L'_2, L_3, and L'_3 are demagnetized. Throughout this mode, diodes D_1, D'_1, D_3, D'_3, D_4, and D'_4 are reverse biased, and diodes D_2, D'_2, D_5, and D'_5 are forward biased.

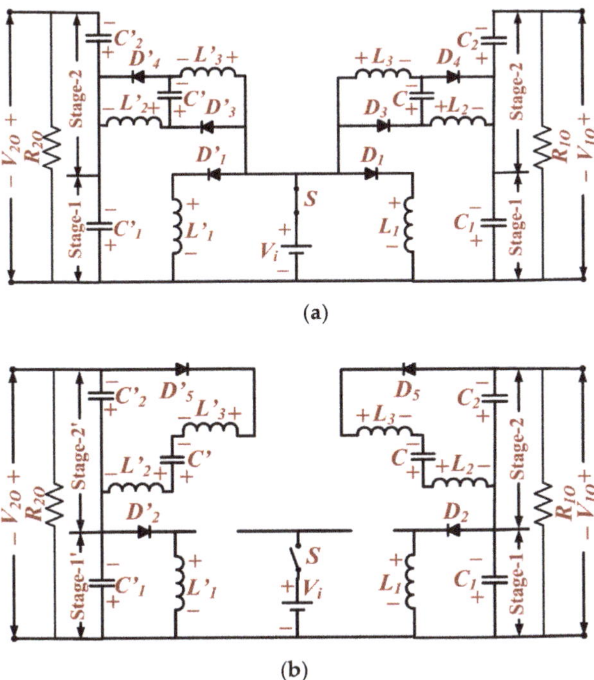

Figure 15. Equivalent circuitry of a DSDO L–2LC$_m$ converter: (a) ON mode; (b) OFF mode.

The voltages across inductors can be obtained as follows,

$$V_{L1} = V_{L'1} = -V_{C1}; (V_{L2} = V_{L'2} = V_{L3} = V_{L3}) = \frac{-(V_{C2} - V_C)}{2} = \frac{-(V_{C2} - V_{C'})}{2} \tag{11}$$

The output voltages V_{1o} and V_{2o} are obtained as follows,

$$\left. \begin{array}{l} \frac{V_{C1}}{V_i} = \frac{V_{C'1}}{V_i} = D(1-D)^{-1}, \frac{V_{C2}}{V_i} = \frac{V_{C'2}}{V_i} = \frac{(1+D)}{(1-D)^2} \\ V_{1o} = V_{2o} = -V_i \times \left(D(1-D)^{-1} + (1+D)(1-D)^{-2} \right) \end{array} \right\} \tag{12}$$

3. Comparative Study

In this section, the new DSDO converter configurations are compared with possible parallel combination of conventional converters and recently addressed DC–DC converters. Table 1 tabulates the comparison in terms of number of components and devices, voltage conversion ratio, and ratio of voltage across switch and input voltage. It is observed that one can achieve multiple output voltages by using conventional converters in parallel. However, the voltage conversion ratio is limited and not suitable for feeding high-voltage loads. Hybrid multiple output converters provide two different voltage levels while using common front-end structure. However, the voltage conversion ratio is not significantly improved by using a hybrid structure. The proposed converter provides a higher voltage conversion ratio compared to parallel combination of the conventional converters. In Figure 16a, the voltage conversion ratios of the converters are compared graphically. It is notable that all proposed converters provide inverting high voltage with medium duty cycle. It is observed that the DSDO L–2LC and DSDO L–2LC$_m$ converters generate higher voltage conversion ratios compared to the other proposed converters and in comparison to recent DC–DC converters. Figure 16b compares the number of diodes, control switches, inductors, and capacitors. It concludes that the DSDO L–2LC$_m$ converter requires fewer diodes than the DSDO L–2LC converter, while both provide the same voltage conversion ratio.

Table 1. Comparison of Converters.

Converter	Switches	Inductors	Diodes	Capacitors	Voltage Conversion Ratio (V_{1o}, V_{2o})	Voltage Stress across Switch (V_s/V_i)
Converter without common front-end structure						
Boost-Boost converter (Figure 1a)	2	2	2	2	1/1-D, 1/1-D	1/1-D
Buck-Boost-Buck-Boost converter (Figure 1b)	2	2	2	2	-D/1-D, -D/1-D	1/1-D
Cuk-Cuk Converter (Figure 1c)	2	4	2	4	-D/1-D, -D/1-D	1/1-D
SEPIC-SEPIC Converter (Figure 1d)	2	4	2	4	D/1-D, D/1-D	1/1-D
ZETA-ZETA Converter (Figure 1e)	2	4	2	4	D/1-D, D/1-D	1/1-D
Converter with common front-end structure						
Boost-Boost converter (Figure 1f)	1	1	2	2	1/1-D, 1/1-D	1/1-D
Buck-Boost-Buck-Boost converter (Figure 1g)	1	1	2	2	-D/1-D, -D/1-D	1/1-D
Cuk-Cuk converter (Figure 1h)	1	3	2	4	-D/1-D, -D/1-D	1/1-D
SEPIC-SEPIC converter (Figure 1i)	1	3	2	4	D/1-D, D/1-D	1/1-D
ZETA-ZETA converter (Figure 1j)	1	3	2	4	D/1-D, D/1-D	1/1-D
Hybrid multi-output converter with common front-end structure						
SEPIC-Cuk converter (Figure 2a)	1	3	2	4	-D/1-D, D/1-D	1/1-D
ZETA–Buck-boost converter (Figure 2b)	1	2	2	3	D/1-D, -D/1-D	1/1-D
Boost-Cuk converter (Figure 2c)	1	2	2	3	-D/1-D, 1/1-D	1/1-D
Proposed DSDO converter configurations						
DSDO L–L converters (Figure 4)	1	4	6	4	$D(1\text{-}D)^{-1} + D(1\text{-}D)^{-2}$	$1+ D(1\text{-}D)^{-1} + D(1\text{-}D)^{-2}$
DSDO L–2L converters (Figure 7)	1	6	12	4	$D(1\text{-}D)^{-1} + 2D(1\text{-}D)^{-2}$	$1+ D(1\text{-}D)^{-1} + 2D(1\text{-}D)^{-2}$
DSDO L–2LC converters (Figure 10)	1	6	14	6	$D(1\text{-}D)^{-1} + (1 + D)(1\text{-}D)^{-2}$	$1+ D(1\text{-}D)^{-1} + (1 + D)(1\text{-}D)^{-2}$
DSDO L–2LC$_m$ converters (Figure 13)	1	6	10	6	$D(1\text{-}D)^{-1} + (1 + D)(1\text{-}D)^{-2}$	$1+ D(1\text{-}D)^{-1} + (1 + D)(1\text{-}D)^{-2}$

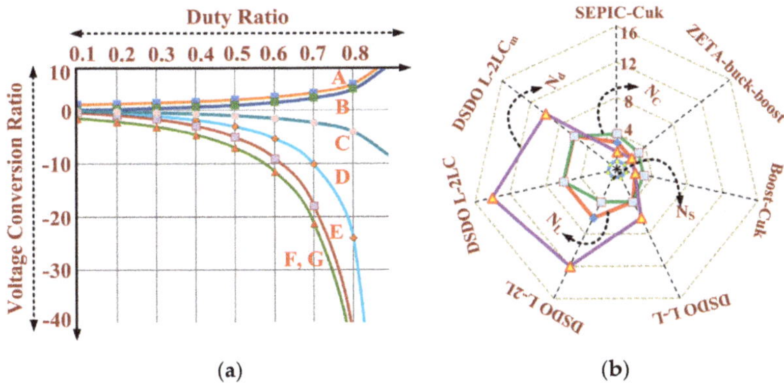

Figure 16. Comparison of (**a**) voltage conversion ratio versus duty cycle and (**b**) number of diodes (N_d), number of capacitors (N_C), number of Switches (N_S), and number of inductors (N_L). A: Boost, B: SEPIC, C: Cuk or Buck-Boost, D: DSDO L–L, E: DSDO L–2L, F: DSDO L–2LC, G: DSDO L–2LC$_m$ converters.

4. Simulation and Experimental Results

The principle and performance of the proposed DSDO converter configurations are validated through numerical simulation software. The converters were designed and tested with two loads, each rated to 100 W with a single input voltage of 20 V and with 25 kHz switching frequency. For simulation and prototype hardware implementation, the values of the reactive components were 220 µF for capacitors and 700 µH for inductors.

DSDO L–L converter: Figure 17a shows the voltage across the two different loads R_{1o} and R_{2o}. Voltage of −105 V was generated across each load with a fixed 60% duty cycle. Figure 17b,c depicts the voltage across capacitors C_1, C'_1, C_2, and C'_2. The voltage magnitude across capacitors C_1 and C'_1 are the same, i.e., 30 V. The voltage magnitudes across capacitors C_2 and C'_2 are both 75 V. Figure 17d shows that the voltage across switch S of a DSDO L–L converter is 125 V. DSDO L–2L converter: Figure 18a shows the voltage across the two different loads R_{1o} and R_{2o}. Voltage of −180 V was generated across each load with a fixed 60% duty cycle. Figure 18b,c depicts the voltage across capacitors C_1, C'_1, C_2, and C'_2, and shows that the voltage magnitude across capacitors C_1 and C'_1 is 30 V. The voltage magnitudes across capacitors C_2 and C'_2 are 150 V. Figure 18d shows that the voltage across switch S of the DSDO L–2L converter is 200 V. DSDO L–2LC converter: Figure 19a depicts the voltage across the two different loads R_{1o} and R_{2o}. Voltage of −230 V is generated across each load with a fixed 60% duty cycle. Figure 19b,c shows the waveforms of the voltage across capacitors C_1, C'_1, C_2, and C'_2.

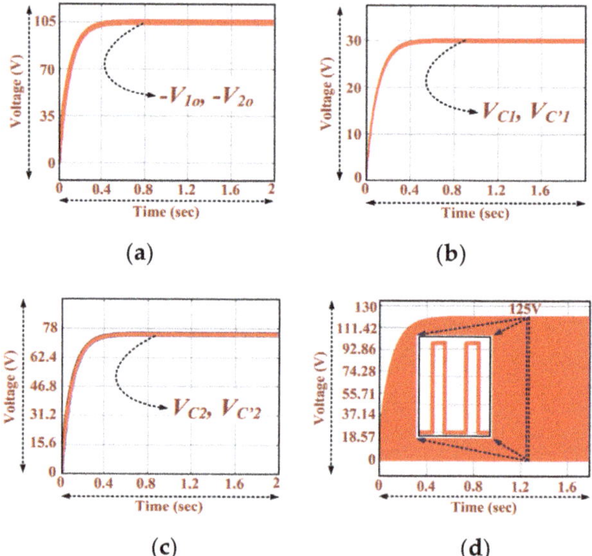

Figure 17. Simulation results of the DSDO L–L converter configuration: (**a**) Output voltages waveforms; (**b**) Voltage waveforms across stage-1 and stage-2; (**c**) Voltage waveforms across stage-1′ and stage-2′; (**d**) Voltage waveform across switch S.

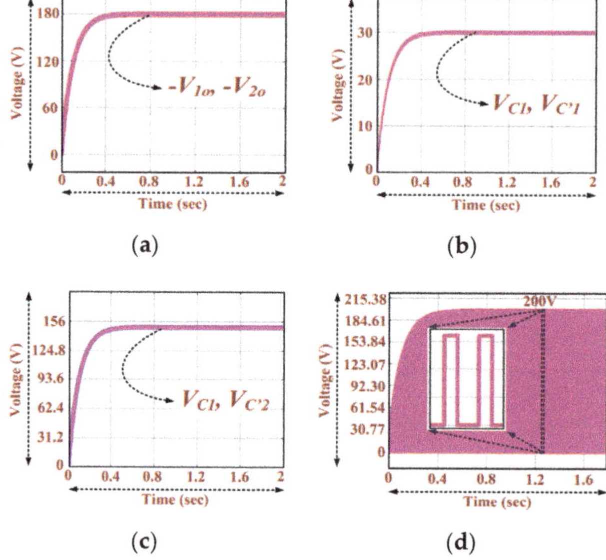

Figure 18. Simulation results of the DSDO L–2L converter configuration: (**a**) Output voltages waveforms; (**b**) Voltage waveforms at stage-1 and stage-2; (**c**) Voltage waveforms at stage-1′ and stage-2′; (**d**) Voltage waveform across switch S.

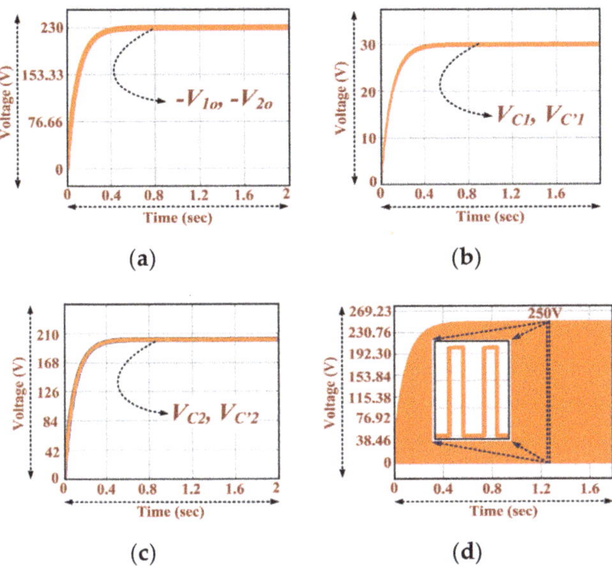

Figure 19. Simulation results of the DSDO L–2LC converter configuration: (**a**) Output voltages waveforms; (**b**) Voltage waveforms across stage-1 and Stage-2; (**c**) Voltage waveforms across stage-1′ and stage-2′; (**d**) Voltage across switch S.

The voltage magnitudes across capacitors C_1 and C'_1 are 30 V and the voltage across capacitors C_2 and C'_2 are 200 V. Figure 19d shows the voltage across switch S of the DSDO L–2LC converter and its magnitude is 250 V.

DSDO L–2LC$_m$ converter: Figure 20a shows the waveforms of the voltage across the two different loads R_{1o} and R_{2o}. Voltage of −230 V is generated across each load with a fixed 60% duty cycle. Figure 20b,c shows the waveforms of the voltage across capacitors C_1, C'_1, C_2, and C'_2. The voltage magnitudes across capacitors C_1 and C'_1 are 30 V, and voltage across capacitors C_2 and C'_2 are 200 V. Figure 20d shows the voltage across switch S of the DSDO L–2LC$_m$ converter and its magnitude is 250 V. Figure 21 shows the preliminary implemented hardware of the DSDO L–L converter. The designed hardware is tested with input voltage of 20 V and the output voltages are controlled at −105 V. Digitally controlled pulses are generated with the help of FPGA (Field Programmable Gate Array). Figures 22a and 22b the voltages across R_{1o} and R_{2o} with the voltage of switch S, respectively. Using a 20 V input supply, −104.2 V is successfully generated across each load and the voltage across switch S is 124.08 V. Figure 22c depicts the voltage and current waveform of inductors L_1 and L_2, respectively. In the ON state, the voltage across inductor L_1 is 20 V, which confirms that inductor L_1 is magnetized with input voltage. In the OFF state, the voltage across inductor L_1 is −30 V, i.e., inductor L_1 is demagnetized to charge the capacitor C_1. In the ON state, the voltage across inductor L_2 is 50 V, which confirms that the inductor L_2 is magnetized by the input voltage and the voltage across capacitor C_1. In the OFF state, the voltage across inductor L_2 is −75 V, i.e., inductor L_2 is demagnetized to charge capacitor C_2. Figure 22d depicts the voltage and current waveforms of inductor L'_1, L'_2 respectively. In the ON state, the voltage across inductor L'_1 is 20 V, which confirms that inductor L'_1 is magnetized by the input voltage. In the OFF state, the voltage across inductor L'_1 is −30 V, i.e., inductor L'_1 is demagnetized to charge capacitor C'_1. In the ON state, the voltage across inductor L'_2 is 50 V, which confirms that inductor L'_2 is magnetized by the input voltage and the voltage across capacitor C'_1. In the OFF state, the voltage across inductor L'_2 is −75 V, i.e., inductor L'_2 is demagnetized to charge capacitor C'_2.

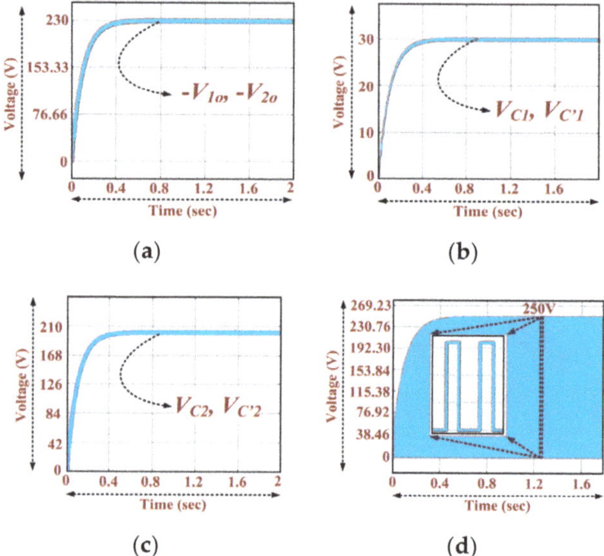

Figure 20. Simulation results of the DSDO L–2LC$_m$ converter configuration: (**a**) Output voltages waveforms; (**b**) Voltage waveforms across stage-1 and Stage-2; (**c**) Voltage waveforms across stage-1′ and stage-2′; (**d**) Voltage across switch S.

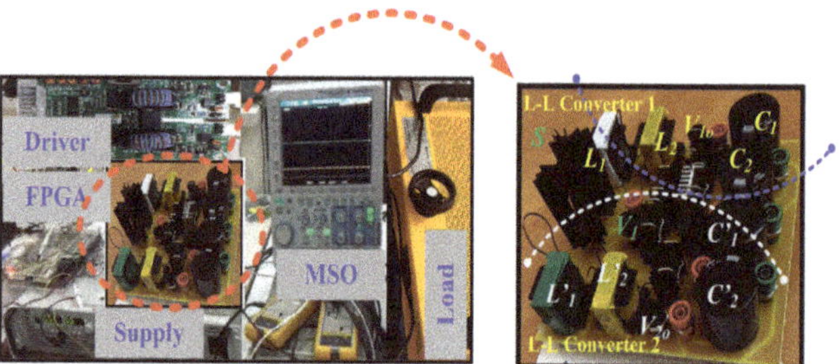

Figure 21. Experimental setup of the DSDO L–L converter configuration.

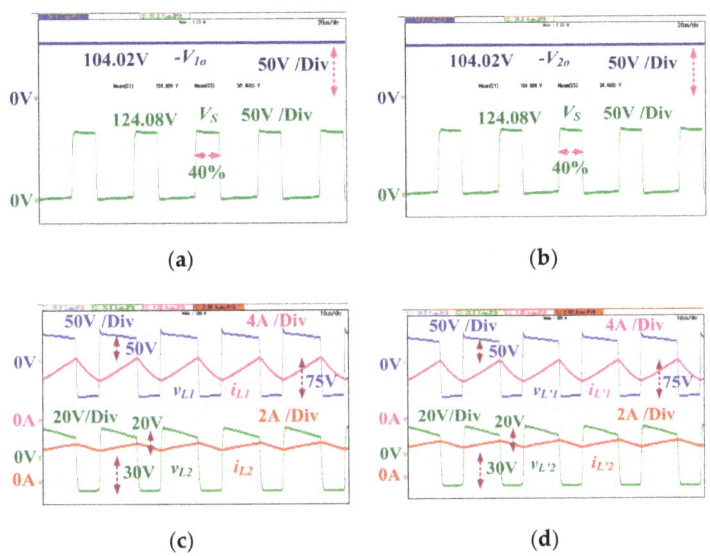

(a)   (b)

(c)   (d)

Figure 22. Experimental results of the DSDO L–L converter configuration: (**a**) voltage at load R_{1o} and voltage across switch S waveforms; (**b**) voltage at load R_{2o} and voltage across switch S waveforms; (**c**) inductor voltages and current of L converter-1; and (**d**) inductor voltages and current of L converter-2.

5. Conclusions

This article developed four DSDO converter configurations for high voltage fuel-cell electric vehicular loads. The proposed converters need a single controlling semiconductor switch and are able to feed two loads with high voltage conversion ratio. The circuitry of the DSDO L–L, DSDO L–2L, DSDO L–2LC, and DSDO L–2LC$_{m}$ converters are developed by merging with two L–L, two L–2L, two L–2LC, and two L–2LC$_{m}$ converters, respectively. The operating principles of the proposed converters are discussed with detailed theoretical analysis and governing equations for the output voltage conversion ratio. Finally, it is concluded that among the proposed converters, the DSDO L–2LC and DSDO L–2LC$_{m}$ converters provide higher output voltage and are effective in comparison with DC–DC converters. Both simulation and experimental results show that the proposed DSDO L–L converters had the expected performance.

Author Contributions: All authors contributed equally for the research work, to present the final manuscript as full research article in current version.

Funding: This research received no external funding.

Conflicts of Interest: The authors declare no conflict of interest.

References

1. Rezaee, S.; Farjah, E. A DC–DC Multiport Module for Integrating Plug-In Electric Vehicles in a Parking Lot: Topology and Operation. *IEEE Trans. Power Electron* **2014**, *29*, 5688–5695. [CrossRef]
2. Mercorelli, P.; Kubasiak, N.; Liu, S. Model predictive control of an electromagnetic actuator fed by multilevel PWM inverter. In Proceedings of the IEEE International Symposium on Industrial Electronics, Ajaccio, France, 4–7 May 2004; pp. 531–535.
3. Mercorelli, P.; Kubasiak, N.; Liu, S. Multilevel bridge governor by using model predictive control in wavelet packets for tracking trajectories. In Proceedings of the IEEE International Conference on Robotics and Automation, New Orleans, LA, USA, 26 April–1 May 2004; pp. 4079–4084.

4.   Faraji, R.; Adib, E.; Farzanehfard, H. Soft-switched non-isolated high step-up multi-port DC-DC converter for hybrid energy system with minimum number of switches. *Intl. J. Electr. Power Energy Sys.* **2019**, *106*, 511–519. [CrossRef]

5.   Mercorelli, P. A Multilevel Inverter Bridge Control Structure with Energy Storage Using Model Predictive Control for Flat Systems. *J. Eng.* **2019**, 1–15. [CrossRef]

6.   Kubasiak, N.; Mercorelli, P.; Liu, S. Model Predictive Control of Transistor Pulse Converter for Feeding Electromagnetic Valve Actuator with Energy Storage. In Proceedings of the IEEE 44th Conference on Decision and Control, Seville, Spain, 12–15 December 2005; pp. 6794–6799.

7.   Lucia, U. Overview on fuel cells. *Renew. Sustain. Energy Rev.* **2014**, *30*, 164–169. [CrossRef]

8.   Peighambardoust, S.J.; Rowshanzamir, S.; Amjadi, M. Review of the proton exchange membranes for fuel cell applications. *Intl. J. Hydrog. Energy* **2010**, *35*, 9349–9384. [CrossRef]

9.   Zhang, Z.; Pittini, R.; Andersen, M.A.E.; Thomsena, O.C. A Review and Design of Power Electronics Converters for Fuel Cell Hybrid System Applications. *Energy Procedia* **2012**, *20*, 301–310. [CrossRef]

10.   Cheng, S.; Lo, Y.; Chiu, H.; Kuo, S. High-Efficiency Digital-Controlled Interleaved Power Converter for High-Power PEM Fuel-Cell Applications. *IEEE Trans. Ind. Electron.* **2013**, *60*, 773–780. [CrossRef]

11.   Bhaskar, M.S.; Al-ammari, R.; Mohammad, M.; Padmanaban, S.; Iqbal, A. A New Triple-Switch-Triple-Mode High Step-Up Converter with Wide Range of Duty Cycle for DC Microgrid Applications. *IEEE Trans. Ind. Appl.* **2019**, 1. [CrossRef]

12.   Maroti, P.K.; Al-Ammari, R.; Bhaskar, M.S.; Meraj, M.; Iqbal, A.; Padmanaban, S.; Rahman, S. New tri-switching state non-isolated high gain DC–DC boost converter for microgrid application. *IET Power Electron* **2019**, *12*, 2741–2750. [CrossRef]

13.   Yang, L.; Liang, T.; Chen, J. Transformerless DC–DC Converters with High Step-Up Voltage Gain. *IEEE Trans. Ind. Electron.* **2009**, *56*, 3144–3152. [CrossRef]

14.   Bhaskar, M.S.; Meraj, M.; Iqbal, A.; Padmanaban, S.; Maroti, P.K.; Alammari, R. High Gain Transformer-Less Double-Duty-Triple-Mode DC/DC Converter for DC Microgrid. *IEEE Access* **2019**, *7*, 36353–36370. [CrossRef]

15.   Muranda, C.; Ozsoy, E.; Padmanaban, S.; Bhaskar, M.S.; Fedak, V.; Ramachandaramurthy, V.K. Modified SEPIC DC-to-DC boost converter with high output-gain configuration for renewable applications. In Proceedings of the IEEE Conference on Energy Conversion CENCON'17, Kuala Lumpur, Malaysia, 30–31 October 2017; pp. 317–322.

16.   Oluwafemi, A.W.; Ozsoy, E.; Padmanaban, S.; Bhaskar, M.S.; Ramachandaramurthy, V.K.; Fedak, V. A modified high output-gain cuk converter circuit configuration for renewable applications—A comprehensive investigation. In Proceedings of the IEEE Conference on Energy Conversion CENCON'17, Kuala Lumpur, Malaysia, 30–31 October 2017; pp. 117–122.

17.   Morales-Saldana, A.; Gutierrez, E.E.C.; Leyva-Ramos, J. Modeling of switch-mode dc-dc cascade converters. *IEEE Trans. Aerosp. Electron. Syst.* **2002**, *38*, 295–299. [CrossRef]

18.   Bhaskar, M.S.; Padmanaban, S.; Blaabjerg, F.; Wheeler, P.W. An Improved Multistage Switched Inductor Boost Converter (Improved M-SIBC) for Renewable Energy Applications: A key to Enhance Conversion Ratio. In Proceedings of the IEEE 19th Workshop on Control and Modeling for Power Electronics COMPEL'18, Padua, Italy, 25–28 June 2018.

19.   De Novaes, Y.R.; Rufer, A.; Barbi, I. A New Quadratic, Three-Level, DC/DC Converter Suitable for Fuel Cell Applications. In Proceedings of the IEEE Power Conversion Conference-Nagoya, Nagoya, Japan, 2–5 April 2007; pp. 601–607.

20.   Iqbal, A.; Bhaskar, M.S.; Meraj, M.; Padmanaban, S. DC-Transformer Modelling, Analysis and Comparison of the Experimental Investigation of a Non-Inverting and Non-Isolated Nx Multilevel Boost Converter (Nx MBC) for Low to High DC Voltage Applications. *IEEE Access* **2018**, *6*, 70935–70951. [CrossRef]

21.   Iqbal, A.; Bhaskar, M.S.; Meraj, M.; Padmanaban, S.; Rahman, S. Closed-Loop Control and Boundary for CCM and DCM of Non-Isolated Inverting Nx Multilevel Boost Converter for High Voltage Step-Up Applications. *IEEE Trans. Ind. Electron.* **2019**, 1. [CrossRef]

22.   Bhaskar, M.S.; Padmanaban, S.; Blaabjerg, F. A Multistage DC-DC Step-Up Self-Balanced and Magnetic Component-Free Converter for Photovoltaic Applications: Hardware Implementation. *Energies* **2017**, *10*, 719. [CrossRef]

23. Ranjana, M.S.B.; Alammari, R.; Meraj, M.; Iqbal, A.; Padmanaban, S. Modified Multilevel Buck-Boost Converter with Equal Voltage across Each Capacitor: Analysis and Experimental Investigations. *IET Power Electron.* **2019**. [CrossRef]
24. Bhaskar, M.S.; Meraj, M.; Iqbal, A.; Padmanaban, S. Non-isolated Symmetrical Interleaved Multilevel Boost Converter with Reduction in Voltage Rating of Capacitors for High Voltage Microgrid Applications. *IEEE Trans. Ind. Appl.* **2019**, 1. [CrossRef]
25. Tahan, M.; Bamgboje, D.; Hu, T. Flyback-Based Multiple Output dc-dc Converter with Independent Voltage Regulation. In Proceedings of the 9th IEEE International Symposium on Power Electronics for Distributed Generation Systems, PEDG, Charlotte, NC, USA, 25–28 June 2018; pp. 1–8.
26. Wai, R.J.; Liaw, J.J. High-Efficiency-Isolated Single-Input Multiple-Output Bidirectional Converter. *IEEE Trans. Power Electron.* **2015**, *30*, 4914–4930. [CrossRef]
27. Kim, J.K.; Choi, S.W.; Kim, C.E.; Moon, G.W. A New Standby Structure Using Multi-Output Full-Bridge Converter Integrating Flyback Converter. *IEEE Trans. Ind. Electron.* **2011**, *58*, 4763–4767. [CrossRef]
28. Ferrera, M.; Litrán, S.; Aranda, E.; Márquez, J.A. A Converter for Bipolar DC Link Based on SEPIC-Cuk Combination. *IEEE Trans. Power Electron.* **2015**, *30*, 6483–6487. [CrossRef]
29. Durán, E.; Litrán, S.P.; Ferrera, M.B.; Andújar, J.M. A Zeta-Buck-Boost converter combination for Single-Input Multiple-Output applications. In Proceedings of the 42nd Annual Conference of the IEEE Industrial Electronics Society, Florence, Italy, 23–26 October 2016; pp. 1251–1256.
30. Pires, V.F.; Foito, D.; Baptista, F.B.; Silva, J. A photovoltaic generator system with a DC/DC converter based on an integrated Boost-Ćuk topology. *Sol. Energy* **2016**, *136*, 1–9. [CrossRef]
31. Liu, X.; Xu, J.; Chen, Z.; Wang, N. Single-Inductor Dual-Output Buck-Boost Power Factor Correction Converter. *IEEE Trans. Ind. Electron.* **2015**, *62*, 943–952. [CrossRef]
32. Sanjeevikumar, P.; Bhaskar, M.S.; Dhond, P.; Blaabjerg, F.; Pecht, M. Non-isolated Sextuple Output Hybrid Triad Converter Configurations for High Step-Up Renewable Energy Applications. In *Power Systems and Energy Management*; Springer: Berlin/Heidelberg, Germany, 2018; pp. 1–12.
33. Patra, P.; Patra, A.; Misra, N. A Single-Inductor Multiple-Output Switcher with Simultaneous Buck, Boost, and Inverted Outputs. *IEEE Trans. Power Electron.* **2011**, *27*, 1936–1951. [CrossRef]
34. Bhaskar, M.S.; Padmanaban, S.; Kulkarni, R.; Blaabjerg, F.; Seshagiri, S.; Hajizadeh, A. Novel LY converter topologies for high gain transfer ratio: A new breed of XY family. In Proceedings of the 4th IET Clean Energy and Technology Conference, IET-CEAT, Kuala Lumpur, Malaysia, 14–15 November 2016; pp. 1–8.
35. Mahajan, S.B.; Sanjeevikumar, P.; Wheeler, P.; Blaabjerg, F.; Rivera, M.; Kulkarni, R. X-Y converter family: A new breed of buck boost converter for high step-up renewable energy applications. In Proceedings of the IEEE International Conference on Automatica IEEE-ICA-ACCA, Curico, Chile, 19–21 October 2016; pp. 1–8.
36. Bhaskar, M.S.; Sanjeevikumar, P.; Iqbal, A.; Meraj, M.; Howeldar, A.; Kamuruzzaman, J. L-L Converter for Fuel Cell Vehicular Power Train Applications: Hardware Implementation of Primary Member of X-Y Converter Family. In Proceedings of the IEEE Power Electronics, Drives and Energy Systems PEDES'18, Chennai, India, 18–21 December 2018.
37. Maroti, P.K.; Padmanaban, S.; Bhaskar, M.S.; Blaabjerg, F.; Ramachandaramurthy, K.V.; Siano, P.; Fedak, V. A novel 2L-Y DC-DC converter topologies for high conversion ratio renewable application. In Proceedings of the IEEE Conference on Energy Conversion CENCON'17, Kuala Lumpur, Malaysia, 30 October 2017; pp. 323–328.
38. Padmanaban, S.; Bhaskar, M.S.; Blaabjerg, F.; Yang, Y. A New DC-DC Multilevel Breed of XY Converter Family for Renewable Energy Applications: LY Multilevel Structured Boost Converter. In Proceedings of the 44th Annual Conference of the IEEE Industrial Electronics Society, Washington, DC, USA, 21–23 October 2018; pp. 6110–6115.

Article

DC Grid for Domestic Electrification

G. Arunkumar [1], D. Elangovan [1,*], P. Sanjeevikumar [2], Jens Bo Holm Nielsen [2], Zbigniew Leonowicz [3] and Peter K. Joseph [1]

[1] School of Electrical Engineering, Vellore Institute of Technology (VIT) University, Vellore, Tamilnadu 632014, India; g.arunkumar@vit.ac.in (G.A.); peterk.joseph@vit.ac.in (P.K.J.)

[2] Center for Bioenergy and Green Engineering, Department of Energy Technology, Aalborg University, 6700 Esbjerg, Denmark; san@et.aau.dk (P.S.); jhn@et.aau.dk (J.B.H.N.)

[3] Department of Electrical Engineering, Wroclaw University of Technology, Wyb. Wyspianskiego 27, I-7, 50370 Wroclaw, Poland; zbigniew.leonowicz@pwr.edu.pl

*   Correspondence: elangovan.devaraj@vit.ac.in; Tel.: +91-989-420-7260

Received: 26 April 2019; Accepted: 27 May 2019; Published: 5 June 2019

Abstract: Various statistics indicate that many of the parts of India, especially rural and island areas have either partial or no access to electricity. The main reason for this scenario is the immense expanse of which the power producing stations and the distribution hubs are located from these rural and distant areas. This emphasizes the significance of subsidiarity of power generation by means of renewable energy resources. Although in current energy production scenario electricity supply is principally by AC current, a large variety of the everyday utility devices like cell phone chargers, computers, laptop chargers etc. all work internally with DC power. The count of intermediate energy transfer steps are significantly abridged by providing DC power to mentioned devices. The paper also states other works that prove the increase in overall system efficiency and thereby cost reduction. With an abundance of solar power at disposal and major modification in the area of power electronic conversion devices, this article suggests a DC grid that can be used for a household in a distant or rural area to power the aforementioned, utilizing Solar PV. A system was designed for a household which is not connected to the main grid and was successfully simulated for several loads totaling to 250 W with the help of an isolated flyback converter at the front end and suitable power electronic conversion devices at each load points. Maximum abstraction of operational energy from renewable sources at a residential and commercial level is intended with the suggested direct current systems.

Keywords: DC grid; microgrid; DC-DC converter; renewable energy

1. Introduction

The World Energy Outlook 2015 statuses that nearly 17% of the total inhabitants in the world lacks access to electric power at homes [1]. As per the reports of International Energy Agency, India has over 237 million citizens belonging to this category. According to CEEW (Council for Energy, Environment, and Water), over 50% of houses in the states of West Bengal, Bihar, Madhya Pradesh, Uttar Pradesh, Jharkhand, Orissa has a shortage of electric power in spite of being grid connected. Regardless of the efforts put in over the years to electrify rural stretches, many households in five out of these six states have not more than 8 h of supply or no supply at all and are regularly subjected to blackouts. The states may perhaps slightly better conditions but the same cannot be said for the unfortunate low-income strata of citizens lacking admission to electricity. Most of these homes use kerosene for their lighting purposes which give poor illumination. It also sends out unhealthy fumes, may cause fire perils, ecologically unfavorable and also too exclusive when bought at market tariffs unless subsidized by the Government [2,3].

It may come as surprise to see that not all houses in a village in India maybe be electrified even though the village is said to be grid connected. By the explanation provided by the Government of

India, "A village is considered electrified when 10% of the homes in a village are connected to the grid." As on May 2016, 18,452 villages are remaining to be electrified [4]. Nevertheless, during recent past numerous of individuals were able to harvest electricity with the help of fleet-footed economic growth and along with numerous sponsored programs, the country has made significant augmentation in its electrical infrastructure. In the energy production sector, the fossil fuels and conventional methods of power generation are losing it demand rapidly since policymakers around the globe are stressing on the effects of global warming and climate changes. The world is advancing towards green energy to meet the ever increasing power demand, thus requiring a mixing up numerous resources both conventional and non-conventional [5]. The unique energy situation in India is making the country's growth objectives to be revised and modified so that it strive to meet its current demands as well as generating energy which is clean, efficient and environmental friendly [6]. This encourages economy to take up more enterprises and initiatives to extract maximum energy from renewable resources.

The abundance of solar radiation and decreasing prices of Photovoltaic components ease of its maintenance and scalability is making Solar PV generation more popular among renewable energy sources [7]. Even though in most parts of the country the sun shines adequately up to 10 to 12 h a day for the most part of a year, the huge impact of distributed solar power has not been amply exploited. In places receiving sunlight higher than 1400 equivalent peak hours annually, the gap in power shortage can be bridged by using solar energy. The government of India has become conscious of these facts and is enhancing the consumption of renewable power sources, specifically the solar power, in meeting the demand-supply gap nationwide. Thus, pertinent strategic guidelines have been created for endorsing solar power usage across the country. It aims at achieving 100 GW PV (photovoltaic) capacity from current statistics of 20 GW by 2022 with the assistance of the Jawaharlal Nehru National Solar Mission (JNNSM). The JNNSM intends to power rural areas, which was deprived of electricity before. Encouraged by the progress made in 4 years it now aims to achieve 100 GW by 2022 with solar rooftops contributing 40 GW of this. In the light of the launch of JNNSM program few states launched their separate solar policies. The Solar Energy policy of Tamil Nadu initiated in 2012 is targeting 5 GW solar energy by 2023. To encourage solar rooftops the Tamilnadu government provides huge incentives and almost 30% subsidies to buildings incorporating solar rooftops. This movement has encouraged the studies and improvement on Low Voltage DC arrangements, as they are suitable for residence applications as well as can be easily integrated with renewable energy sources and storage systems [8,9]. The rising demand for aggregating renewable energy resources is bringing back DC into the energy distribution frame since it is easy to integrate renewable sources into the grid in such case. Most loads at the utilization terminal these days are DC or non-sinusoidal. As a result, many types of exploration have been going on dc dissemination systems and their prospective uses in residential applications [10]. The DC also enjoys numerous other advantages over the AC system, one of the significant being the reduced number of converters at each power conversion legs and better efficiency in comparison with AC grid. If the solar DC output voltage is fed straight to these device appliances, the conversation stages are reduced from three stage DC-AC-DC to two stage DC-DC when solar PVs and fuel cells are interconnected with DC microgrids [11–15]. Additionally, the lack of reactive power decreases the current required to pass the equal magnitude of energy [16] and also mitigates the issues of skin effect, power factor and harmonics [17].Studies conducted on DC distribution systems in various residences located in various locations and different topologies in the United States showed that energy savings estimation can be up to 5% in case of a non-storage system and up to 14% for a system using storage [18,19]. There are more optimistic researchers that aim to pull of energy savings of 25–30% [20]. Exchanging the prevailing alternating current delivery grids with direct current is impracticable and economically non-viable. This is why DC can become the idyllic pick when it comes to energizing remote areas which are not connected to the main power grid, also known as "island areas". This type of areas can be made self-sustainable by harnessing non-conventional energy resources.

This article proposes one direct current microgrid which uses Solar PV to facilitate a domestically situated appliance in an isolated area to energize itself. Microgrids are entities that can be self-controlled and operated in island or grid connected mode when interconnected with the local distribution systems [21]. They mainly refer to small-scale power network having voltage levels lesser than 20 kV and power rating up to 1 MW. Using this system, unsolicited energy changeover steps and losses accompanying by the same are eliminated. The major advantage of DC Microgrid is its ability to comply easily with DC loads and Distributed Energy Resources (DERs). For example, only a DC-DC conversion stage is required in a DC Microgrid when it is connected to solar PV and a battery storage, thus provides a simple and cost cutting structure with the better control strategy. This boosts the general system efficiency and makes the system less complicated. It also supplements the performance and the life of components [22,23]. The absence of normalization, instruction and improvement of protection devices for DC-DC converters are few of the major problems that DC power systems must solve, before being regarded as an appropriate option that supersedes AC power systems in rural and island areas [17].

Figure 1 illustrates the methodology outline of this research works. The works done and the outcome of this methodology is explained in coming sessions.

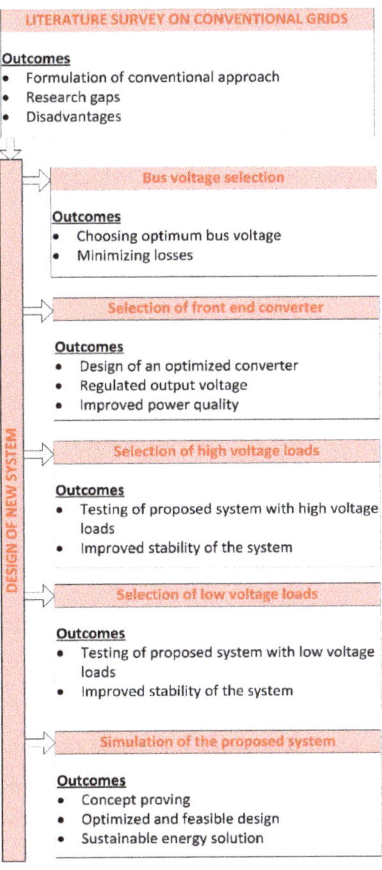

Figure 1. Methodology outline.

Existing Works in DC Microgrids

An analysis has been performed on a 48 V DC microgrid integrated with PV panels using very effective DC loads utilized in a multi-storied building in India [24,25]. The findings show that the DC microgrid is far efficient in bringing cost savings, thereby dropping the electricity invoices. A practical employment of low power solar system designed to supply the basic power needs of a low-income family in India has been studied [9]. The system supplies a cumulative load of 125 W from PV Panel. This system may not be able to meet up to the power requirement of a fully electrified and digital household but is able to show that when minimizing the system cost is priority a low voltage DC distribution system have no challengers.

Works on larger test beds such as 5 kW with high DC link voltage of 380 V has also been conducted to study the feasibility of DC as distribution system [15]. A solar hybrid system of grid connection along with solar array panel feeding 220 V DC link powering up an entire household is realized in [5]. An effective Maximum Power Point Tracking (MPPT) algorithm to obtain constant DC voltage of 12 V or 24 V using a PI controller is studied in [6]. An Off Grid Home (OGH) which is inverter less system to power lighting loads are deployed in [8].

Green Office and Apartments (GOA) technology is a solution offered to ensure all day power using an integration of grid and batteries charged from solar PV [26]. A DC microgrid consisting of 250 W solar panel and charge controllers to regulate battery charging has been proposed in [27,28]. Suggest a novel reconfigurable inverter topology which can perform DC to DC, DC to AC and grid connection at the same time. An experimental prototype of a power balancing circuit to solve mismatching problems while connecting various renewable to a DC link is proposed in [29,30]. Elaborates the concepts of DC house and Null Net Energy (NNE) buildings which supply DC to residential buildings. Various Multiple Input Multiple Output (MIMOCs) DC-to-DC converters that can be used as front end converter for a DC distribution in future homes is discussed in [31–42].

Apart from the conventional microgrid works, some researches are done in the field of advanced aspects of microgrid implementation. Researches [43–45] discusses about the consumers with distributed storage capacity. In this case, the demand sharing and power quality improvement will be much easier. Refs. [46,47] considers renewable energy sharing mechanism of multiple consumers, rather than the individual renewable energy harvesting topology. Since this research work discusses about the implementation of a DC microgrid in rural domestic area, this advanced techniques are neglected for the initial phase. In addition, since solar energy is weather dependent, to ensure a regulated supply irrespective of the weather or time, storage devices or weather independent renewable energy sources like fuel cells need to be integrated to the microgrid [48,49]. This part also neglected from the simulation, as the outcome will be the same.

2. Selection of Bus Voltage

The DC grid distribution system having several practical challenges in distributing a regulated power supply [32]. The DC microgrid supplying low voltage and higher currents requires high gauge cables, which leads to an increase in overall losses [32,33]. Thus in order to reduce the losses and save the installation cost, the DC microgrid voltage must be sufficiently high enough. As a paradox, if the link voltage is too high, it leads to the occurrence of sparks, arcing and electric shock. Many research works have been done in order to reduce the arcing and spark phenomenon in order to optimize the DC distribution system. However, this paper deals with loads not requiring more than 240 V voltage and 3.42 A current. Hence the DC link voltage is taken as 72 V [34–37].

Since the majority of the domestic electrical appliances internally needs DC voltage for its operation, which is obtained conventionally by stepping down of rectified AC voltage supply [38]. Renewable energy resources can directly produce this low value of DC voltage [39]. Hence the rectification stage can be avoided if the load is powered with DC. A customary magnitude for DC grid voltage is not fixed for a microgrid. The chosen loads for this research has rated voltage varies in the range from 5 V to 230 V. For ensuring a coherent transition from grid voltage to rated load voltage, an optimum value of grid voltage of 72 V is chosen [40,41].

3. Front End Isolated DC to DC Converter

The input voltage V_{dc} of the DC microgrid is considered to be a solar panel whose output is expected to be 24 V. A 20 W, 12 V solar panel (54 × 46 cm) is used for implementing the solar array. To make the rated input to the grid, seven parallel connections of two series connected panels are used. The distribution losses in the microgrid can be reduced to a low value by stepping up the input voltage to a DC voltage of 72 V by a Flyback converter. A flyback converter is chosen for the proposed system as the primary side DC-DC converter for the purpose that it can facilitate galvanic seclusion in amongst the input and the DC microgrid. The specifications of the selected flyback converter are input voltage as 24 V, output voltage as 72 V and output power as 250 W. The simplicity of its topology compared to other isolated SMPS topologies is an added advantage. It also has the lesser component count and lowers cost, making it popular. This will function for an extensive difference of the source voltage, as well as, it can facilitate numerous secluded DC voltage outcomes.

L, C and R denotes inductor, capacitor and resistor respectively. Lm denotes the mutual inductance. For an input voltage V_{dc} of 24 V and grid voltage V_{grd} of 72 V, duty cycle ratio of the flyback converter is 0.42. The front end converter is designed to energize a cumulative device power of 250 W. Isolation transformer of turn's ratio 1:4 is chosen for the proposed topology. The magnetizing inductance Lm of the isolation transformer is 85 μH. Switching frequency is selected as 50 KHz, and for a 1% voltage-ripple, capacitor C1 of 50 μF is used. A clamping circuit is also connected to the isolation transformer to absorb the energy stored in the inductor and provide a path for its dissipation to avoid high surge voltage. The capacitance Ccl of 1 μH is used in the clamping circuit. The microgrid voltage is fed to various devices by Point of Load (POL) converters [42]. Depending upon load specifications POL converters can be Buck-boost, Buck or Boost. Table 1 show various loads utilized by the proposed system. Figure 2 shows the proposed topology for the DC microgrid including the front end converter, bus and loads.

Table 1. Loads selected for the analysis of work.

Device	Wattage (W)	Rated Voltage(V)
DC FAN	90	48
LAPTOP	65	19
LED	5 × 18 (90)	240

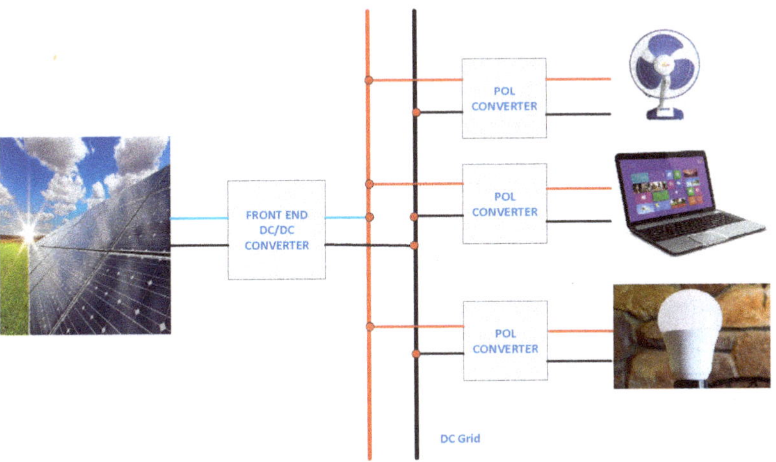

Figure 2. Proposed Schematic for the proposed circuit topology.

The circuit topology of the complete system including the front end converter, high voltage loads and low voltage loads are as shown in Figure 3. Here M1, M2, M3 and M4 are the controlled switches.

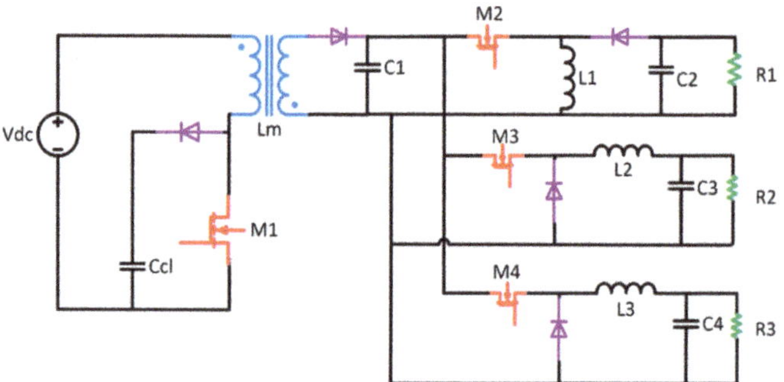

Figure 3. Complete circuit topology including front end Flyback converter for 0–240 V loads.

4. Loads with 24 V to 240 V Rating

The voltage essential for these loads is provided by a buck-boost voltage converter. For a home illuminating application, we are considering five 9 W Syska B22 LED bulb with 240 V DC voltage ratings. A buck-boost converter premeditated for a 1% peak voltage ripple and 10% current ripple of the rated voltage and current respectively. The proposed arrangement of the 24–240 V loads are illustrated in Figure 4. The designed values of inductor and capacitor is tabulated in Table 2.

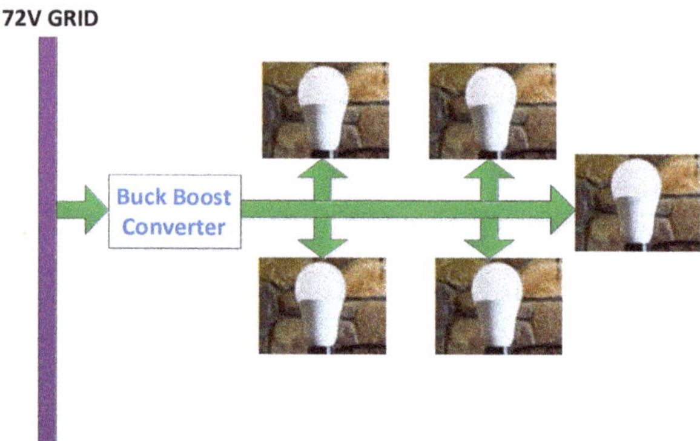

Figure 4. Proposed arrangement for 24–240 V Devices.

Table 2. Calculated values of various converter parameters for energizing LV devices.

Device	Duty Formula, (D)	Duty	Converter	Inductance	Capacitance
LED LIGHTS (R1)	$\frac{V_{grd}+V_{R1}}{V_{R1}}$	0.769	Buck-Boost	L1: 24.3 mH	C2: 5 μF
DC FAN (R2)	$\frac{V_{R2}}{V_{grd}}$	0.667	Buck	L2: 1.69 mH	C3: 1 μF
LAPTOP CHARGER (R3)	$\frac{V_{R3}}{V_{grd}}$	0.263	Buck	L3: 1 mH	C4: 5μF

5. Loads with <24 V Rating

Low voltage loads like laptop and DC fan are considered in this section. Figure 5 above illustrates the designed connection diagram for the <24 V devices. These loads essentially need a ripple-free DC output voltage which is usually acquired by using high-efficient DC conversion stages followed by a stepping up PFC (power factor correction) circuit. This setup contributes bulkiness to the system [41]. By replacing the above mentioned circuitry with a steeping down buck converter, the power quality of the grid can be maintained with a minimized space consumption. This will reduces the development cost, dimensions and enhances the lifespan of the device [39]. In rural areas, usually the application side dispersal transformer having a 20% to 25% reduced voltage than the general fixed values. Operation of conventional induction motor based devices like household fans with such voltage variation from the general fixed values may results in higher iron losses, which may leads to the permanent damage of motor [8]. The calculated values of various converter parameters for energizing <24 V devices are tabulated in Table 2. Here VR1, VR2 and VR3 represents the voltage drop across <24 V loads R1, R2 and R3 respectively.

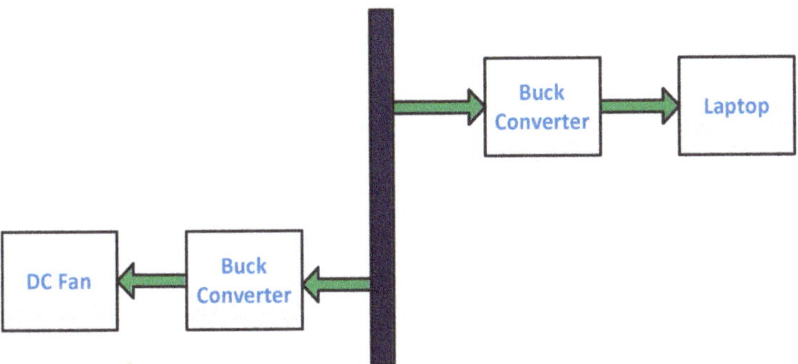

Figure 5. Proposed arrangement for <24 V devices.

To mitigate these effects, modern brushless DC motors for DC fans can be used instead of conventional fans having less ripple percentage. In addition the reduction in losses, various advantages like improved power density, enhanced torque, higher life-span, and easy control and reduced maintenance cost.

6. Numerical Simulation Results

The PSIM Professional Version 9.1.1.400 (Vellore, Tamilnadu, India) was used to simulate the proposed system. The equivalent models of real time loads, devices and sources are used for the simulation. The simulated values of each loads are tabulated in Table 3. In this initial phase of research, simulated results are considered for formulating the conclusion. The flyback converter illustrated in Figure 3 is simulated to formulate the parameters. The system in total has four DC-DC converters. 72 V DC bus voltage used for the electrifying DC microgrid is obtained from front end converter as illustrated in Figure 6a. The magnetizing current waveforms from the flyback converter is shown in Figure 6b. Figure 6a illustrates the magnetizing current of the flyback transformer. The current ripple in the inductor magnetizing current is obtained as 9.64%. Figure 6b shows the DC grid voltage waveform. As the grid input is given by flyback converter, the grid voltage is 72 V as per the rating. The voltage ripple is obtained as 0.9%, which is feasible for a domestic power network. Figure 7a illustrates the LED input voltage waveform and Figure 7b shows the LED input current waveform. The voltage ripple and current ripple is obtained as 0.83% and 1.06% respectively. This reduced ripple denotes the high power quality of the microgrid. Figure 8a,b represents the laptop input voltage and current respectively from the grid. The waveforms are of high power quality. The voltage ripple and current ripple are obtained as 0.7% and 0.58% respectively. Similarly Figure 9a,b illustrates the DC fan input voltage and input current respectively. The voltage ripple and current ripple are found as 2.37% and 0.337% accordingly. From the waveforms of these loads, it is clear that the power quality of the proposed microgrid topology is very high compared to other conventional topologies.

Table 3. Output voltages and currents of each load.

Load	V_{bus} (V)	Duty Cycle Formula, (D)	Duty Cycle (D)	V_{out} (V)	It (A)
LED LIGHTS	71.66	$\frac{Vgrd+VR1}{VR1}$	0.769	239.7	0.375
DC FAN	71.66	$\frac{VR2}{Vgrd}$	0.667	47.9	1.87
LAPTOP CHARGER	71.66	$\frac{VR3}{Vgrd}$	0.263	19.8	3.41

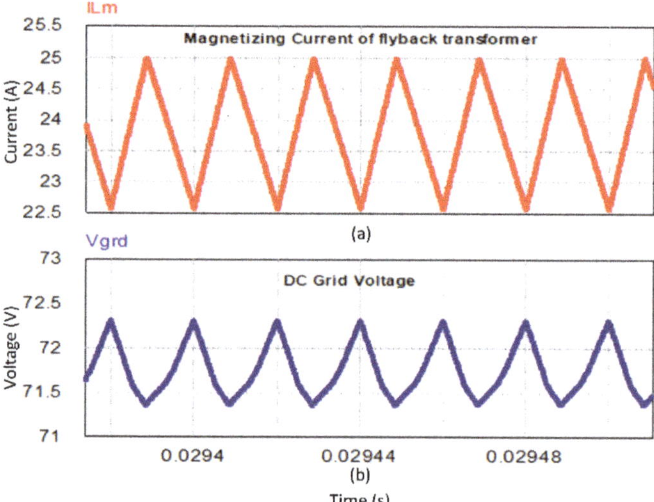

Figure 6. Output of Simulation DC-Grid Voltage. Average Voltage = 71.6 V, voltage ripple = 0.9% and output of Simulation—magnetizing inductor current. Average inductor current 24.39 A, current ripple = 9.64%. (**a**) Magnetizing current of flyback converter; (**b**) DC grid voltage.

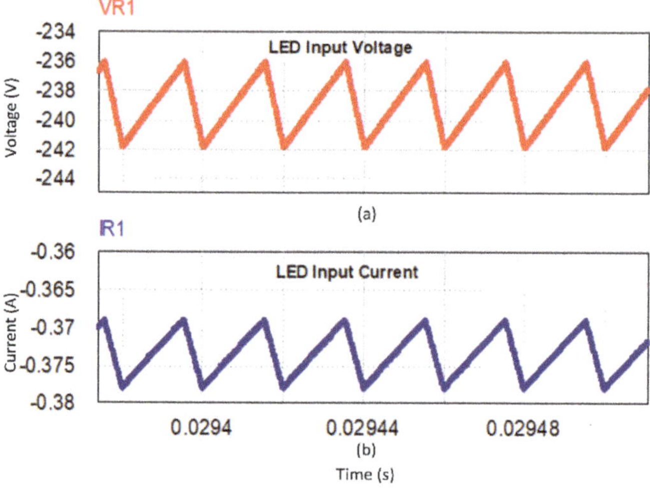

Figure 7. LED Input Voltage and Current Waveforms. LED Input Voltage Ripple: 0.83% Current Ripple: 1.06%. (**a**) LED input voltage; (**b**) LED input current.

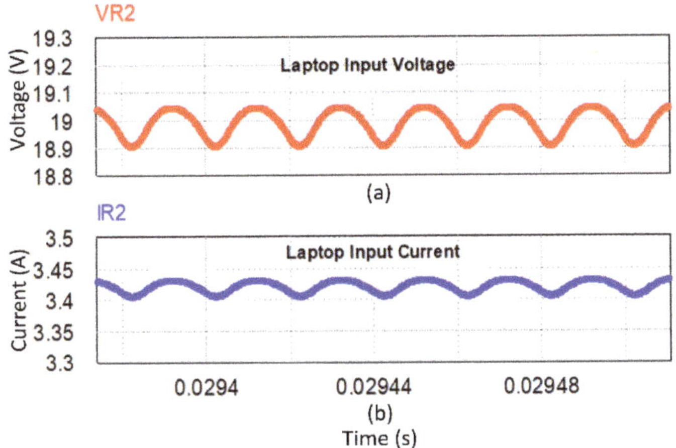

Figure 8. Laptop Input Voltage and Current Waveform. Laptop Input Voltage Ripple: 0.7%; Current Ripple: 0.58%. (**a**) Laptop input voltage; (**b**) Laptop input current.

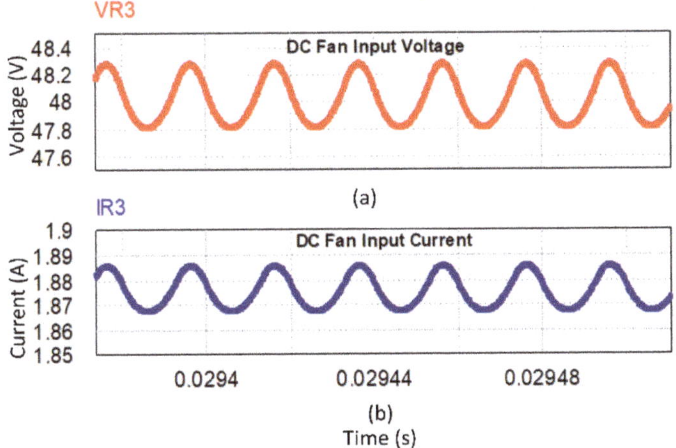

Figure 9. DC Fan Load Input Voltage and Current Waveform. DC Fan Input Voltage Ripple: 2.37%; Current Ripple: 0.337%. (**a**) DC fan input voltage; (**b**) DC fan input current.

7. Conclusions

This research proposes to design and simulation of a DC microgrid that facilitates standalone powering of a rural household which uses less than 250 W load from Solar PV array. The DC to DC POL conversion systems were effectively connected to a DC bus of 72 V. This grid is able to highlight the benefits that a DC grid arrangement have above the traditional AC grids, with a supreme advantage of reduced converter count at the device end. DC grid is designed to be 72 V as no fixed standards are available for this topology. The grid voltage on simulation is obtained as 72 V. The conversion circuitry for each device were developed and the prerequisite voltage values were attained from the systems. On observing the simulation results, it can be inferred that the designed DC grid can supply the rated power desirable for each load. By realizing on a higher scale, the scope of this project can be commercialized to power individual homes in an island area thereby achieving the goal of 0% unpowered villages.

Author Contributions: All authors contributed equally to the final dissemination of the research investigation as a full article.

Funding: This research activity received support from EEEIC International, Poland.

Acknowledgments: The authors would like to acknowledge the technical assistance received from the Center for Bioenergy and Green Engineering, Department of Energy Technology, Aalborg University, Denmark.

Conflicts of Interest: The authors declare no conflict of interest.

References

1. Sieminski, A. *Annual Energy Outlook 2015*; US Energy Information Administration: Washington, DC, USA, 2015.
2. Loomba, P.; Asgotraa, S.; Podmore, R. DC solar microgrids—A successful technology for rural sustainable development. In Proceedings of the IEEE PES Power Africa, Livingstone, Zambia, 28 June–2 July 2016; pp. 204–208.
3. Jhunjhunwala, A.; Aditya, L.; Prabhjot, K. Solar-dc microgrid for Indian homes: A transforming power scenario. *IEEE Electrif. Mag.* **2016**, *4*, 10–19. [CrossRef]
4. Chandel, S.S.; Shrivastva, R.; Sharma, V.; Ramasamy, P. Overview of the initiatives in renewable energy sector under the national action plan on climate change in India. *Renew. Sustain. Energy Rev.* **2016**, *54*, 866–873. [CrossRef]
5. Makarabbi, G.; Gavade, V.; Panguloori, R.B.; Mishra, P.R. Compatibility and performance study of home appliances in a DC home distribution system. In Proceedings of the IEEE International Conference on Power Electronics, Drives and Energy Systems (PEDES), Mumbai, India, 16–19 December 2014.
6. Rajesh, M.P.; Pindoriya, N.M.; Rajendran, S. Simulation of DC/DC converter for DC nano-grid integrated with solar PV generation. In Proceedings of the IEEE Innovative Smart Grid Technologies-Asia (ISGT ASIA), Bangkok, Thailand, 3–6 November 2015.
7. Panguloori, R.B.; Mishra, P.R.; Boeke, U. Economic viability improvement of solar powered Indian rural banks through DC grids. In Proceedings of the Annual IEEE India Conference (INDICON), Hyderabad, India, 16–18 December 2011.
8. Kaur, P.; Jain, S.; Jhunjhunwala, A. Solar-DC deployment experience in off-grid and near off-grid homes: Economics, technology and policy analysis. In Proceedings of the IEEE First International Conference on DC Microgrids (ICDCM), Atlanta, GA, USA, 7–10 June 2015; pp. 26–31.
9. Global Buildings Performance Network. *Residential Buildings in India: Energy Use Projections and Savings Potentials*; Global Buildings Performance Network: Ahmedabad, India, 2014.
10. Rodriguez-Diaz, E.; Vasquez, J.C.; Guerrero, J.M. Intelligent DC homes in future sustainable energy systems: When efficiency and intelligence work together. *IEEE Consum. Electron. Mag.* **2016**, *5*, 74–80. [CrossRef]
11. Nilsson, D.; Sannino, A. Efficiency analysis of low and medium-voltage DC distribution systems. In Proceedings of the Power Engineering Society General Meeting, Denver, CO, USA, 6–10 June 2004; pp. 2315–2321.
12. Rodriguez-Otero, M.A.; O'Neill-Carrillo, E. Efficient home appliances for a future DC residence. In Proceedings of the IEEE Energy 2030 Conference, Atlanta, GA, USA, 17–18 November 2008; pp. 1–6.
13. Otero, R.; Angel, M. Power quality issues and feasibility study in a DC residential renewable energy system. *Mast. Abstr. Int.* **2009**, *47*.
14. Sustainable Energy Program Report. The Use of Direct Current Output from PV Systems in Buildings. Available online: http://www.berr.gov.uk/files/file17277.pdf (accessed on 7 July 2008).
15. Jeon, J.Y.; Kim, J.S.; Choe, G.Y.; Lee, B.K.; Hur, J.; Jin, H.C. Design guideline of DC distribution systems for home appliances: Issues and solution. In Proceedings of the IEEE International Electric Machines & Drives Conference (IEMDC), Niagara Falls, ON, USA, 15–18 May 2011; pp. 657–662.
16. Center for Decentralized Power Systems. *Technological Comparative Study of Solar Lighting Systems for Homes*; Indian Institute of Technology: Chennai, India, 2015.
17. Rodriguez-Diaz, E.; Savaghebi, M.; Vasquez, J.C.; Guerrero, J.M. An overview of low voltage DC distribution systems for residential applications. In Proceedings of the 5th International Conference on Consumer Electronics-Berlin (ICCE-Berlin), Berlin, Germany, 6–9 September 2015; pp. 318–322.

18. Vossos, E. *Optimizing Energy Savings from "Direct-DC" in US Residential Buildings*; Ernest Orlando Lawrence Berkeley National Laboratory: Berkeley, CA, USA, 2011.
19. Vossos, V.; Garbesi, K.; Shen, H. Energy savings from direct-DC in US residential buildings. *Energy Build.* **2014**, *68*, 223–231. [CrossRef]
20. Savage, P.; Nordhaus, R.R.; Jamieson, S.P. Dc Microgrids: Benefits and Barriers. In *From Silos to Systems: Issues in Clean Energy and Climate Change*; Yale School of Forestry & Environmental Studies: New Haven, CT, USA, 2010; pp. 51–66.
21. Koutroulis, E.; Kalaitzakis, K.; Voulgaris, N.C. Development of a microcontroller-based, photovoltaic maximum power point tracking control system. *IEEE Trans. Power Electr.* **2001**, *16*, 46–54. [CrossRef]
22. Tong, Y.; Shan, Z.; Jatskevich, J.; Davoudi, A. A nonisolated multiple-input-multiple-output dc-dc converter for dc distribution of future energy efficient homes. In Proceedings of the IECON 2014—40th Annual Conference of the IEEE Industrial Electronics Society, Dallas, TX, USA, 30 October–1 November 2014; pp. 4126–4132.
23. Weiss, R.; Ott, L.; Boeke, U. Energy-efficient low-voltage DC-grids for commercial buildings. In Proceedings of the IEEE First International Conference on DC Microgrids (ICDCM), San Francisco, CA, USA, 30 March–3 April 2015; pp. 154–158.
24. Tidjani, F.S.; Chandra, A. Integration of renewable energy sources and the utility grid with the Net Zero Energy Building in the Republic of Chad. In Proceedings of the IECON 2012—38th Annual Conference on IEEE Industrial Electronics Society, Montreal, CA, USA, 25–28 October 2012; pp. 1025–1030.
25. Rajaraman, V.; Jhunjhunwala, A.; Kaur, P.; Rajesh, U. Economic analysis of deployment of DC power and appliances along with solar in urban multi-storied buildings. In Proceedings of the IEEE First International Conference on DC Microgrids (ICDCM), Atlanta, GA, USA, 24–27 May 2015; pp. 32–37.
26. Momose, T.; Osaka Gas Company Limited, Japan. Nano-grid: Small scale DC Microgrid for Residential Houses with Cogeneration System in Each House. In Proceedings of the International Gas Union Research Conference, Paris, France, 8–10 October 2008; Currans Associates, Inc.: Red Hook, NY, USA, 2008.
27. Sasidharan, N.; Singh, J.G. A Novel Single-Stage Single-Phase Reconfigurable Inverter Topology for a Solar Powered Hybrid AC/DC Home. *IEEE Trans. Ind. Electr.* **2017**, *64*, 2820–2828. [CrossRef]
28. Sun, K.; Wang, X.; Qiu, Z.; Wu, H.; Xing, Y. A PV generation system based on the centralized-distributed structure and cascaded power balancing mechanism for DC microgrids. In Proceedings of the IEEE 2nd International Future Energy Electronics Conference (IFEEC), Taipei, Taiwan, 1–4 November 2015; pp. 1–6.
29. Shwehdi, M.H.; Mohamed, S.R. Proposed smart DC nano-grid for green buildings—A reflective view. In Proceedings of the International Conference on Renewable Energy Research and Application (ICRERA), Milwaukee, WI, USA, 19–22 October 2014; pp. 765–769.
30. Stieneker, M.; De Doncker, R.W. Medium-voltage DC distribution grids in urban areas. In Proceedings of the IEEE 7th International Symposium on Power Electronics for Distributed Generation Systems (PEDG), Vancouver, BC, Canada, 27–30 June 2016; pp. 1–7.
31. Friedman, M.M.; van Timmeren, A.; Boelman, E.; Schoonman, J. The concept for a dc low voltage house. *Smart Sustain. Built Environ.* **2008**, 85–94.
32. Li, W.; Mou, X.; Zhou, Y.; Marnay, C. On voltage standards for DC home microgrids energized by distributed sources. In Proceedings of the 7th International Power Electronics and Motion Control Conference (IPEMC), Harbin, China, 2–5 June 2012; Volume 3, pp. 2282–2286.
33. Starke, M.; Tolbert, L.M.; Ozpineci, B. AC vs. DC Distribution: A loss comparison. In Proceedings of the IEEE/PES Transmission and Distribution Conference and Exposition, Chicago, IL, USA, 21–24 April 2008; pp. 1–7.
34. Pratt, A.; Kumar, P.; Aldridge, T.V. Evaluation of 400V DC distribution in telco and data centers to improve energy efficiency. In Proceedings of the INTELEC 2007—29th International Telecommunications Energy Conference, Rome, Italy, 30 September–4 October 2007; pp. 32–39.
35. Ammerman, R.F.; Gammon, T.; Sen, P.K.; Nelson, J.P. DC arc models and incident energy calculations. In Proceedings of the Record of Conference Papers Industry Applications Society 56th Annual Petroleum and Chemical Industry Conference, Anaheim, CA, USA, 14–16 September 2009; pp. 1–13.
36. Baran, M.E.; Mahajan, N.R. DC distribution for industrial systems: Opportunities and challenges. *IEEE Trans. Ind. Appl.* **2003**, *39*, 1596–1601. [CrossRef]

37. Manandhar, U.; Ukil, A.; Jonathan, T.K.K. Efficiency comparison of DC and AC microgrid. In Proceedings of the IEEE PES Innovative Smart Grid Technologies Conference—Asia (ISGT Asia), Bangkok, Thailand, 3–6 November 2015; pp. 1–6.

38. Jagadish Kumar Patra, H.M.; Tania, D.E.; Arunkumar, G. A Review on Advancements in DC Microgrid Technology. *Proc. Today* **2016**, *9*, 1265–1279.

39. Wunder, B.; Ott, L.; Szpek, M.; Boeke, U.; Weiß, R. Energy efficient DC-grids for commercial buildings. In Proceedings of the IEEE 36th International Telecommunications Energy Conference (INTELEC), Vancouver, BC, Canada, 28 September–2 October 2014; pp. 1–8.

40. Umanand, L. *Power Electronics—Essentials and Applications*; Wiley India Pvt. Ltd.: New Delhi, India, 2009.

41. Rykov, K.; Duarte, J.L.; Szpek, M.; Olsson, J.; Zeltner, S.; Ott, L. Converter impedance characterization for stability analysis of low-voltage DC-grids. In Proceedings of the IEEE PES Innovative Smart Grid Technologies Asia (ISGT Asia), Washington, DC, USA, 19–22 February 2014; pp. 1–5.

42. Carli, R.; Dotoli, M. Energy scheduling of a smart home under nonlinear pricing. In Proceedings of the 53rd IEEE Conference on Decision and Control, Los Angeles, CA, USA, 15–17 December 2014; pp. 5648–5653.

43. Sperstad, I.B.; Korpås, M. Energy Storage Scheduling in Distribution Systems Considering Wind and Photovoltaic Generation Uncertainties. *Energies* **2019**, *12*, 1231. [CrossRef]

44. Hosseini, S.M.; Carli, R.; Dotoli, M. Model Predictive Control for Real-Time Residential Energy Scheduling under Uncertainties. In Proceedings of the 2018 IEEE International Conference on Systems, Man, and Cybernetics (SMC), Miyazaki, Japan, 7–10 October 2018; pp. 1386–1391.

45. Wu, Y.; Lau, V.K.; Tsang, D.H.; Qian, L.P.; Meng, L. Optimal energy scheduling for residential smart grid with centralized renewable energy source. *IEEE Syst. J.* **2013**, *8*, 562–576. [CrossRef]

46. Carli, R.; Dotoli, M. A decentralized resource allocation approach for sharing renewable energy among interconnected smart homes. In Proceedings of the 54th IEEE Conference on Decision and Control (CDC), Osaka, Japan, 15–18 December 2015; pp. 5903–5908.

47. Sun, L.; Wu, G.; Xue, Y.; Shen, J.; Li, D.; Lee, K.Y. Coordinated control strategies for fuel cell power plant in a microgrid. *IEEE Trans. Energy Convers.* **2017**, *33*, 1–9. [CrossRef]

48. Patterson, M.; Macia, N.F.; Kannan, A.M. Hybrid microgrid model based on solar photovoltaic battery fuel cell system for intermittent load applications. *IEEE Trans. Energy Convers.* **2014**, *30*, 359–366. [CrossRef]

49. Farrokhabadi, M.; König, S.; Cañizares, C.A.; Bhattacharya, K.; Leibfried, T. Battery energy storage system models for microgrid stability analysis and dynamic simulation. *IEEE Trans. Power Syst.* **2017**, *33*, 2301–2312. [CrossRef]

Article

A Three-Phase Transformerless T-Type- NPC-MLI for Grid Connected PV Systems with Common-Mode Leakage Current Mitigation

P. Madasamy [1], V. Suresh Kumar [2], P. Sanjeevikumar [3,*], Jens Bo Holm-Nielsen [3], Eklas Hosain [4] and C. Bharatiraja [5,*]

[1] Department of Electrical and Electronics Engineering, Alagappa Chettiar college of Engineering and Technology, Karaikudi 630 003, Tamilnadu, India; mjasmitha0612@gmail.com

[2] Department of Electrical and Electronics Engineering, Thiagarajar College of Engineering, Madurai 625 015, Tamilnadu, India; vskeee@tce.edu

[3] Center for Bioenergy and Green Engineering, Department of Energy Technology, Aalborg University, 6700 Esbjerg, Denmark; jhn@et.aau.dk

[4] Oregon Renewable Energy Center (OREC), Department of Electrical Engineering & Renewable Energy, Oregon Tech, Klamath Falls, OR 97601, USA; eklas.hossain@oit.edu

[5] Department of Electrical and Electronics Engineering, SRM University, Chennai 603 203, Tamilnadu, India

* Correspondence: san@et.aau.dk (P.S.); bharatiraja@gmail.com (C.B.); Tel.: +91-904-270-1695 (C.B.)

Received: 4 June 2019; Accepted: 21 June 2019; Published: 24 June 2019

Abstract: DC to AC inverters are the well-known and improved in various kinds photovoltaic (PV) and gird tied systems. However, these inverters are require interfacing transformers to be synchronized with the grid-connected system. Therefore, the system is bulky and not economy. The transformerless inverter (TLI) topologies and its grid interface techniques are increasingly engrossed for the benefit of high efficiency, reliability, and low cost. The main concern in the TL inverters is common mode voltage (CMV), which causes the switching-frequency leakage current, grid interface concerns and exaggerates the EMI problems. The single-phase inverter two-level topologies are well developed with additional switches and components for eliminating the CMV. Multilevel inverters (MLIs) based grid connected transformerless inverter topology is being researched to avail additional benefits from MLI, even through that are trust topologies presented in the literature. With the above aim, this paper has proposed three -phase three-level T type NP-MLI (TNP-MLI) topology with transformerless PV grid connected proficiency. The CM leakage current should handle over mitigating CMV through removing unwanted switching events in the inverter pulse width modulation (PWM). This paper is proposes PV connected T type NP-MLI interface with three-phase grid connected system with the help of improved space vector modulation (SVM) technique to mitigate the CM leakage current to overcome the above said requests on the PV tied TL grid connected system. This proposed the SVM technique to mitigate the CM leakage current by selecting only mediums, and zero vectors with suitable current control method in order to maintain the inverter current and grid interface requirements. The proposed PV tied TNP-MLI offering higher efficiency, lower breakdown voltage on the devices, smaller THD of output voltage, good reliability, and long life span. The paper also investigated the CM leakage currents envisage and behavior for the three-phase MLI through the inverter switching function, which is not discussed before. The proposed SVM on TL-TNP-MLI offers the reliable PV grid interface with very low switching-frequency leakage current (200mA) for all the PV and inverter operation conditions. The feasibility and effectiveness of the TLI and its control strategy is confirmed through the MATLAB/Simulink simulation model directly as compared with 2kW roof top PV plant connected TL-TNP-MLI experimentation, showing good accordance with theoretical investigation. The simulation and experimental results are demonstrated and presented in the good stability of steady state and dynamics performances. The proposed inverter reduces the cost of grid interface transformer, harmonics filter, and CMV suppressions choke.

Keywords: Hybrid Microgrid; Battery Electric Vehicle; Energy management strategy; Vehicle-to-Vehicle Charging; Energy Storage Unit

---

1. Introduction

Photovoltaic (PV) based power generation is an unavoidable segment in the electrical power generation system (PGS) to meet the world power demand. The most recent renewables 2018 global status report indicates that 450GW of new PV plants have been set up worldwide in 2017, a 125% proliferation when compared to 2016 [1]. The reported global electricity generations from renewable energy sources alone in 2017 has been estimated around 26.5%, out of which 1.9% is from PV power generations.

The leading PV power producers were China, European Union, and the United States of America, with 131.1 GW, 108 GW, and 51 GW, respectively [2,3]. Countries like Japan, India, United Kingdom (UK), and Australia are the next pioneers in generating more power from PV planets.

Even though the price of PV panel has been mostly dropped, the complete cost of both the components investment and the generation of grid-tied PV system are quiet high. Hence, the PV tied grid-tied voltage source inverters (VSIs) are need to be prudently designed for accomplishing the low cost, high efficiency, and small size, in addition to low weight. The PV grid-tied VSIs are associated with the line-frequency transformers (LFTs) to afford galvanic-isolation in commercial PV inverter system structures. Nevertheless, LFTs are heavy and large, building the complete system bulky as well as inflexible to install. The use of LFTs for line-frequency isolation in grid-connected inverters, high-frequency isolation transformers are considered for their smaller size, lower cost, as well as total system weight. On the other hand, these high-frequency transformers connect inverter system have different power conversion stages, which affect the system and diminish the system overall efficiency and straightforwardness [4,5]. As a result, the PV grid-tied transformerless inverters (TLIs) are broadly used and installed in the low-power PV distributed generation systems (particularly 5kW and less). However inappropriately, TLIs are producing the common-mode leakage currents, which cause the leakage current flow due to the present PV panel parasitic capacitances [6]. This leakage current leads to serious safety concern and electro magnetics interference issues [7]. Hence, parasitic capacitances leakage current must be able to mitigate within a recommended choice [8]. Furthermore, the PV inverters tied grid-connected system should fulfill the grid interfacing standards and codes, leakage current detection, grid frequency protection, active and reactive control, and power quality standards [9–13]. Hence, the PV large manufacturers companies, such as Fronius international, Sunny boy, and SMA solar technology, are strict with PV standards and codes to satisfy the system safety and reliability requests. These companies are following German codes (VDE0126-1-1, IEC 60755 issued time, 2006 and VDE-AR-N-4105 issued time, 2011) for leakage current (less than 300mA) and grid frequency (47.5 < f < 51.5). Hence, over 300 mA leakage current, the PV inverter should be trigger to breakdown within 0.3 sec, as per the VDE 0126-1-1 code standard.

From the above-mentioned investigation, the important concerns in the PV tied transformerless inverters should be considered for efficiency, reliability, and cost. In industry practices, the commercial TLIs are designed with the total efficiency of more than 97% to 98 %. The total efficiency of the distributed PV tied TLI systems are manufactured by Steca's Grid and Good We Technology's, with 97.7% and 96.9%, respectively [14,15]. In view of PV inverters installation, 10%–20% overall installation cost is the initial system cost [16], hence the price drops for PV inverters installations certainly promote the PV product affordability. The U.S. energy department demonstrated that the TLI system for the residential systems (≤ 10 kW) had dropped from nine U.S. dollars to 4.8 U.S. dollars per Watt, since 2009 to 2014 [17].

Most of the installed PV plants are erected by single-phase type two-level voltage source inverter (VSI) topologies [2], which generate pulsating AC output power. Hence, the inverters require large

DC-link capacitors; high collector–emitter voltage based switching devices and bulky filters ensure the grid standard codes. Henceforth for the last quarter decades, three-phase three-level NPC (3L-NPC) inverters are preferred for the alternate of two-level VSIs, which has been the hub of various research studies on PV applications [3–7]. Due to their small DC-link capacitors and the constant ac power on the output, NP-MLIs gave higher efficiency, lower breakdown voltage on the devices, smaller THD of output voltage, good reliability, and long life span [3,18–24]. Three-phase three-level NPC-MLIs are separated into two types (I and T) based on their leg shapes. After the arrival of T-type-NPC-MLI, I-type-NPC-MLI were evaded due to their long current commutation paths, higher stray inductance, and high conduction loss. The T-type 3L-NPC is being explored more to progress the better system efficiency [25–32]. The research on NPC-MLI has been extensively researched in PV and DC Distribution Network due its influencing of the stability, lower total harmonic distortion (THD), and lower electromagnetic noise [32–34]. These improvements mainly focused on DC-link balancing and common-mode voltage mitigations [35,36].

Most of the PV-tied grid connected inverter topologies are transformer based, which isolates the PV panels from the grid. The line-frequency transformer is avoided and high-frequency transformer is preferred due to the size and weight issues. However, the high-frequency transformer degrades the efficiency and makes the system more complex [37,38]. As a result, transformerless (TL) PV tied inverter–grid connection structure is greatly appreciated and studied by the power engineers due its less size, weight, and cost [39,40]. Although the TL inverters are able to gain system efficiencies of up to 97%, due to the absence of galvanic isolation and large stray capacitances, higher common mode voltage (CMV) is generated by VSI, which escorts the higher CM leakage current on the PV modules. Therefore, TL inverters suffer higher THD, ripple in grid current, degrading power quality, and warns of human safety, along with forming worsening EMI issues [41,42]. Figure 1 shows the leakage current bath for the Transformerless PV connected system.

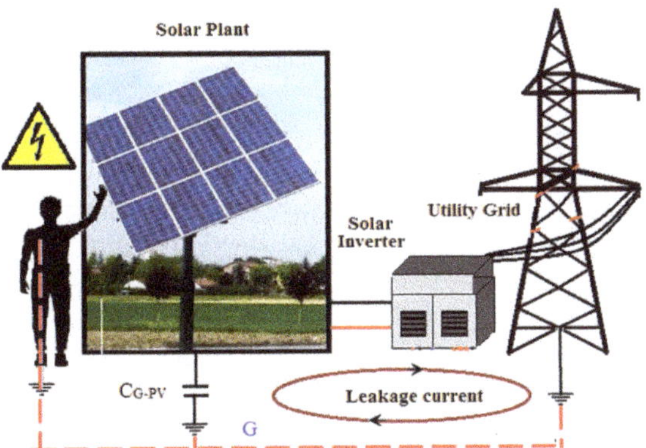

Figure 1. Leakage current for Transformerless photovoltaic (PV) connected system.

The PV commissioning companies recommended limiting the leakage current, for example, in German standard 300mA, is used as the reference value. In order to keep this recommendation, most of the inverter manufacturers maintain expertise in inverter topologies, special controllers, and novel pulse width modulation (PWM) techniques. There are techniques for the TL system to reduce the leakage current through adding up devices in the inverter, such as diodes, switches, and passive elements [43–47]. However, changing the PWM switching pattern is more efficient than adding the devices and increasing the cost and reducing trust on continuous operation [37], which are mainly changing the pulses based on CMV development. When compared to other PWM methods, space vector modulation (SVM) is a pledge PWM to creating different PWM in a binary way. The methods use three medium vectors (3MV), keeps minimal CMV as the constant value, and it does not root less leakage current. Nevertheless, the maximum fundamental output voltage of the inverter is restricted $V_{DC}/2$. A different method in [45] uses two MVs and one zero vector (ZV) and it does not cause the leakage current. However, similar to 3MV, its limited output voltage, and hence the better dc-link voltage, is compulsory to interface with the grid. Lee. J.S et.al has developed modulation techniques for reducing the leakage current while balancing the DC-link voltages. The method using the medium, zero, and large vectors reduces the CMV voltage that significantly roots the leakage current with better DC-link unitization [47].

Based on the aforementioned recognitions, a novel three-phase three-level TL T-type- NPC-MLI (3L-TL-TNPC-MLI) is proposed for the PV grid connected system with leakage current reduction. The novel TL-TNPC-MLI topology, together with the SVM strategy, have been proposed, aiming to improve system efficiency through the reduction of the commutation and conduction loss on the inverter, and leakage current reduction. The validity of the proposed inverter and its SVM control algorithm is verified for 1.5 kW PV grid connected system via simulation and experiment study. From the results, it is verified that the proposed inverter and PMW algorithm is well set for the perfect grid interface with reducing the leakage current.

The paper is outlined, as follows: nomenclature deals with the List of symbols and abbreviations. Section 2 accomplishes the analytical model and derivation of the leakage current on the three-phase transformerless MLI PV-grid connected system. Section 3 discusses the proposed PV tied three-phase three-level TL-TNP-MLI for th grid connected system. Sections 4 and 5 accomplishes the MATLAB/Simulink simulations and hardware prototype experimentations. In conclusion, the advantages of the proposed system are presented in Section 6.

2. Leakage Current Analysis Transformerless PV Inverter System

In the PV tied grid connected systems, the isolation transformer is connected between the PV inverters and the grid, as shown in Figure 1. The parasitic capacitances (C_{G-PV}) is connected in the 1 (dc$^+$ terminal) and 2 (dc$^-$terminal) of the PV array. The mid-point of terminals 1 and 2 is called the neutral point (N). The voltage difference between terminals 1 and 2 is related to ground (G). The L_A, L_B, and L_C are line the inductances of each phase, L_{CA}, L_{CB}, and L_{CC} are the line inductances of each phase with respect to grid and, L_G is the grid inductance. When the variations of the 1 and 2 terminals would source leakage currents from the PV panel to the ground. The leakage current value is depending on the amplitude and frequency of the voltage fluctuations and, in addition to the value of the parasitic capacitance (leakage capacitance) [26]. The leakage capacitance value depends on many factors, such as PV panel and frame structure, dust or salt covering the PV panel, and weather conditions, and so on.

Due to the fact of C_{G-PV} and inverter topology, common-mode voltage (CMV) is generated, which causes the common-mode current (CMC). The CMC is a danger, and it falloffs the PV lifetime [36]. In a general PV tied grid connected system with isolation transformer, the CMC can only find its path through the stray capacitances of the transformer. Generally, this is the reason why galvanic isolation (with a transformer) based PV systems are not affected due to the low frequency leakage current behavior, irrespective of the converters and their modulation techniques. As per German standard-DIN VDE0126-1-1, when the ground current goes beyond 300 mA (peak), the inverter needs to disconnect within 3 sec, and it furnishes the needs for restrictions concerning ground leakage and fault currents [18]. However, in TLI, due to the transformer absence, the converter and modulation methods can only do the CMC elimination. With this aim, the leakage current model is derived related with three-phase inverter switching. In order to illustrate the path for the common-mode current (CMC), the stray elements are added to the system, as in Figure 2.

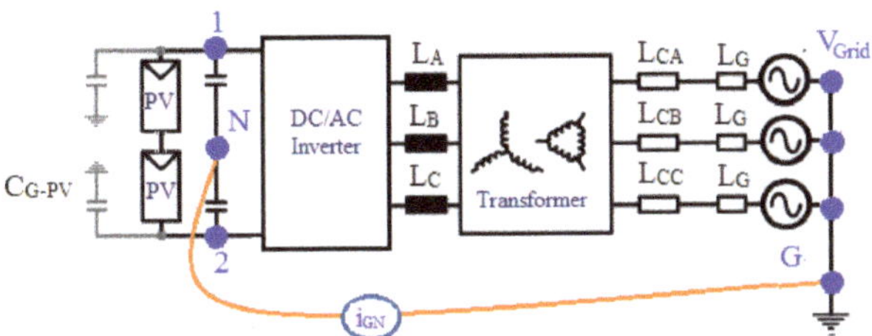

Figure 2. Grid connected PV system including the parasitic capacitance to ground of the PV array.

Figure 3 shows the three-phase PV-grid connected inverter model with $C_{G\text{-}PV}$ and CMV source that individually calculate the CMV and differential mode voltage (DMV), and these can be related with the leakage current model.

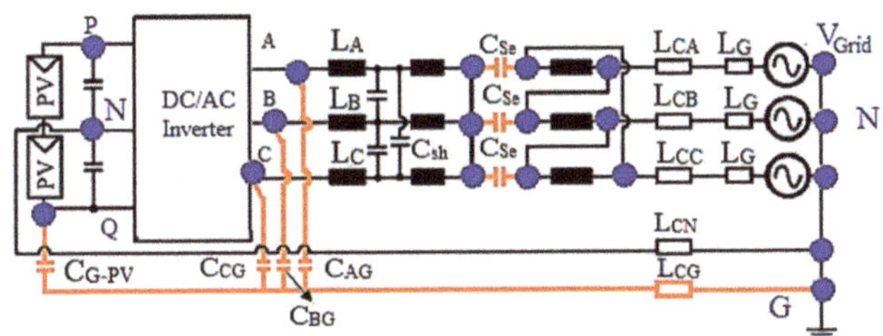

Figure 3. Three-phase PV-grid connected inverter model with stray elements.

Where the parameters are described, as follows; C_{sh} is shunt capacitance; C_{se} is series capacitance; L_{CN} is inductance between inverter neutral point and grid; L_{CG} is inductance between PV terminals and grid; and, C_{AG}, C_{BG}, and C_{CG} are the capacitances between each phase with respect to grid ground (see Figure 4).

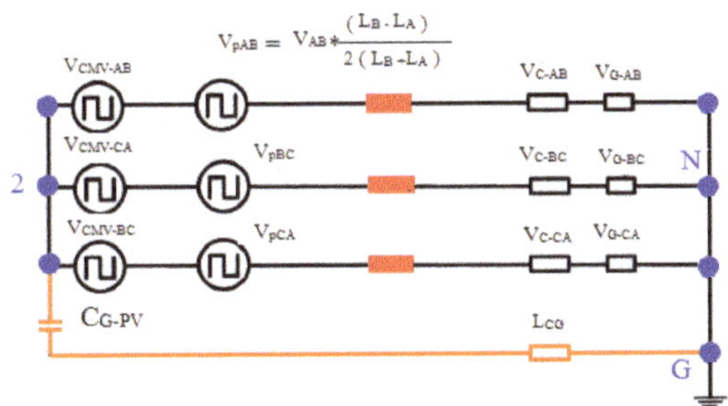

Figure 4. Three-phase PV-grid connected inverter model [5].

In a three-phase system, CMV and differential mode voltage (DMV) calculations between A to B or B to C or C to A phases are same. Thus, only A and phase B investigated in this paper. They calculated as [5],

$$\text{CMV between A and B phases, } V_{CM-AB} = \frac{V_{AN} + V_{BN}}{2} \tag{1}$$

$$\text{DMV between A and B phases, } V_{DM-AB} = V_{AN} - V_{BN} \tag{2}$$

From (1) and (2), the out voltage expressed as

$$V_{AN} = \frac{V_{DM-AB}}{2} + V_{CM-AB} \tag{3}$$

$$V_{BN} = -\frac{V_{DM-AB}}{2} + V_{CM-AB} \tag{4}$$

By observing the above equations, the CM model for A and B phases are shown in Figures 5 and 6. Inverter stray capacitances are significantly identical, since the output inductances are identical for all three phases; the model can be simplified, as presented in Figure 7a. The final CM equivalent circuit model for the three-phase system is based on the designed two-phase circuit CM model, as shown in Figure 7b.

The CMV for all three phases expressed as [20],

$$V_{CM} = \frac{V_{AN} + V_{BN} + V_{CN}}{3} \tag{5}$$

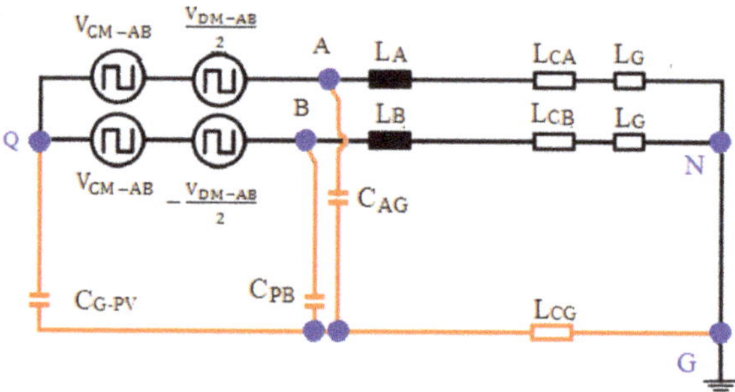

Figure 5. PV-grid connected inverter model only A and B phase [5].

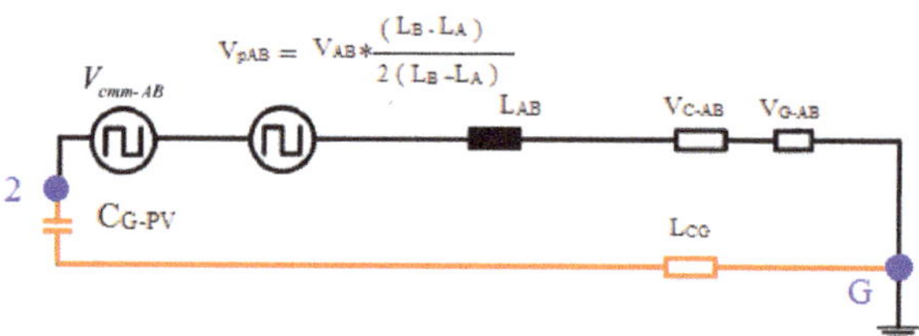

Figure 6. Single line model of total common mode voltage (CMV) [5].

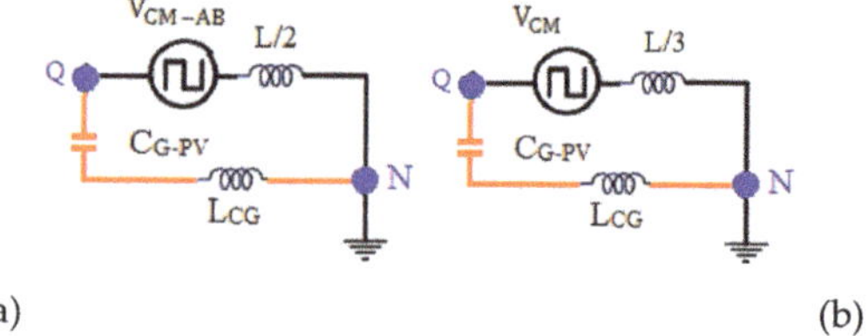

(a)   (b)

Figure 7. CMV model (a) two-phase model, and (b) three-phase model.

The output voltages of the three-level VSI are determined, based on the inverter's legs connection with one point among P, N, and Q, as in Figure 3. If the grid ground (G) is the standard, these three connection points' output voltages are $V_{DC} + V_N$, $V_{DC}/2 + V_N$, and V_N, respectively.

The grid currents is expressed as [5],

$$\frac{V_{AN} + V_N - E_a}{z_L} + \frac{V_{BN} + V_N - E_b}{z_L} + \frac{V_{CN} + V_N - E_c}{z_L} = 0 \tag{6}$$

z_L is the impedance of the L-filter

Sum of grid currents is zero,

$$E_a + E_b + E_c = 0 \tag{7}$$

$$V_N = \frac{V_{AN} + V_{BN} + V_{CN}}{3} \tag{8}$$

From (5) and (8), the V_{CM} is related to the output voltages of inverter, which means that V_{CM} can be influenced by the inverter switching state.

Consider that S_a, S_b, and S_c are the switching states of the inverter legs. Hence, the switching function is defined as,

$$S_{A,B,C}(V_{AN},\ V_{BN},\ V_{CN})_{1,\frac{1}{2},0} = \begin{array}{ccc} V_{DC} & V_{DC}/2 & 0 \\ V_{DC} & V_{DC}/2 & 0 \\ V_{DC} & V_{DC}/2 & 0 \end{array} \tag{9}$$

From (6) and (9)

$$V_N \text{ or } V_{CM} = \frac{V_{DC}(S_a) + V_{DC}(S_b) + V_{DC}(S_c)}{3} \tag{10}$$

From (10), the possibilities of the CMV in the inverter switching are $-V_{DC}$, $-5V_{DC}/6$, $-2V_{DC}/3$, $-V_{DC}/2$, $-V_{DC}/3$, $-V_{DC}/6$, and 0. These ac-components of V_N cause the leakage current through the line, including C_{PV}.

Further, the leakage current is calculated, as [5],

$$|I_{CM}| = \frac{|V_{CM}|}{|Z_{PV}|} \tag{11}$$

Due the square wave CMV nature, the total CM leakage current is the addition of odd multiples of switching frequency current components. Hence, the total leakage current (I_{TCM}) is obtained, as [5],

$$I_{TCM} = \sum_{n=1,3,5,7,\dots}^{\infty} |I_{CM}(nf)| \tag{12}$$

From (12), if the switching frequency increases, there is an increase in I_{TCM}. However, it is not proportional to the switching frequency due to the damping resistor, the leakage resistor on the PV, R_{PV}. From the above discussion, I_{TCM} is influenced by leakage capacitor C_{PV}, and the switching action of the inverter. Hence, inverter switching based leakage current elimination is the best method, since C_{PV} is decisive by the environmental factors.

3. Proposed PV Tied Three-Phase Three-Level TL-TNPC-MLI for Grid Connected System with Leakage Current Reduction

This section discussed the proposed PV tied grid connected TL-TNPC-MLI, including CM leakage current elimination by using the full CM elimination switching technique with grid interfacing.

3.1. TL-TNPC-MLI Operation for Zero CMV.

Figure 8 shows the three-phase 3L-TL-TMLI power circuit, which has 12-IGBTs (S_{a1}-S_{c4}) and 12-anti-parallel free-wheeling diodes (FWD; D_{a1}-D_{c4}) three-phase 3L-TL-TMLI power circuit. It involves three-phase leg with four IGBTs (S_{a1}-S_{a4}) and four FWD (D_{a1}-D_{a4}) in a leg. Here, the top and bottom switches IGBTs/diodes (S_{a1}/D_{a1} and S_{a4}/D_{a4}) that were employed with 1200V and middle switches (S_{a2}

/ D_{a2} and S_{a3} / D_{a3}) operated by two 600V. Hence, due to small blocking voltage, middle switches consume very less switching and conduction losses, even though there are two devices that were connected in series [40]. The inverter includes two dc-link identical capacitors ($C_1 = C_2 = V_{DC}/2$) and six equal value split inductors ($L_{a1} = L_{a2} = L_{b1} = L_{b2} = L_{c1} = L_{c2} = L$). In view of T-MLI with TL concept, the ground connection is the essential one for any TL inverter topologies, while the centre of PV cluster does not ground, the proposed TL-TMLI output terminals can be abundantly associated to the grid; else, mid-point (N) of both the dc-link capacitors (V_{C1} and V_{C2}) must be connected to the PV cluster mid-point and grid neutral (G). Table 1 illustrates the switching operation of TL T-MLI, the modes of operation is situated based on the dc-link mid-point (N) connection to DC$^+$, N and DC$^-$. There are 27 possible switching states of operation on 3L-TL-TMLI for the full cyclic operation of one grid cycle. Here, the modes are created based on the switching ON and OFF position on each inverter leg. These modes are categorized in four groups (G-1 to G-4). The possible switching state and its switching groups are tabulated in Table 2.

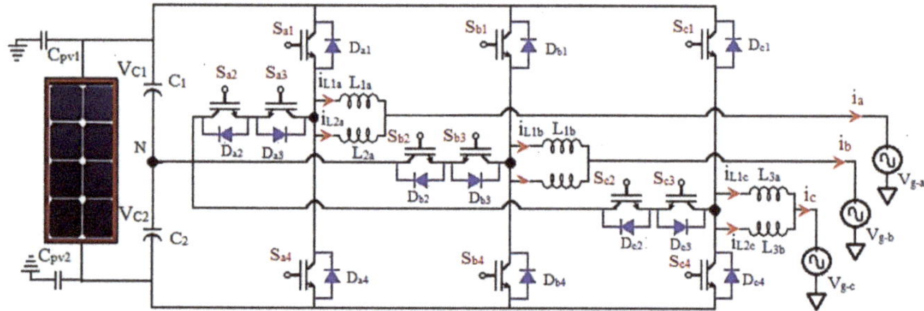

Figure 8. PV tied Three-level TL T- MLI power circuit.

Table 1. PV connected Three-level T-type MLI power circuit.

State	V_{out}	S_{a1}	S_{a2}	S_{a3}	S_{a4}
DC$^+$	$+V_{dc}/2$	ON	ON	OFF	OFF
N	0	OFF	ON	ON	OFF
DC$^-$	$-V_{dc}/2$	OFF	OFF	ON	ON

Table 2. Mode of operation of 3L-TL-TMLI.

Groups	Groups	Switching Mode
G-1	Zero Vector (ZV) switching	(000),(111),(-1-1-1)
G-2	Small Vector (SV) switching	{(100),(011)},{(110),(001),{(010),(101)}{(011),(100),{(001), (110)},{(101),(0-10)}
G-3	Medium Vector (MV) switching	(10-1), (01-1), (-110), (-101), (0-11), (1-10)
G-4	Large Vector (LV) switching	(1-1-1), (11-1), (-11-1), (-111), (-1-11), (1-11)

By applying the switching states in equation (10), each switching state has different possibilities of the CMV in the inverter switching are $V_{dc}/6$, $V_{dc}/3$, $V_{dc}/2$, 0, $-V_{dc}/6$, $-V_{dc}/3$, and $-V_{dc}/2$ as shown in Table 3. For example, for the -1-1-1 switching state, the obtained CMV is $V_{com_{[-1-1-1]}} = \frac{1}{3} \cdot (-3) \cdot \frac{V_{dc}}{2} = -\frac{V_{dc}}{2}$. Similarly, the MV 10-1 switching state CMV is $V_{com_{[1\,0-1]}} = \frac{1}{3} \cdot (0) = 0$. Table 1 illustrates the CMV for the different switching state inverter. For all 27 switching states, the CMVs are summarized as $+V_{dc}/2$, 0, $+V_{dc}/3$, $+V_{dc}/6$. Figure 8 shows the space vector diagram (SVD) for three-level TMLI with their switching and ITS corresponding CMV. The SVD consists of six sectors, 27 switching states, and 24 sub-triangles Here, every switching states has its own CMV, which is tabulated as seven groups and the same represented in three-level SVD, as shown in Figure 8. Here, expect group-D (six MVs and one ZV) all other group switching states are producing CMV and these CMV components cause the leakage current through the line, including C_{PV}.

Table 3. Switching states and CMV.

Group	Switching States of 3-Level Inverter	CMV Generated
A	(111)	$V_{dc}/2$
B	(110), (101), (011)	$V_{dc}/3$
C	(1-11), (11-1),(-111), (001), (010), (100)	$V_{dc}/6$
D	(000), (01-1), (0 –11), (10-1), (1-10), (–101), (–110)	0
E	(– 1-11),(–11 –1), (1 –1 –1), (00 –1), (0–10), (–1 00)	$-V_{dc}/6$
F	(– 10 –1), (0 –1 –1), (0 –1 –1)	$-V_{dc}/3$
G	(–1 –1 –1)	$-V_{dc}/2$

While operating SVM by selecting D –group-switching sequence, the CMV is fully eliminated, and, as a result, leakage current is fully eliminated. Figure 9 shows the refined SVD with using only D-group switching. When considering the case of V^* moving in sector-1 in order to zero CMV, the suggested switching structure uses zero CMV vectors, as V_Z [000], V_{M1} [10-1], and V_{M2} [01-1]. The uses of MVs with ZV facilitated by fixing/switching at the middle of the sector are 30, 90, 150, 210, 270, 330, (-30). Hence, the proposed scheme trundles on the principle of reference phase angle discrete hoping. Thus, sector-1 is reordered between -30 to 30 ($-30^0 < \theta > 30^0$). Similarly, sector-2 is 30 to 90 ($30^0 < \theta > 90^0$), and so on. This angle determination procedure has expanded for all six sectors in the SVD and each sector consists of the three zero CMV switching vectors (two MVs and one ZV switching state) as shown in Figure 10. Of course, due the absence of LVs, the full CMV elimination limited m_a is to 0.7. Therefore, the inverter output voltages reach up to 70% from the full linear m_a range. The dc-link voltage of the inverter must be increased in order to increase the voltage performance.

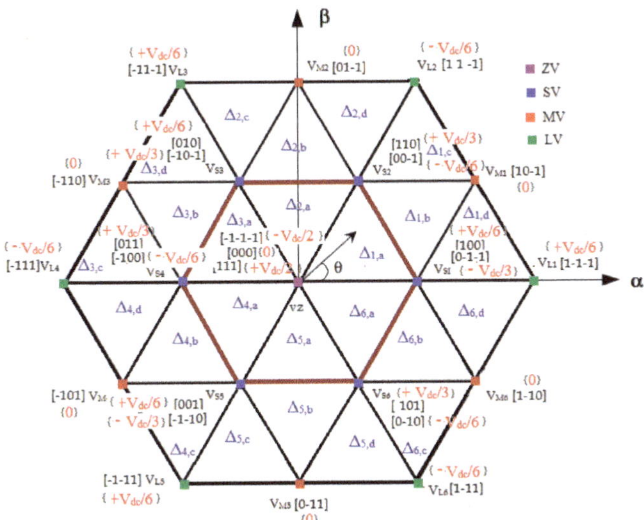

Figure 9. Three-phase space vector diagram (SVD) for three-level MLI.

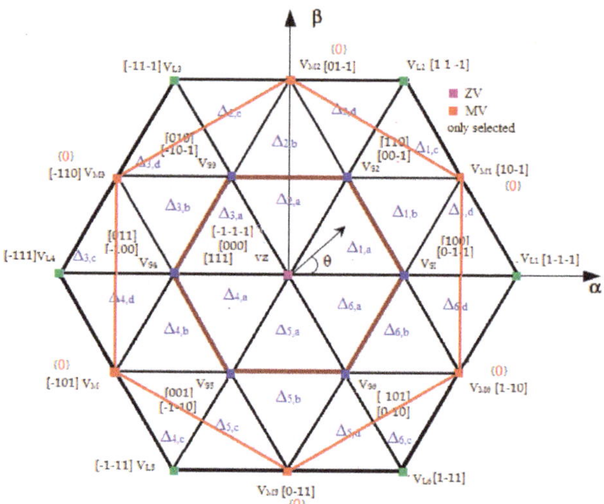

Figure 10. Proposed full mitigation CMV SVD.

3.2. Proposed Close Loop Grid Connected TL-TNPC-MLI

Figure 11 shows the proposed control system block. As presented in closed loop system, the PV panels are and controlled by P&O—maximum power point tracking (MPPT) method and the control output drive the boost converter to meet the desired dc-link voltage for the inverter. To connect the inverter power with utility grid, the phase lock loop (PLL) that is used to measure the phase locked angle of the grid voltage (θ) and its nominal frequency (w_r) from the V_g. The three-phase inverter (i_{inv}) and load current (i_L) are converted from abc to dq to match the PV output V_{dc}.

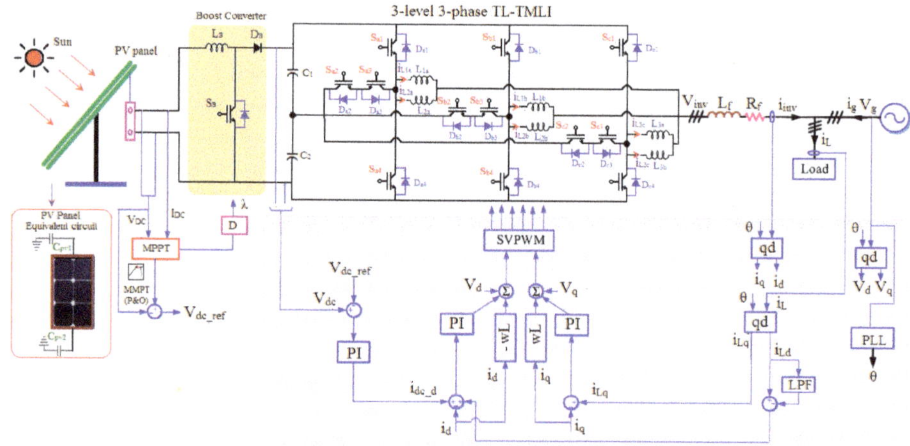

Figure 11. Proposed closed control PV tied grid connected TL-TMLI schematic control block.

By using (13) and (14), according to [20], reference V_α and V_β are generated, which control the SVM to achieve the inverter grid interface.

$$V_\alpha = -wLi_d + V_d + i'_{dc\_d} \tag{13}$$

$$V_\beta = wLi_q + V_q + i'_{L\_q} \tag{14}$$

Finally, the predetermined component (V_α and V_β) is given to the SVM block, which gives the voltage control reference (V^*) with desired modulation index (m_a) to match the grid voltage.

The hysteresis current control (HCC) is used for controlling the inductor current based on the grid current reference. The proposed HCC is giving suitable switching sequence to SVM for controlling inverter. In any HCC, it is necessary to retain the actual load current near the reference load current within the hysteresis boundaries. The main mathematical challenge in current controllers is to generate the reference current components. Figure 12a illustrates the building block of HCC of an inverter. Here, the reference current (i_{ref}) is related with the actual current (i_{act}) to produce the error current, $\vec{\varepsilon_i}$= (i_{Lref} - i_{Lact}), which will produce the error gain (control input u(t)) between the two hysteresis bands B_U and U_L (upper and lower hysteresis bands) [24]. This control signal generates the control input u (t) to drive the inverter.

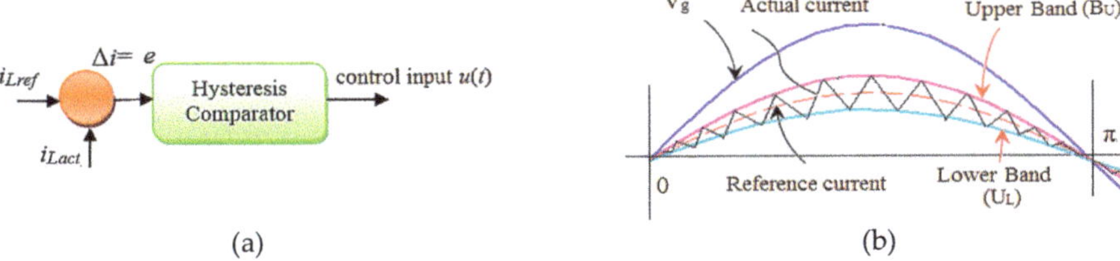

(a)

(b)

Figure 12. Hysteresis current control (HCC) control strategy [24]. (a). HCC comparator, (b). HCC band.

The Sinusoidal PWM (SPWM) based conventional HCC (CHCC) involves three individual references. Therefore, due to the non-interference between the commutation phases, the CHCC deviates from the variable switching frequency over the fundamental frequency cycle, which results in an irregular inverter operation. This happens, because each phase current not only depends on the corresponding phase voltage, but is also affected by the other voltages [23]. However, SVM-based CCs control the current using a single reference vector. Here, the tip of the reference current vector moves around the hexagonal-coordinate system in SVD [20]. Some of the research on HCC for MLI has been attempted through SVM [24,35]. However, these controllers have difficulties in attempting to make a direct current control and they use more numbers of hysteresis bands in order to achieve the results.

Unlike the reported CHCC and SVCC, the proposed HSVCC not only maintains the current, but it also takes care of dc-link capacitor balancing for the inverter, which assures the capacitor balancing within the recommended significance and produces the effective output voltage with current control to the grid. These switching options enables the CC with capacitors balancing throughout the operating condition (change in PV, Inverter, and grid parameters).

Here, when the reference current (i_{Lref}) and actual current (i_{Lact}) lie in SVD, and it is expressed as [24],

$$\vec{i_{Lref}} = i_{Lref\_\alpha} + j i_{Lref\_\beta} \tag{15}$$

$$\vec{i_{Lact}} = i_{Lact\_\alpha} + j i_{Lact\_\beta} \tag{16}$$

Subsequently, the error vector, $\vec{\varepsilon_i}$ is calculated from (2) and (3),

$$\vec{\varepsilon_i} = \vec{i_{Lref}} - j \vec{i_{Lact}} \tag{17}$$

Now, the $\vec{\varepsilon_i}$ expressed in the $\alpha\beta$-vector components frame is

$$\vec{\varepsilon_i} = \vec{\Delta_{i\alpha}} + j\vec{\Delta_{i\beta}} \tag{18}$$

From Equation (5), the position of the $\vec{\varepsilon_i}$ in SVD is calculated and the switching vectors are then selected to keep the $\vec{i}_{Lact}$ close to the $\vec{i}_{Lref}$.

It is assumed that the inverter has unity power factor (PF), i.e., the inverter current i_L is in phase with the grid voltage V_g. This assumption is true for PV grid-connected inverter (unity PF between the in-grid current and the grid voltage). The high frequency inverter switching pulse (f_s) must be stopped at the time $x\pi$ in order to confirm the zero-inverter current beforehand the zero crossing of the V_g, hence the inverter current is forced to zero by the grid voltage, as shown in Figure 13.

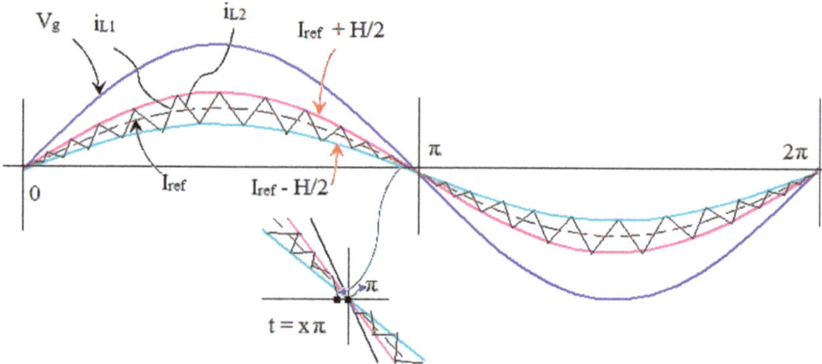

Figure 13. Phase control between grid voltage V_g and inductor current i_{L12}.

The time $x\pi$ to turn OFF the high-frequency inverter switching signals is calculated from the inverter switching action for an interval between x to x.

$$\frac{\int_{x\pi}^{\pi}\left[V_g.\sin(wt) + V_D(ON) + V_S(ON)\right]}{L} \geq I_{ref.}\sin(x\pi) + \frac{H}{2} \quad (19)$$

Here, V_D (ON) and V_S (ON) are the ON-state voltages of clamping diodes and the power switches, respectively; 'H' is the bandwidth of current hysteresis, i.e., the ripple of inductor current, and x is the percent of high-frequency driving signal continuance. HCC is a nonlinear controller, which has high performance, high stability, and fast dynamic response. Nevertheless the output current is widely distributed in the energy of spread-spectrum due to the unfixed switching frequency. Therefore, the filter design is highly difficult. There are many studies on hysteresis band calculation [37]; among them, the real-time regulation method that was based on the switching frequency is the effective method for the calculation of the H value. Fixed frequency HCC can be obtained by calculating the hysteresis band H while using the transient value of the V_g and output voltage of PV array. Based on that, the hysteresis band 'H' is calculated as,

$$H = \frac{V_g\left(V_{PV} - 2.V_g\right)}{V_{PV}.L.f_s} \quad (20)$$

Here, the variant parameters are V_{PV}, V_g, f_S, and fixed one is filter inductor (L)

4. Simulation Validation

The performance of the proposed PV connected three-phase three-level TL-T-MLI simulation models are performed while using MATLAB/Simulink 14.b. Here, the 2 kW PV array module is configured (series model) to feed the power to inverter with each module rated at 152.5W. The T-MLI maintains the dc-link voltage that is close to 456V by the help of two (C1 and C_2) 470µF capacitors. The inverter is connected to 230V/50 Hz utility grid via 4 mH quality split inductors, L_{x1} and L_{x2}

(where. x = a,b,c). The inverter switching frequency (f_{sw}) is maintained throughout the operation of the inverter as 5 kHz. Simulation studies are carried out in two main directions: verifying the validity of the current controller for controlling inductors current and validating the grid connection of PV tied TL inverter with zero circulating current.

Initially, the PV is set to obtain 456V DC-link voltage for T-MLI. The simple boost converter with the P&O Maximum power point tracking (MMPT) method is used to maintain the MLI DC-link as constant. Figure 14 shows the model of the PV array with the MPPT algorithm results, which is taken at various irradiations and constant cell temperature in increasing and decreasing irradiation with the oscillation of MPP voltage. From the results, it could be seen that the MPPT algorithm was able to follow the MPP of PV array within 1.5 times grid period. The inverter operation tested for variable and fixed band hysteresis. In Figures 15 and 16, the current error (Δ_{ei}) trajectory and controlled current trajectory in α-β plane for variable and fixed band hysteresis are shown. These results indicate the HCC characteristic operation and its error minimization are well worked for both bands. When the TL T-MLI operates in the HCC with a fixed band and variable band, the waveforms of i_{Lref} and i_{Lact} are shown in Figure 17. From the results, it can be viewed that i_{Lact} visibly tracks i_{Lref} and minimizes the error by using the proposed HCC for both the fixed band and variable hysteresis band, which ensure the perfect inverter current synchronous with grid voltage. Figure 18a,b shows the inverter leg inductor currents (i_{L1} to i_{L6}) for variable and fixed band hysteresis. It shows that all of the inductors are properly and identically charged. Hence, three-phase inverters current are equal to each other.

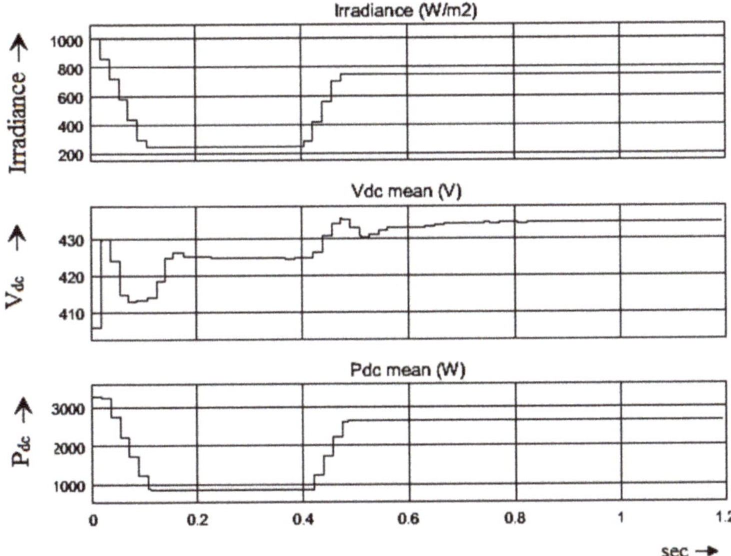

Figure 14. PV array performance for different irradiance.

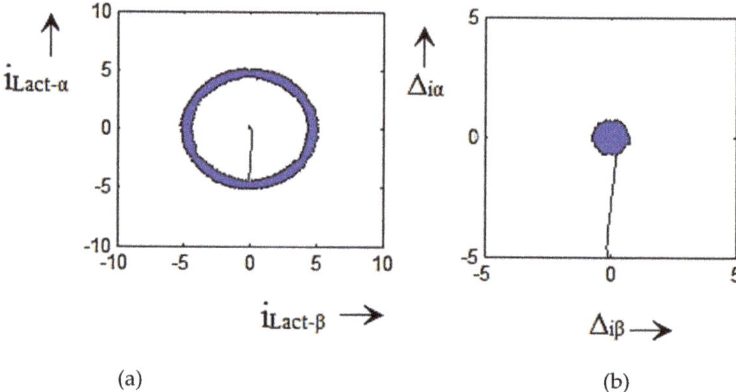

Figure 15. Current error (Δ_{ei}) trajectory and controlled current trajectory in α-β plane for variable and fixed band hysteresis. (**a**). $i_{Lref\text{-}\alpha}$ and $i_{Lact\text{-}\beta}$ Trajectory (**b**). $\Delta_{i\ \alpha}$ and $\Delta_{i\ \beta}$ Trajectory.

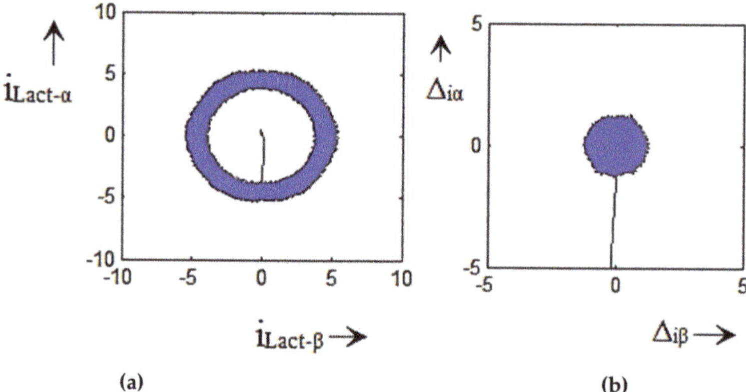

Figure 16. Current error (Δ_{ei}) trajectory, and controlled current trajectory in α-β plane for variable and variable band hysteresis. (**a**). $i_{Lref\text{-}\alpha}$ and $i_{Lact\text{-}\beta}$ Trajectory (**b**). $\Delta_{i\ \alpha}$ and $\Delta_{i\ \beta}$ Trajectory.

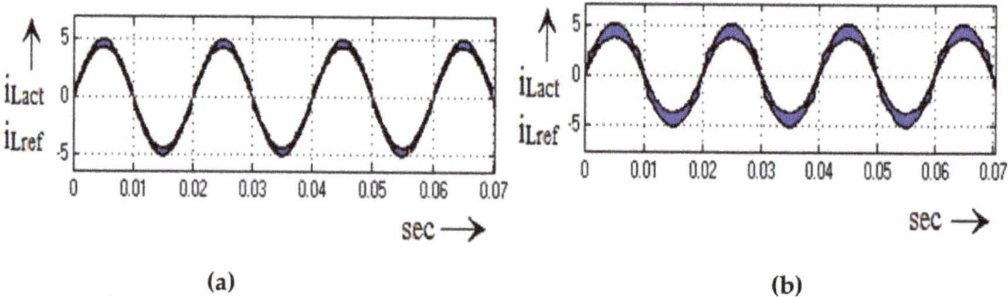

Figure 17. i_{Lact} and i_{Lref} tracking; (**a**) variable hysteresis band, (**b**) variable and variable band Hysteresis.

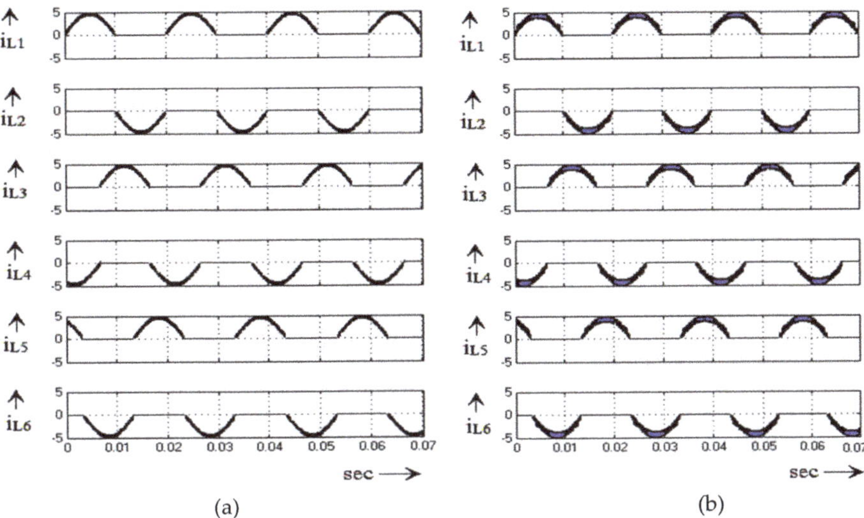

(a)   (b)

Figure 18. Inverter leg inductor currents (i_{L1} to i_{L6}); (**a**) while applying fixed hysteresis band, (**b**) while applying variable hysteresis band.

Figure 19 shows the waveforms of grid voltage V_a,V_b,V_c and inverter current i_a,i_b,i_c. The one cycle V_g and i_a is given in Figure 20. Based on Figure 20 (zoomed waveform), it could realize that V_g and I_{inv} are meeting together in the zero crossing point for ensuring the perfect grid synchronization. The harmonics spectra of the inverter current and inverter voltage are carried out up to the 500th order (25 kHz/50Hz), as shown Figure 21; it can be seen the THD percentages for both are less, as 14.52% and 9.14%, respectively.

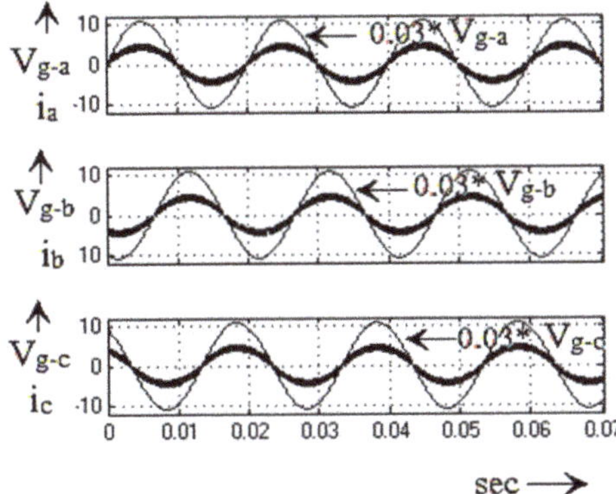

Figure 19. Waveforms of grid voltage V_a,V_b,V_c, and inverter current i_a,i_b,i_c.

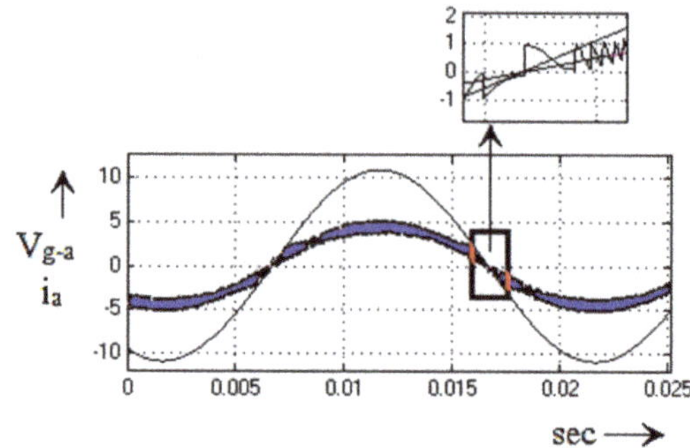

Figure 20. V_g and I_{inv} waveforms of proposed PV-grid connected TL-TNPC-MLI.

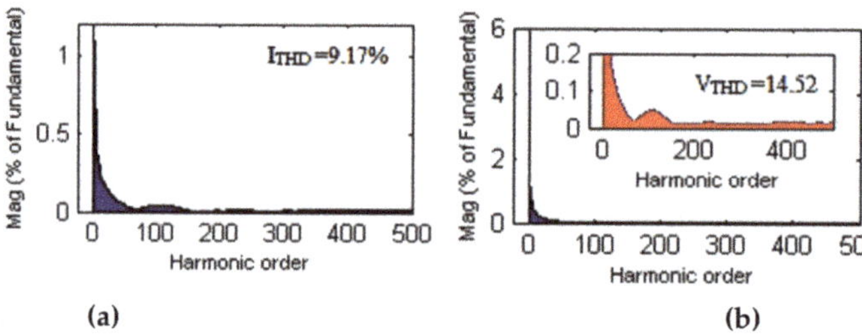

(a) (b)

Figure 21. Harmonics spectra of the inverter current and inverter voltage. (**a**). Current THD (**b**). Voltage THD.

Figure 22 shows the performance of the inverter, when there is a change in grid frequency and current magnitudes, here the actual current reaches the reference current by using constant-frequency in 1/5 of the grid frequency interval itself. Figure 23 shows the leakage current of the inverter is observed as very low (100mA), which is close to zero. Based on the waveform, it could clearly understand that the proposed inverter fully mitigates the CM leakage current. Figure 24 shows the performance of the PV tied grid-connected inverter in different grid variations.

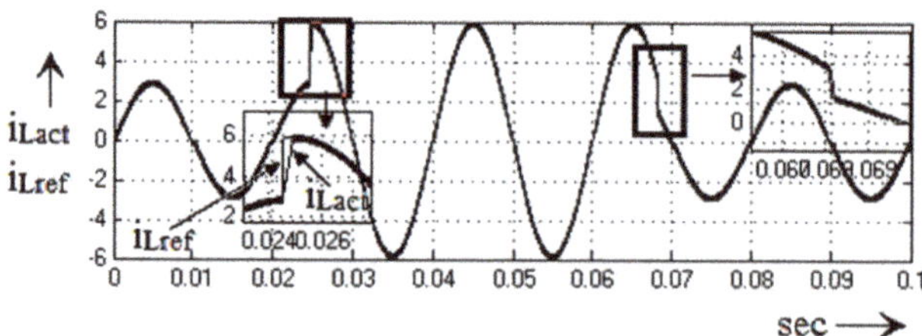

Figure 22. Performance of the PV tied grid connected TL-TMLI by change in grid current i_{Lref}.

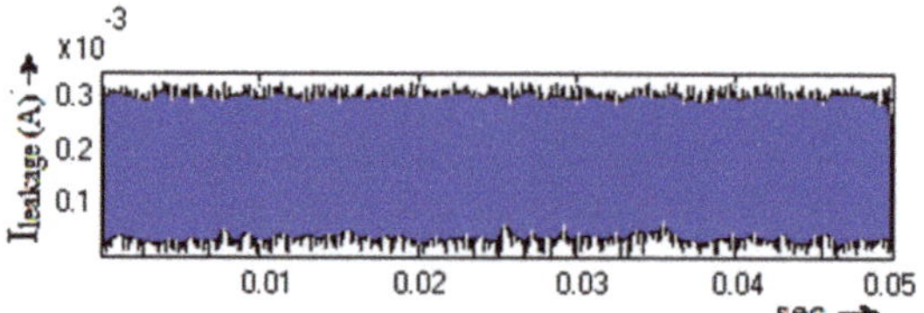

Figure 23. Common-mode, $i_{leakage}$.

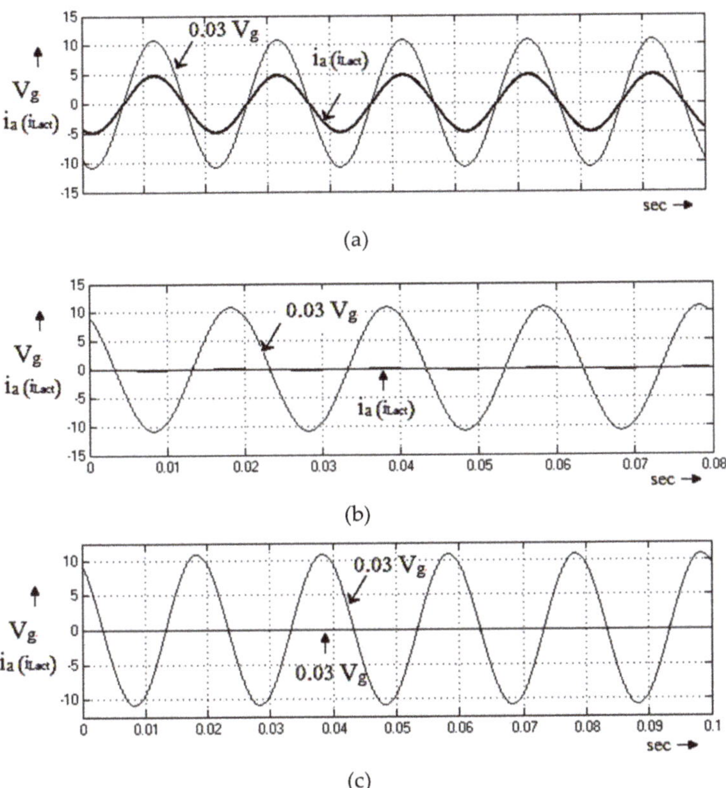

(a)

(b)

(c)

Figure 24. Steady operation waveform with different grid-current command under variable-band constant-frequency HSVCC; (**a**) rated current, (**b**) 5% of rated grid current, and, (**c**) Zero-grid-current reference.

Next, the proposed system is investigated under dynamic condition, when there is an abrupt rise and drop on the grid-current, grid voltage, and PV input voltage, as shown in Figure 25 and 26. Figure 25 illustrates the performance of the PV tied grid connected TL-TMLI by a change in grid voltage V_g. Here, the grid voltage rises at 2.2ms from 230V to 250 V. Figure 26 illustrates the PV voltage variation at 3 msec from 400V to 500V. The proposed controller operated well for both the grid voltage rise and the PV voltage rise. The same way when there are sudden changes in V_g and PV input voltage, the inverter interfacing manipulations on HCC react soon to obtain the actual current matching with the reference current.

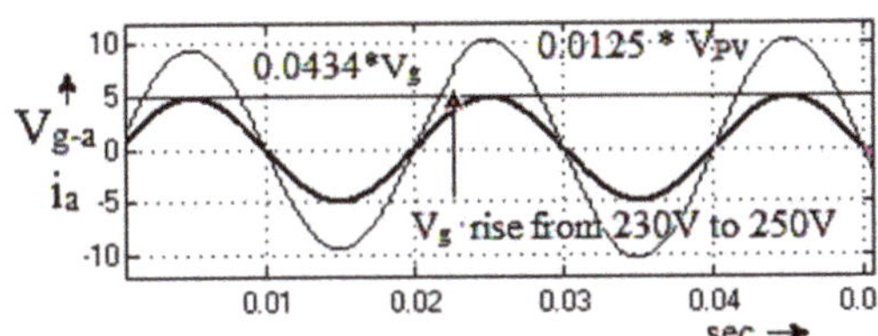

Figure 25. Performance of the PV tied grid connected TL-TMLI by change in grid voltage V_g.

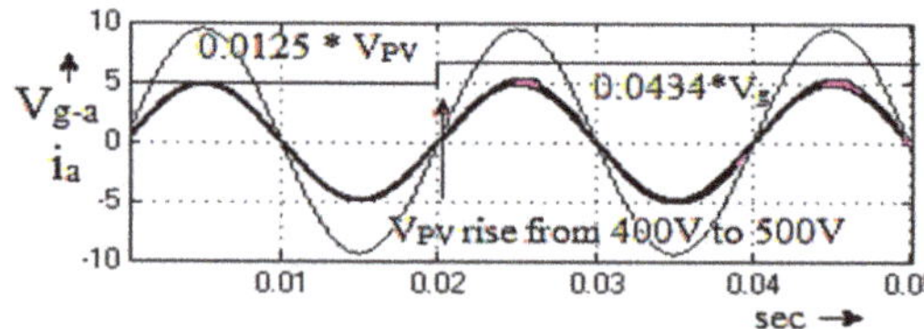

Figure 26. Performance of the PV tied grid connected TL-TMLI by change in PV input voltage V_{PV}.

Next, the inverter performance is tested with HSVCC for different operation frequencies. Figure 27 presents the transient response of current for step change (at 0.023sec from 50Hz to 60Hz) in the frequency of the reference current. From the tested results, it is confirmed that the proposed HSVCC also exhibits superior performance in transient state.

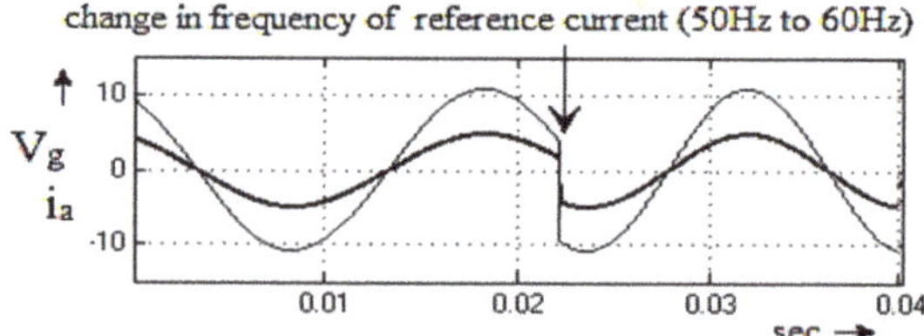

Figure 27. Transient response of the PV tied grid-connected inverter by change grid frequency.

5. Experimental Result

A 1.5kW prototype (shown in Figure 28) is built and CM leakage current elimination SVM signals are produced and tested on a Spartan-III field-programmable gate array (FPGA) board in order to confirm the performance of the proposed topology based 2kW roof-top PV grid connected system, which was programmed through the MATLAB/Simulink-system generated Xilinx tool. The specifications laboratory prototype proposed system parameters are listed; PV output based dc-link voltage = 400, DC-link capacitance $C_1 = C_2$ = 100mF, stray capacitance (used a thin-film capacitor) =100nF, maximum m_a = 0.9, output active, three phase filter inductance L_f=1mH, filter resistance R_f = 10Ω, fundamental frequency f =50Hz, switching frequency f_s = 10kHz, dead time t_D= 4µs. All of the collaborative results were measured by a digital oscilloscope (DSO)) Tektronix MDO4034B-3 and Yokogawa power analyzer. Figure 29 shows the phase-A inverter leg inductor currents (i_{L1a} and i_{L2a}). It shows that the both inductors are charged properly and identically. Similarly, other inverters phase legs -B and C inductors i_{L1b}, i_{L2b} and i_{L1c}, i_{L2c} also maintains identical nature, like phase-A inductors. Hence, the three-phase inverters currents are equal to each other. Figure 30 shows the inverter current (i_{inv}) and Figure 31 shows the inverter voltage. The inverter current harmonics profile is measured and is given in Figure 32. Here, the percentage current THD is observed as 10.251% and all higher harmonics are eliminated. Figure 33 shows the numerical harmonics values for inverter voltage and current. From the results, it could observe that the percentage voltage THD is about 18.051. Here, the third order harmonics is predominant and all other higher orders are very low. Figure 34 shows the i_L and V_g, and inverter i $_{leakage}$. In order to show that the interface between V_g and i_L, the V_g and I_L are measured in the same scope window as 600V/div and 5A/div, respectively. Based on the waveform, it can be seen that V_g and I_L are meeting together in the same zero crossing point for ensuring the grid synchronization. The second scale of Figure 34 showing leakage current of the inverter is observed as very low (200mA), which is close to zero. This value is less than the threshold value referred to VDE0126-1-1, IEC 60755, and VDE-AR-N-4105 PV tied grid connected TLI codes. Figure 35 illustrates the phasor angle of inverter and the grid voltage and current, which confirms the perfect phase sequence of inverter grid interface. With respect to PV input, the proposed PV tied gird connected T type NP-MLI transformerless topology maintains that the efficiency is more 95% in all PV conditions. Based on the above discussed experimental results of the proposed TL PV tied grid connected system, not only the removing the transformer and inverter leg leakage current. It is also ensuring the enhanced performance such as accuracy in voltage and current waveform, quicker converging speed with smaller amount of response time, no steady-state oscillation around the PVs maximum power point values.

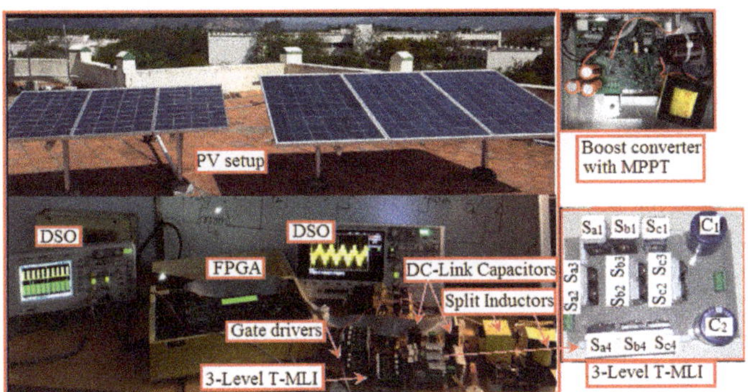

Figure 28. Experimental setup of PV tied reconnected tranformerless T-MLI.

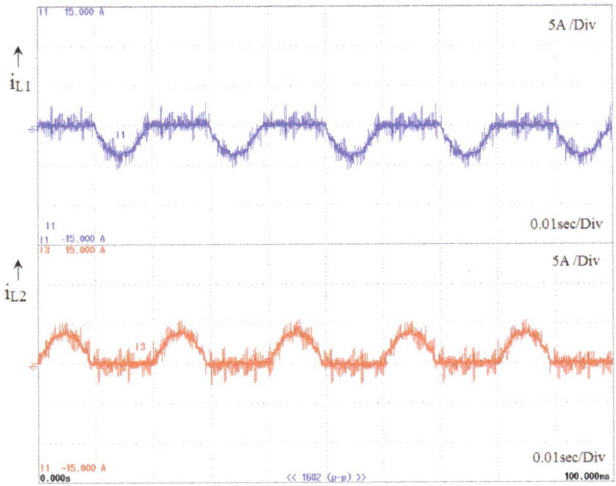

Figure 29. Inverter leg inductor currents (i_{L1a} and i_{L2a}).

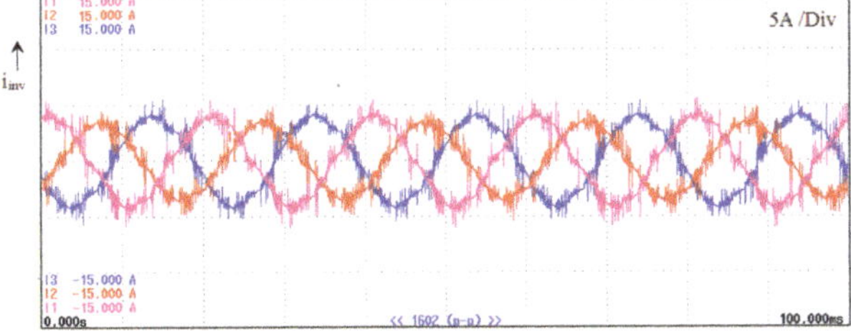

Figure 30. Three phase inverter current (I_{inv}).

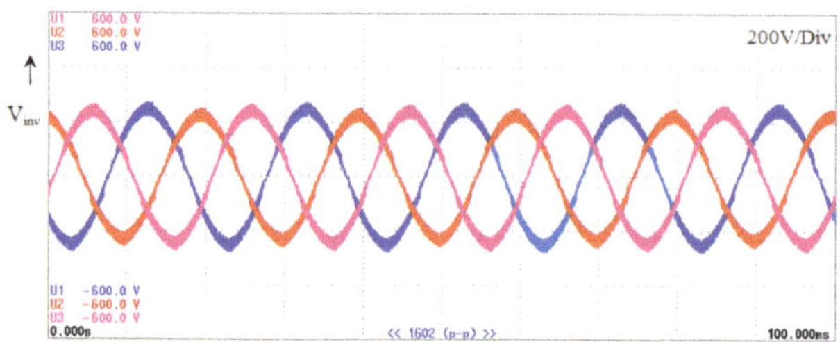

Figure 31. Three phase inverter voltage.

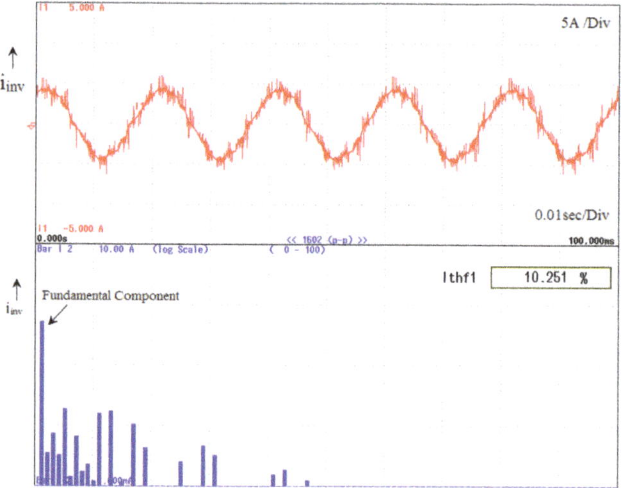

Figure 32. Inverter current (i_L) and harmonics spectra.

			Order	U1 [V]	hdf [%]	Order	U1 [V]	hdf [%]
fPLL1:U1	49.902	Hz	Total	293.51		dc	----------	--------
fPLL2:U2	49.894	Hz	1	288.69	98.357	2	0.54	0.183
			3	3.14	1.068	4	0.40	0.138
Urms1	297.76	V	5	43.20	14.717	6	0.19	0.064
Irms1	0.6413	A	7	21.29	7.253	8	0.41	0.141
P1	123.89	W	9	0.55	0.188	10	0.31	0.107
S1	190.95	VA	11	4.45	1.515	12	0.06	0.020
Q1	145.30	var	13	1.56	0.531	14	0.11	0.037
λ1	0.6488		15	0.52	0.176	16	0.15	0.050
Φ1	G49.55	°	17	1.19	0.406	18	0.19	0.066
			19	2.36	0.806	20	0.20	0.069
Uthd1	18.051	%	21	0.40	0.138	22	0.11	0.037
Ithd1	10.251	%	23	7.02	2.391	24	0.18	0.061
Pthd1	2.340	%	25	8.38	2.855	26	0.24	0.083
Uthf1	11.581	%	27	0.25	0.084	28	0.22	0.075
Ithf1	2.517	%	29	9.50	3.236	30	0.23	0.079
Utif1	---O F---		31	8.71	2.968	32	0.25	0.085
Itif1	---O F---		33	0.07	0.025	34	0.11	0.038
hvf1	7.286	%	35	5.69	1.939	36	0.30	0.101
hcf1	4.419	%	37	3.95	1.344	38	0.36	0.121
Kfact1	1.6793		39	0.10	0.033	40	0.14	0.047
PAGE 1/11							PAGE 1/13	

Figure 33. Numerical harmonics spectra of inverter voltage and inverter current.

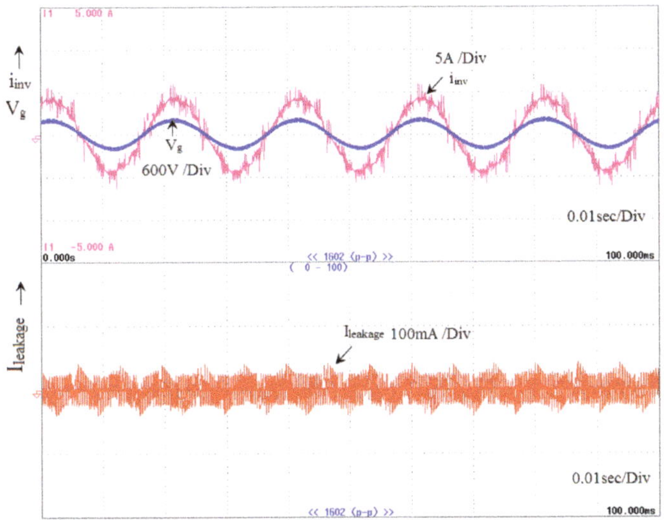

Figure 34. Inverter current (I_L) and grid voltage (V_g), and inverter leakage current.

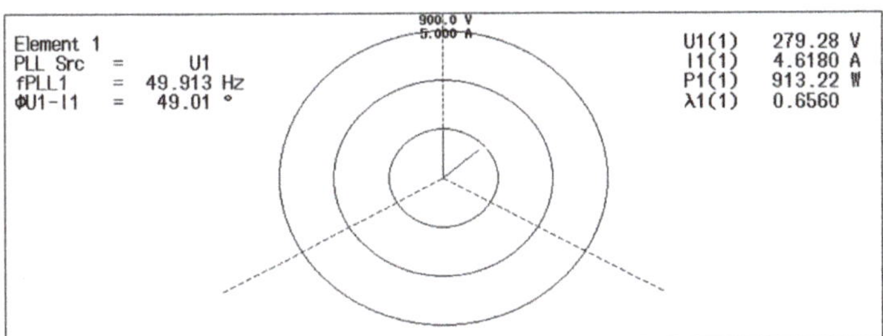

Figure 35. Phasor angle of inverter and grid voltage and current.

6. Conclusions

The interested on photovoltaic tied transformerless inverter topologies and its grid interface techniques are increasingly engrossed for the benefit of high efficiency, reliability, and low cost. The main issue in the transformerless inverter is CMV, which causes the switching-frequency leakage current. The single-phase inverter and two-level topologies are well developed with additional switches and components for eliminating the CMV. Even through there are trust topologies presented in literature, MLIs based grid connected transformerless inverter topology is being researched to avail the additional benefits from MLI, such as lower device breakdown voltage, smaller harmonics effect in the output voltage, good reliability, and long life span.

With above aim, this paper has proposed three-phase three-level T type NP-MLI topology with transformerless PV grid connected proficiency. The proposed MLI is successfully connected with the PV and grid connected system with the help of improved space vector modulation technique to mitigate the CM leakage current.

First, the paper analyzed the three-phase PV tied grid connected transformerless inverter in detail and derived the mathematical model for CMV, DMV, and common mode leakage current. Subsequently, followed by CMV model, the inverter switching function equation is derived for the opportunity of leakage current mitigation. Second, the paper proposed the PV gird interfaced closed model for three-phase three-level T type NP-MLI topology with help of improved space vector modulation technique for controlling the current and mitigating the CM leakage current. The combined space vector and hysteresis current control has been proposed based on the behavior and requirements of the grid connected standards. The proposed arrangements are confirmed by MATLAB/Simulink simulations and FPGA based 1.5 kW PV tied TL-TNP-MLI grid connected experimentations. The simulation and experimental results significantly recognized that the proposed PV tied gird connected T type NP-MLI transformerless topology is observed as very low CM leakage current (200mA) for all PV and inverter operation conditions. The results are demonstrated and presented in the good stability of steady state and dynamics performances. The proposed inverter reduces current as well as voltage harmonics, and reduction leakage (current near to zero), which makes a significant reduction in harmonics filter and the complete elimination of CMV suppressions choke receptivity.

Author Contributions: All authors are involved developing the concept, simulation and experimental validation and to make the article error free technical outcome for the set investigation work.

Funding: This research received no external funding.

Conflicts of Interest: The authors declare no conflict of interest.

Abbreviations

$C_{G\text{-}PV}$	parasitic capacitances
V_{CM}	Common mode voltage
V_{DC}	DC-link voltage
I_{CM}	Leakage current
V^*	SVM Reference voltage
V_Z	Zero voltage vector
V_S	Small voltage vector
V_M	Medium voltage vector
V_Z	Large voltage vector
I_{TCM}	Total leakage current
V_α and V_β	α and β SVM coordinated voltage
i_{ref}	Reference current
i_{act}	Actual current
$\vec{\varepsilon_i}$	Error current
$u(t))$	Control input
m_a	Modulation index
f_s	Frequency inverter switching pulse
f	Grid frequency inverter switching pulse
V_g	Grid voltage
Δ_{ei}	current error
TL	Transformerless inverter
TNP-MLI	Neutral point multilevel inverter
SVM	Space vector modulation
DMV	Common mode voltage
SVD	Space vector diagram
HCC	hysteresis current control
HSVCC	hysteresis Space vector current control
MMPT	Maximum power point tracking

References

1. Global Status Report. *Renewables*; REN21: Paris, France, 2018; ISBN 978-3-9818911-3-3.
2. Zsiborács, H.; Baranyai, N.H.; Csányi, S.; Vincze, A.; Pintér, G. Economic Analysis of Grid-Connected PV System Regulations: A Hungarian Case Study. *Electronics* **2019**, *8*, 149. [CrossRef]
3. Zsiborács, H.; Baranyai, N.H.; Csányi, S.; Vincze, A.; Pintér, G. Economic and Technical Aspects of Flexible Storage Photovoltaic Systems in Europe. *Energies* **2018**, *11*, 1445. [CrossRef]

4. Sahan, B.; Vergara, A.N.; Henze, N.; Engler, A.; Zacharias, P. A single stage PVmodule integrated converter based on a low-power current source inverter. *IEEE Trans. Ind. Electron.* **2008**, *55*, 2602–2609. [CrossRef]

5. Li, Q.; Wolfs, P. A review of the single phase photovoltaic module integrated converter topologies with three different dc link configuration. *IEEE Trans. Power Electron.* **2008**, *23*, 1320–1333.

6. Lopez, O.; Freijedo, F.D.; Yepes, A.G.; Fernandez Comesana, P.; Malvar, J.; Teodorescu, R.; Doval Gandoy, J. Eliminating ground current in atransformerless photovoltaic application. *IEEE Trans. Energy Convers.* **2010**, *25*, 140–147. [CrossRef]

7. Xiao, H.; Xie, S. Leakage current analytical model and applicationin single-phase transformerless photovoltaic grid-connected inverter. *IEEE Trans. Electromagn. Compat.* **2010**, *52*, 902–913. [CrossRef]

8. VDE-AR-N. *4105: Power Generation Systems Connected to the Low-Voltage Distribution Network—Technical Minimum Requirements for the Connection to and Parallel Operation with Low-Voltage Distribution Networks*; DIN_VDE Normo: Berlin, Germany, 2011.

9. DIN VDE V 0126-1-1. *Automatic Disconnection Device between a Generator and the Public Low-Voltage Grid*; DIN VDE V 0126-1-1. Berlin, Germany, 2006. Available online: https://www.hylaw.eu/database/national-legislation/germany/din-vde-v-012611-automatic-disconnection-device-between-a-generator-and-the-public-lowv (accessed on 24 June 2019).

10. IEEE Application Guide for Std. *1547, IEEE Standard for Interconnecting Distributed Resources with Electric Power Systems*; IEEE Standard 1547.2-2008; IEEE: Piscataway, NJ, USA, 2009.

11. IEC 61727. *Photovoltaic (PV) Systems—Characteristics of the Utility Interface*; IEC 61727; IEC: Geneva, Switzerland, 2004.

12. UL 1741. *Standard for Inverters, Converters, Controllers, and Interconnection System Equipment for Use with Distributed Energy Resources*; UL 1741; UL: Northbrook, IL, USA, 2010.

13. EN 61000-3-2. *Electromagnetic Compatibility (EMC)—Part 3-2: Limits—Limits for Harmonic Current Emissions (Equipment Input Current Under 16 A Per Phase)*; EN 61000-3-2: Brussels, Belgium, 2006.

14. Steca Grid. Available online: www.steca.com (accessed on 24 June 2019).

15. Goodway. Available online: www.goodwe.com (accessed on 24 June 2019).

16. Goodrich, A.; James, T.; Woodhouse, M. *Residential, Commercial, Utility-Scale Photovoltaic (PV) System Prices in the United States: Current drivers and Cost-Reduction Opportunities*; Nat. Renewable Energy Lab.: Golden, CO, USA, 2012.

17. Feldman, D.; Margolis, R.; Bolinger, M.; Chung, D.; Fu, R.; Seel, J.; Wiser, R. *Photovoltaic System Pricing Trends: Historical, RecentNear-Term Projections*, 2013th ed.; U.S. Dept. Energy: Oak Ridge, TN, USA, 2013.

18. Koutroulis, E.; Blaabjerg, F. Design optimization of transformerless grid-connected PV inverters including reliability. *IEEE Trans. Power Electron.* **2013**, *28*, 325–335. [CrossRef]

19. Rodriguez, J.; Bernet, S.; Steimer, P.K.; Lizama, I.E. A survey on neutral-point-clamped inverters. *IEEE Trans. Ind. Electron.* **2010**, *57*, 2219–2230. [CrossRef]

20. Bharatiraja, C.; Jeevananthan, S.; Munda, J.L.; Latha, R. Improved SVPWM vector selection approaches in OVM region to reduce common-mode voltage for three-level neutral point clamped inverter. *Int. J. Electr. Power Energy Syst.* **2016**, *79*, 285–297. [CrossRef]

21. Selvaraj, R.; Chelliah, T.R.; Khare, D.; Bharatiraja, C. Fault Tolerant Operation of Parallel-Connected 3L-Neutral-Point Clamped Back-to-Back Converters Serving to Large Hydro-Generating Units. *IEEE Trans. Ind. Appl.* **2018**, *54*, 5429–5443. [CrossRef]

22. Maheshwari, R.; Munk-Nielsen, S.; Busquets-Monge, S. Design of neutral-point voltage controller of a three-level NPC inverter with small DC-link capacitor. *IEEE Trans. Ind. Electron.* **2013**, *60*, 1861–1871. [CrossRef]

23. Bharatiraja, C.; Jeevananthan, S.; Latha, R. FPGA based practical implementation of NPC-MLI with SVPWM for an autonomous operation PV system with capacitor balancing. *Int. J. Electr. Power Energy Syst.* **2014**, *61*, 489–509. [CrossRef]

24. Bharatiraja, C.; Jeevananthan, S.; Latha, R. Vector Selection Approach-based Hexagonal Hysteresis Space Vector Current Controller for a Three-phase Diode Clamped MLI with Capacitor Voltage Balancing. *IET Power Electron.* **2016**, *9*, 1350–1361. [CrossRef]

25. Yong Wang, W.W.; Shi, N.X.; Wang, C.M. Diode-Free T-Type Three-Level Neutral-Point-Clamped Inverter for Low-Voltage Renewable Energy System. *IEEE Trans. Ind. Electron.* **2014**, *61*, 6168–6174. [CrossRef]

26. Schweizer, M.; Kolar, J.W. Design and implementation of a highly efficient three-level T-type converter for low-voltage applications. *IEEE Trans. Power Electron.* **2013**, *28*, 899–907. [CrossRef]

27. Chokkalingam, B.; Bhaskar, M.S.; Padmanaban, S.; Vigna, K.R.; Iqbal, A. Investigations of Multi-Carrier Pulse Width Modulation Schemes for Diode Free Neutral Point Clamped Multilevel Inverters. *J. Power Electron.* **2019**, *19*, 702–713.

28. Bharatiraja, C.; Sanjeevikumar, P.; Blaabjerg, F. Critical Investigation and Comparative Analysis of Advanced PWM Techniques for Three-Phase Three-Level NPC-MLI Drives. *Electr. Power Compon. Syst.* **2018**, *46*, 258–269.

29. Bharatiraja, C.; Jeevananthan, S.; Munda, J.L. Timing Correction Algorithm for SVPWM Based Diode-Clamped MLI Operated in overmodulation Region. *IEEE J. Sel. Top. Power Electron. Appl.* **2018**, *6*, 233–245. [CrossRef]

30. Santhakumar, C.; Shivakumar, R.; Bharatiraja, C.; Sanjeevikumar, P. Carrier shifting algorithms for the mitigation of circulating current in diode clamped MLI fed induction motor drive. *Int. J. Power Electron. Drive Syst.* **2017**, *8*, 844–852. [CrossRef]

31. Bharatiraja, C.; Raghu, S.; Rao, K.R.S. Comparative analysis of different PWM techniques to reduce the common mode voltage in three-level neutral-point- clamped inverters for variable speed induction drives. *Int. J. Power Electron. Drive Syst.* **2013**, *3*, 105–116.

32. Lyu, J.G.; Wang, J.D.; Hu, W.B.; Wu, Z.F. Research on the Neutral-Point Voltage Balance for NPC Three-Level Inverters under Non-Ideal Grid Conditions. *Energies* **2018**, *11*, 1331. [CrossRef]

33. Wang, Y.; Xu, Q.; Ma, Z.; Zhu, H. An Improved Control and Energy Management Strategy of Three-Level NPC Converter Based DC Distribution Network. *Energies* **2017**, *10*, 1635. [CrossRef]

34. Chen, Z.; Yu, W.; Ni, X.; Huang, A. A new modulation technique to reduce leakage current without compromising modulation index in PV systems. In Proceedings of the 2015 IEEE Energy Conversion Congress and Exposition (ECCE), Montreal, QC, Canada, 20–24 September 2015.

35. Zhang, L.; Sun, K.; Feng, L.; Wu, H.; Xing, Y. A Family of Neutral Point Clamped Full-Bridge Topologies for Transformerless Photovoltaic Grid-Tied Inverters. *IEEE Trans. Power Electron.* **2013**, *28*, 730–739. [CrossRef]

36. Hashempour, M.M.; Yang, M.Y.; Lee, T.L. An Adaptive Control of DPWM for Clamped-Three-Level Photovoltaic Inverters With Unbalanced Neutral-Point Voltage. *IEEE Trans. Ind. Appl.* **2018**, *54*, 6133–6148. [CrossRef]

37. Cavalcanti, M.C.; Farias, A.M.; Oliveira, K.C.; Neves, F.A.S.; Afonso, J.L. Eliminating leakage currents in neutral point clamped inverters for photovoltaic system. *IEEE Trans. Ind. Electron.* **2012**, *59*, 435–443. [CrossRef]

38. Araujo, S.; Zacharias, P.; Mallwitz, R. Highly efficient single-phase transformerless inverters for grid-connected photovoltaic systems. *IEEE Trans. Ind. Electron.* **2010**, *57*, 3118–3128. [CrossRef]

39. Albalawi, H.; Zaid, S.A. An H5 Transformerless Inverter for Grid Connected PV Systems with Improved Utilization Factor and a Simple Maximum Power Point Algorithm. *Energies* **2018**, *11*, 2912. [CrossRef]

40. Selvamuthukumaran, R.; Garg, A.; Gupta, R. Hybrid Multicarrier Modulation to Reduce Leakage Current in a Transformerless Cascaded Multilevel Inverter for Photovoltaic Systems. *IEEE Trans. Power Electron.* **2015**, *30*, 1779–1783. [CrossRef]

41. Lopez, O.; Teodorescu, R.; Doval-Gandoy, J. Multilevel transformerless topologies for single-phase grid-connected converters, 32nd Annual Conf. onIEEE Industrial Electronics. *IECON* **2006**, *1*, 5191–5196.

42. Kwon, J.M.; Nam, K.H.; Kwon, B.H. Photovoltaic power conditioning system with line connection. *IEEE Trans. Ind. Electron.* **2006**, *53*, 1048–1054. [CrossRef]

43. Bharatiraja, C.; Padmanaban, S.; Siano, P.; Leonowicz, Z.; Iqbal, A. A hexagonal hysteresis space vector current controller for single Z-source network multilevel inverter with capacitor balancing. In Proceedings of the 2017 IEEE International Conference on Environment and Electrical Engineering and 2017 IEEE Industrial and Commercial Power Systems Europe, Milan, Italy, 6–9 June 2017.

44. Wang, J.; Zhai, F.; Wang, J.; Jiang, W.; Li, J.; Li, L.; Huang, X. A Novel Discontinuous Modulation Strategy with Reduced Common Mode Voltage and Removed DC Offset on Neutral Point Voltage for Neutral Point Clamped Three Level Converter. *IEEE Trans. Power Electron.* **2018**, *6*, 1966–1978. [CrossRef]

45. Bae, Y.; Kim, R.Y. Suppression of Common-Mode Voltage Using a Multicentral Photovoltaic Inverter Topology With Synchronized PWM. *IEEE Trans. Ind. Electron.* **2014**, *61*, 4722–4733. [CrossRef]

46. Hu, X.; Ma, P.; Gao, B.; Zhang, M. An Integrated Step Up Inverter Without Transformer and Leakage Current for Grid-Connected Photovoltaic System. *IEEE Trans. Power Electron.* **2019**. [CrossRef]
47. Lee, J.S.; Lee, K.-B. New Modulation Techniques for a Leakage Current Reduction and a Neutral-Point Voltage Balance in Transformerless Photovoltaic Systems Using a Three-Level Inverter. *IEEE Trans. Power Electron.* **2014**, *29*, 1720–1732. [CrossRef]

Article

A Modified High Voltage Gain Quasi-Impedance Source Coupled Inductor Multilevel Inverter for Photovoltaic Application

Madasamy Periyanayagam [1,*], **Suresh Kumar V** [2], **Bharatiraja Chokkalingam** [3,5,*], **Sanjeevikumar Padmanaban** [4], **Lucian Mihet-Popa** [6] **and Yusuff Adedayo** [5]

[1]   Department of Electrical and Electronics Engineering, Alagappa Chettiar College of Engineering and Technology, Karaikudi 630003, India

[2]   Department of Electrical and Electronics Engineering, Thiagarajar College of Engineering, Madurai 625015, India; vskeee@tce.edu

[3]   Department of Electrical and Electronics Engineering, SRM Institute of Science and Technology, Chennai 603203, India

[4]   Department of Energy Technology, Aalborg University, 6700 Esbjerg, Denmark; san@et.aau.dk

[5]   Department of Electrical Engineering, University of South Africa, Pretoria 003, South Africa; yusufaa@unisa.ac.za

[6]   Faculty of Engineering, Østfold University College, 1671 Kråkeroy-Fredrikstad, Norway; lucian.mihet@hiof.no

*   Correspondence: mjasmitha0612@gmail.com (M.P.); bharatiraja@gmail.com (B.C.); Tel.: +91-904-270-1695 (B.C.)

Received: 19 January 2020; Accepted: 12 February 2020; Published: 17 February 2020

Abstract: The quasi-impedance source inverters/quasi-Z source inverters (Q-ZSIs) have shown improvement to overwhelmed shortcomings of regular voltage-source inverters (VSIs) and current-source inverters (CSIs) in terms of efficiency and buck-boost type operations. The Q-ZSIs encapsulated several significant merits against conventional ZSIs, i.e., realized buck/boost, inversion and power conditioning in a single power stage with improved reliability. The conventional inverters have two major problems; voltage harmonics and boosting capability, which make it impossible to prefer for renewable generation and general-purpose applications such as drive acceleration. This work has proposed a Q-ZSI with five-level six switches coupled inverter. The proposed Q-ZSI has the merits of operation, reduced passive components, higher voltage boosting capability and high efficiency. The modified space vector pulse width modulation (PWM) developed to achieve the desired control on the impedance network and inverter switching states. The proposed PWM integrates the boosting and regular inverter switching state within one sampling period. The PWM has merits such as reduction of coupled inductor size, total harmonic reduction with enhancing of the fundamental voltage profile. In comparison with other multilevel inverters (MLI), it utilizes only half of the power switch and a lower modulation index to attain higher voltage gain. The proposed inverter dealt with photovoltaic (PV) system for the stand-alone load. The proposed boost inverter topology, operating performance and control algorithm is theoretically investigated and validated through MATLAB/Simulink software and experimental upshots. The proposed topology is an attractive solution for the stand-alone and grid-connected system.

Keywords: impedance source; multilevel inverter; coupled inductors; space vector pulse PWM; photovoltaic connected inverter

1. Introduction

Photovoltaic (PV) energy, extracted through solar cells is a mandatory power generation technology to meet out the global power demand [1]. The most prominent merits in PV generation are reduction of fossil fuel usage, less impact on the environment and reduction in power generation cost [2,3]. The large-scale deviation occurs in the floating power generations that abide by climatic conditions. Unfortunately, the main drawback of the PV array panels is a wide range of drop-in voltage. However, the power electronics devices overcome the voltage drops and floating power generation [4]. The power electronics converter and inverter combinations make PV power transfer an efficient process. The traditional power electronics inverters; voltage-source inverter (VSI), and current-source inverter (CSI) defeats options in the PV power generation with the addition of DC-DC converters. This two-stage power conversion needs more semiconductor switches and passive components; hence they may cause the abrupt disturbance on voltage profile [5]. To prevail over the demerits of traditional inverters, single-stage power conversion is introduce named as Z (impedance) source inverter (ZSI). Impedance networks offer an effective power transformation between the source (input) and an extensive range of loads with high efficiency. However, the ZSI lags in the performance like producing discontinuous nature in the input current; inductors do not withstand high current, and voltage stress on the capacitors [6]. The quasi-Z source has been expected to inherit the merits of the ZSI with reduced passive components, continuous input current, constant DC rail voltage for the inverter, and so on. Before embarking on the investigation of the Z source inverter, it is helpful to look into its evolution. Impedance source topology-based researches have proliferated, from the time it was proposed by Peng et al. in 2003; the variety of alterations and novel Z-source topologies has matured exponentially [7]. In the advancement of ZSI, it finds variety of applications such as; variable-speed electrical drives [8,9], uninterruptible power supply [10], in distributed power generations (such as photovoltaic (PV), fuel cell, and wind, etc.), energy storage system (such as battery and supercapacitor), and electric vehicles [11–13]. An earned mark of the Z-source inverter has lost its capability in the form of the input and output voltage ratio profile, switching stress and utilization of higher modulation. These lagging structure needs to be remarked with a quasi (Q)-Z source. The development in the Q-Z source with coupled inductor achieves most compromising effect towards upright of power quality, lower switching dv/dt, better electromagnetic influence, and negligible switching losses [14].

Owing to the advantages and challenges, power electronics researchers have given much interest in impedance-source topology development. The first Z-source was proposed during 2003 (Peng et al.); the variety of alterations, as well as novel topological inventions that have been developed exponentially, are chosen based on the applications and requirements. Concerning power conversion, it is separated into four groups: AC to DC (rectifier), AC to AC (AC voltage regulator), DC to DC (DC chopper), and DC to AC (inverter). This classification is further divided into two-level (conventional VSI) and multi-level DC to AC, AC to AC matrix converters [15] and DC to DC converters in isolated and non-isolated arrangement [16]. Based on the input source (current or voltage), the impedance source topology is further divided into the voltage source and current source Z-source [17]. Besides, considering the impedance components (inductors arrangements), this group can be distributed into coupled inductor (magnetically coupled) or transformer-based [18] and non-transformer (non-magnetically coupled) based [19]. There are selected limitations present in non-magnetically coupled topologies such as lower modulation proportion and lesser, output gain. Therefore, non-transformer topology needs higher boosting inductor and DC-link voltage rating, which may increase the converter switching stress superfluously. Besides, the circuit cost and size for these converters are undesirable. To minimize these concerns, the uses of magnetically coupled inductors or transformers are attractive to increase the operating range and output voltage gain concurrently. The Q-ZSIs proposed in [19,20] offer additional improvement of traditional X shape network topology. Besides the rewards inherited from X shape ZSIs, Q-ZSIs also has its own merits such as continuous current mode operation, reduction of component selection ratings, structured with common DC-rail in the middle of the input and inverter. There are modified Q-ZSI topologies comprehensibly investigated within their improvement to provide continuous input current [21]. Yang et al. propose the current-fed Q-ZSI., including the benefits of the combined buck-boost operation, enhanced reliability, reduced component ratings, as well as single-stage regeneration capability. This topology provides consistency with the input current control and performs better than Q-ZSI.

In the current era, the interests of multilevel inverters (MLIs) have increased than the conventional VSIs, due to their merits [22]. Particularly the development of power semiconductor technology makes it easy to tradeoff the selection of power devices. New MLIs have been recommended in a hybrid approach by involving MOSFETs, IGCTs or GTOs and IGBTs [23–27]. Recently, the interleaved converter topology using coupled-inductors is proposed and extensively applied in low-power claims. These topologies are mostly used to increase the output current with lesser current ripple.

Additionally, the interleaved converter reduces the size of passive components (inductors and capacitors), and the output voltage harmonic profile is considerably increased [28–39]. Banaei et al. discussed the switching stress reduction in Z source-based MLI [36]. Alexandre Battiston et al. deliberated the withdrawal of the ripples in the input current with the suitable coupled inductors arrangement [37]. Li et al. proposed the use of the cascading magnetic cells to obtain a high voltage gain [38] and investigated the voltage gain achievement against smaller duty time. Followed by the authors, Lei et al. offered optimized pulse width modulation (PWM) technique with reduction of switching loss, current ripple, low total harmonic distortion (THD) and high boost gain [39]. The space vector PWM is extensively reviewed for impendence sourced VSIs and MLIs [40–49]. The creation of shoot-through (ST) and placing them in the active switching states, these PWM methods perform better than other carrier-based PWM methods. The ZSI space vector PWM is also studied with two-level MLIs for altering the active and ST switching vector in the space vector diagram for the enhancement of inverter output [44]. However, when connecting a coupled inverter with an impedance network, the PWM methods need to be modified.

Based on this technical background and coupled MLI and control scheme requirements of ZSI, this paper suggests a single-phase MLI coupled inverter topology with five-level output for PV application. This PV tied five-level coupled inverter topology is connecting the modified Q-ZSI with a six-switch coupled inverter, making a single stage DC to AC conversion topology. The proposed Q-ZSI has the merits of operation, namely reduced passive components, voltage boosting capability and high efficiency. The modified space vector PWM is proposed to realize the desired control in impedance network shoot-through and regular inverter switching state to make five-level output. The proposed PWM is integrating the boosting and regular inverter switching state within one sampling period. The PWM has the merits like a reduction of coupled inductor size and triplen harmonic reduction with the enhancement of the fundamental voltage profile.

In comparison with other MLIs, it utilizes only half of the power switch, lower modulation index to obtain high voltage gain. The proposed inverter is simulated and experimentally validated for single phase induction motor load with off-grid fashion. Nevertheless, the proposed inverter topology is a suitable version for both off-grid and on-grid applications.

The paper flow is organized in this way. Section 2 deals with the review of traditional Z-source inverter. The proposed modified Q-impedance converter fed coupled inductor multilevel inverter is explained in Section 3. The modified Space Vector PWM concept and control methods of the proposed inverter are shown in Sections 4–7 explain the simulation and experimentation, respectively. The conclusion is given in Section 8.

2. Review of Traditional Z-source Inverter

2.1. Review of X-Z Source Topology

Figure 1 illustrates the general impedance-source configuration with the impedance-source network for VSI. The basic Z-source network structure is generalized as a necessary X shaped two-port network using two L and C (linear energy storage elements) (Peng et al. 2003). Perhaps designing different Z-source network configurations to expand the converter performance, non-linear elements (switches, diodes, or/and combination of both) is added in the shape two-port network (Peng et al. 2003).

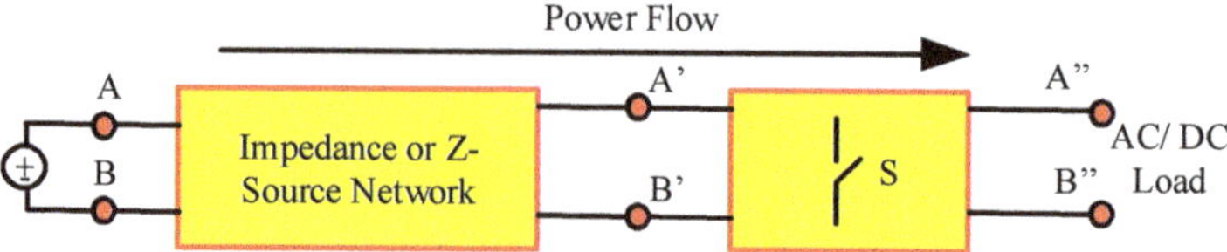

Figure 1. Impedance source inverter basic block diagram.

According to Figure 2, the impedance-source network has two modes of work. During the mode-1, inverter switch is short-circuited to provide the charging loop for an inductor with input DC source. This state is called the 'shoot-through (ST) state'. The second mode is the active mode, where energy stored in the inductors is supplied the load via inverter active mode switching. To connect this ST state with regular inverter active switch, there are a variety of modulation methods available. The modulation methods are distributing the ST states equally based on the modulation index of the inverter (M_a). The upper limit of the modulation index is $1-D_S$. For the maximum boost modulation method, the modulation index is operated in extensive range ($M_a \geq 1$); however, the state involves larger (high rating), passive (L and C) elements [8], and correspondingly distributing the active states, the M_a has its maximum $M_a \leq 1-D_S$ [14]. Therefore, $V_{C1} = V_{C2} = V_{PL}$. During this time, high-frequency generation in the active states causes higher frequency ripples, as shown in Figure 2a. According to the characteristics of Z-source inverter boost mode operation, the performance of inductor and capacitor components and DC-link voltage estimation related to input voltage function are shown in Figure 2b. From the characteristics curve, it is understood that the inverter matches the maximum DC-link voltage with the least input voltage. The inductor curve has a challenging dependency with ST time, and the quick capacitance value raise with the decreasing input voltage is essential. Hence, it is noted that passive elements (inductor and capacitor) are carefully chosen based on the input source voltage and power profile. The voltage rating capacitors are required since the voltage across the capacitors is always greater than the input voltage. Similarly, starting current and voltage surge happens because of the enormous inrush input current. Due to the resonance conditions, these surges are not avoidable, and the absence of bidirectional and soft-start capability limits the application for conventional Z-source topology.

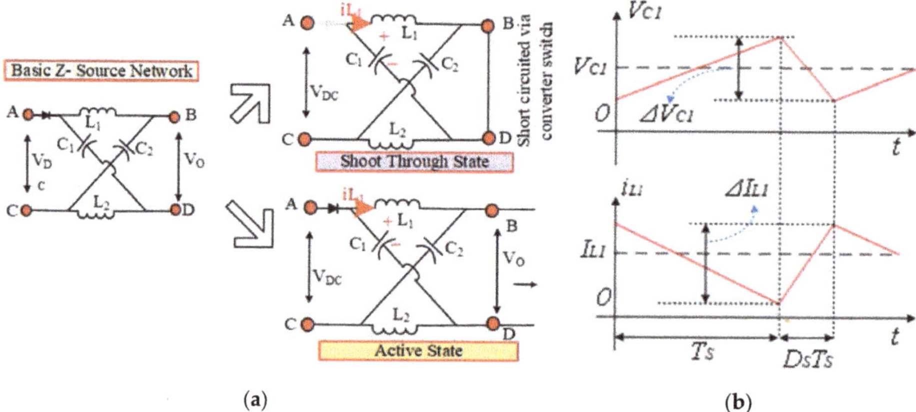

(a)   (b)

Figure 2. Impedance source inverter operation: (**a**) impedance source circuit, shoot through (ST) state, an active state, (**b**) inductor current and capacitor voltage idealized condition waveforms.

To get improvement in the conventional ZSIs, Q-ZSIs are significantly improved to make use of single-stage power conversions. Considering the circuit operations and boosting, characterizes Q-ZSIs as having similar belongings with conventional ZSIs. Figure 3 illustrates the boost-mode characteristics of ZSIs. The Q-ZSIs differ from the conventional X shape-ZSIs by providing continuous input current through lower capacitor C_2 [10]. Regardless of the control strategy of Q-ZSIs, during ST states, the duty ratio is limited concerning inductor rating directly. Hence, there is a possibility of continuous conduction with reduced input inductor current ripples. Besides, Q-ZSI provides the cross conduction switching states to provide the boosting by mutual sharing of inverter switching (top and bottom), which improves the inverter reliability [11]. The Q-ZSI has advantages of reduced component (L and C) ratings, lower switching stress, and continuous current mode operation. The standard ground-sharing option in the Q-ZSI is an additional feature, which is high indeed for PV modules [19]. Even though Q-ZSI has several advantages over conventional X-shape ZSI; it has low DC-link utilization in constant boost operation. To overcome this weakness, Q-ZSIs modified with additional passive components have been proposed [19,20].

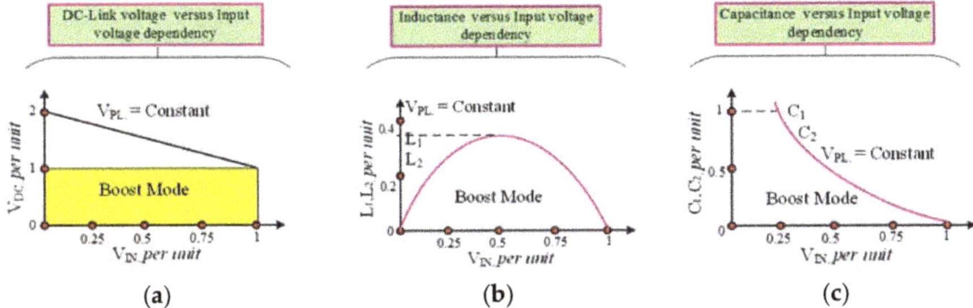

Figure 3. Boost-mode characteristics of Z-source inverter; (**a**) DC-link voltage versus input voltage, (**b**) inductance versus input voltage, and (**c**) input voltage response versus change in capacitances.

2.2. Different Advanced Z-Source Topologies

Due to the advantages and challenges, power electronics researchers have given much interest in impedance-source topology development. These topologies are categorized into two ways; 1. Transformer based and 2. Without transformer (non-magnetically coupled). Figure 4a–f shows the basic different impedance source topologies.

Selected limitations are present in non-magnetically coupled topologies such as low modulation ratio and lesser input-to-output gain. Therefore, this topology needs higher DC-link voltage, which may increase the semiconductors stress needlessly. Besides, the circuit cost and size for these converters are undesirable. Even though the magnetically coupled converters are attractive for their boosting performance, due to the higher DC ripple, the inverter suffers from high harmonics. The disadvantages of magnetically coupled topologies are; (1) raising the shoot-through and magnetizing current while switching; (2) the tightly coupled transformer leads to low leakage impedance; and (3) the need for the snubber circuit is mandatory when the coupling is not fulfilled. Hence, non-magnetic type converters are not highly preferred for PV fed applications [6]. Considering the non-magnetic type converters group, X-shape and Q- impedance converters offer low passive components rating, less duty cycle conversion ratio, and direct conversion. However, it has the poor performance to maintain the input current, reverse blocking capability for the switch, buck-boost operation, absence of bidirectional conversion and direct current control ability. Hence, the paper proposes a coupled inverter connected modified Q-impedance converter to provide single conversion mode power flow operation with multi-level output.

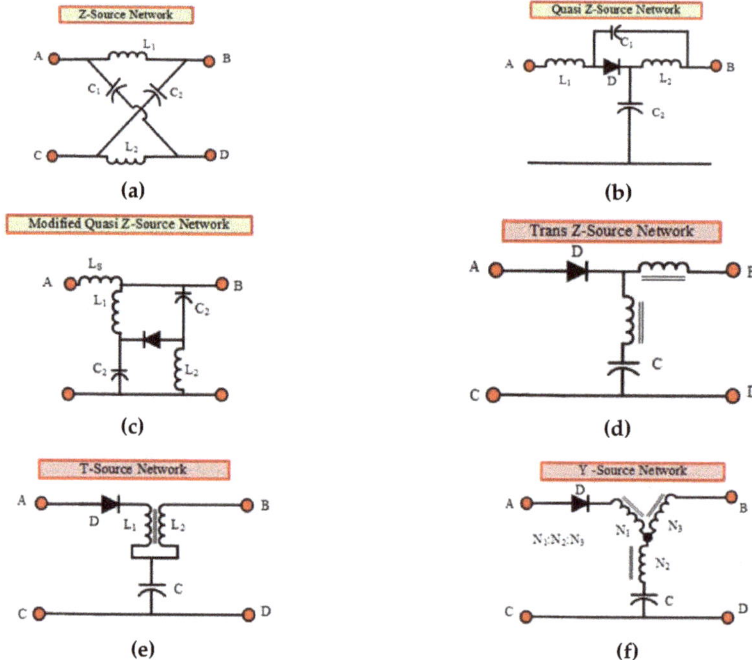

Figure 4. Different Z-source topologies; (**a**) Z-source network, (**b**) Quasi-source network, (**c**) Modified Quasi-source network, (**d**) Trans Z-source network, (**e**) T-source network, (**f**) Y-source network.

3. Proposed Modified Q- Impedance Converter Fed Coupled Inductor Multilevel Inverter.

The traditional Z source inverters can only allow unidirectional power conversion flow with boosting operation. Nevertheless, the proposed topology is different from the conventional Z source inverter or Q-ZSI family to exchange the diode for bi-directional power flow; the proposed Q-ZSI achieve the boosting capability with single-stage conversion associated with the inverter switching scheme as shown in Figure 5. The first suggestion of a proposed modified Q-impedance network is to obtain the continuous input current is controlled possibility. It consists of the three operating states; (1) active state, (2) shoot-through state, and (3) open-zero states. When related to the conventional X-Z/Q-Z topology, the proposed structure streams the minimum DC voltage on capacitor C_2 as well as deliver continuous input current [11,19]. Acknowledging straightforwardness in the control strategy for the proposed quasi-Z-source MLI functions with two operating modes; (1) shoot-through (ST) and, (2) non-shoot-through (NST). In ST mode, among all inverter phase leg, only one leg is conducting for providing ST. In NST mode, all the three leg switches in the inverter are forming the switching states to make a level in MLI. Hence, the inverter is working similar to a standard inverter. The current Z-source inverter switching states are premeditated with two zero states, and six active states. Figure 6 shows the Proposed modified Q-impedance network. The proposed inverter possesses the two open-zero, six active, and one shoot-through states.

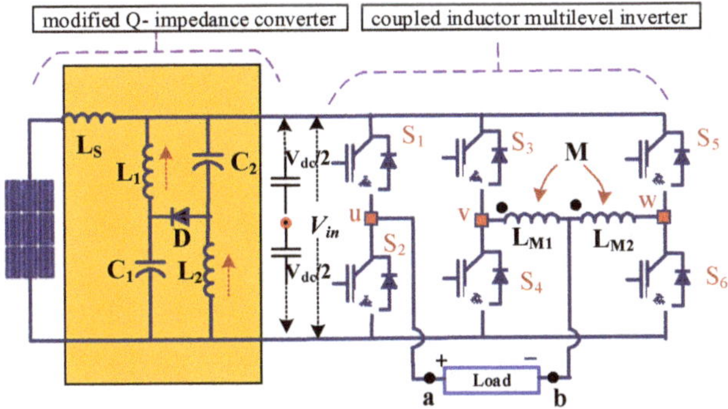

Figure 5. Proposed modified Q-impedance converter fed coupled inductor multilevel inverter (MLI).

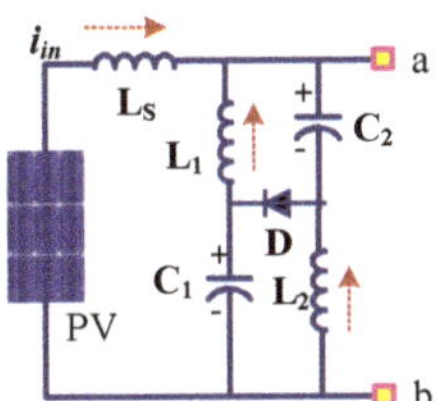

Figure 6. Proposed modified Q-impedance network.

State-1: Figure 7a shows the active states of the inverter. This mode assumes inductors, L_1 and L_2 and capacitors, C_1 and C_2 with chosen identical values as $V_{C1} = V_{C2} = V_C$, $V_{L1} = V_{L2} = V_L$ to maintain a symmetrical output nature. The Kirchhoff's voltage law is applied in Figure 7a,

$$V_{C1} + V_{L1} = V_{C2} + V_{L2} = V_{PV} + V_{Ls} = V_{ab} \tag{1}$$

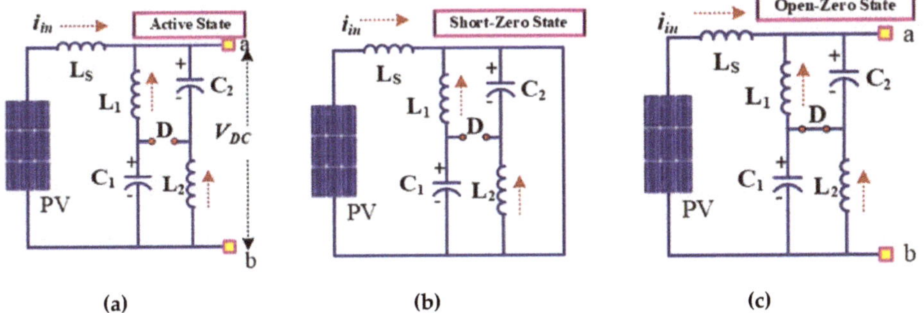

Figure 7. States of proposed Q- impedance network topology; (**a**) State-1: active state, (**b**) State 2: shoot-through states, (**c**) State-3: open-zero states.

Considering the steady-state average voltage of the inductors in one switching event must be zero. $V_{L1} = V_{L2} = V_{Ls} = V_L = \frac{1}{T}\int_0^T V_L(t)dt = 0$. From Equation (1), the steady-state voltage average of inductors is zero at one single switching period. Hence,

$$V_{C1} = V_{C2} = V_C = V_{PV} \tag{2}$$

State-2: In this shoot-through state, when any inverter leg is shorted, the PV array voltage is zero, and the diode is off. Thus, the inverter output voltage is zero due to short circuit as shown in Figure 7b, from the equivalent circuit of Figure 7b the KCL equation is given below,

$$V_{C1} = V_{C2} = V_C = V_{PV} = 0 \tag{3}$$

State-3: This is considered to be an open-zero state, where the inverter is switching legs act as an open circuit, and therefore the added capacitor voltages (V_{C1} and V_{C2}) appear across the V_{ab}. Now, the inductor (L_1 and L_2) currents flow through diode D_1 to charge the capacitors (C_1 and C_2) as exposed in Figure 7c.

From the circuit operation of the proposed inverter, the total switching period is classified as T_A (active state switching time), T_{sh} (shoot-through state switching time), and T_{op} (open-zero state switching time) within one switching cycle, $T_A + T_{sh} + T_{op} = 1$. Considering any of the three inductors voltage, V_L in different states; state-1: $V_L = V_{in} - V_{out}$, state-2: $V_L = V_{in}$, state-3: $V_L = V_{in} - 2V_C$.

Since, $V_L = \frac{1}{T} \int_0^T V_L(t) dt = 0$

$$V_L = T_A(V_{in} - V_{out}) + T_{sh} V_{in} - T_{op} V_{in} = 0 \tag{4}$$

$$V_{boost} = \frac{\left(T_A + T_{sh} - T_{op}\right)}{T_A} V_{PV} \tag{5}$$

From $T_A + T_{sh} = 1 - T_{op}$
Hence,

$$V_{boost} = \frac{\left(1 - 2T_{op}\right)}{T_A} V_{PV} \tag{6}$$

The converter DC output voltage boosting (V_{boost}) has two control degrees of choice (T_A *and* T_{sh}). Figure 8a shows the graphical understanding between V_{boost} and input voltage (V_{PV}) concerning duty cycle D_A. Here, the operation is split into two modes as (1) Mode-1: without open-zero states, and (2) Mode-2: towards short-zero states to open-zero states.

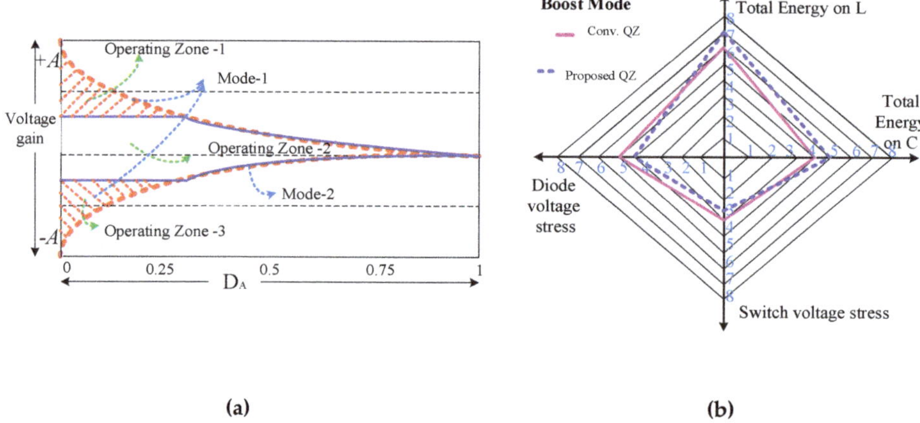

(a)   (b)

Figure 8. (a) Converter operation in different modes, V_{boost} / input voltage (V_{PV}) for duty cycle D_A, (b) energy storage capability and voltage stress radar graph of conventional QZ and proposed QZ.

In mode 1 the $T_{op} = 0$ output voltage depends on the parameter $T_A = 1 - (T_{op} + T_{sh})$. The entire range of the operation depends on the T_A active state switching time. The controlling of the parameter T_A decides the boosting capability of the MLI.

In Mode-2, the converter is in shoot-through states, where $T_{op} = 1 - T_A$. Hence the converter provides minimum voltage gain with a specified duty ratio of T_A. When converter operation moves from shoot-through states to the open-zero states, the output gain would be situated in the middle of the Mode-1 and Mode-2. The marked red area in Figure 8a shows the mode shift of the converter. Therefore, the converter voltage output can be adjusted to the desired values with two control degrees of freedom T_{op} and T_A. Figure 8b shows the radar graph of energy storage capability on C, L, diode voltage stress and switch voltage stress in boost mode for conventional QZ and proposed QZ. It shows the proposed QZ have better energy storage capability and lesser diode and switching voltage stress than conventional QZ.

4. Modes of Operation of Coupled Inductor MLI

The proposed coupled inductor MLI contains a structure having six switches in three legs (u, v and w). The leg v and leg w are connected with identical turns coupled inductor (L_{M1} and L_{M2}). The ST operation is done through any leg. If the leg u is used for ST, then the switch S_1 and S_2 are turned ON simultaneously. Since the ST is allowed in any of the MLI legs, the switching reliability is significantly improved. During the Non-ST (active state), the MLI is functioning with either one upper switch and two-lower switches/two-upper switches, and one-lower switch. Hence, during the active state, the inverter is operating with eight modes of operation to produce five-level voltages $(-V_{dc}, -V_{dc}/2, 0, V_{dc}/2, and V_{dc})$.

During the regular inverter operating conditions, the boosted voltage is appearing across the inverter and provides pure DC current. For the period of the ST time, the inductor is maintaining its voltage precisely equal to capacitor voltage V_C and hence current increases through the inductors leniently and limits the inductor current ripples. The eight modes of operation and their corresponding equivalent circuits of the proposed coupled inductor connected MLI is illustrated in Figure 9.

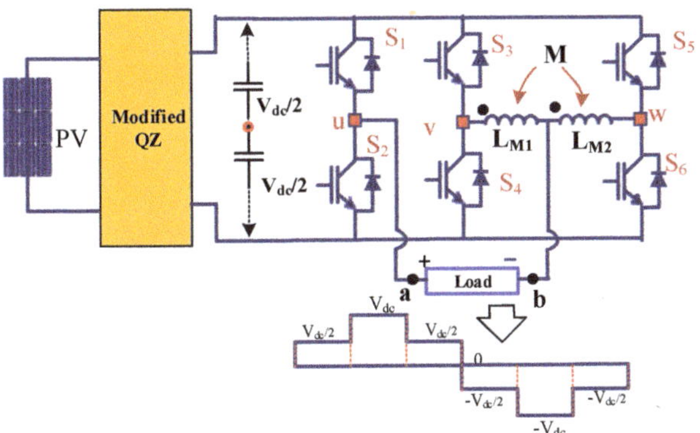

Figure 9. Coupled inductor MLI.

It is a single-phase circuit with two legs associated with coupled circuits. The quasi fed MLI provides $V_{pv} + V_S$ as input voltage to the coupled inductors (L_{M1} and L_{M2}). The mutual inductance (M) of the coupled inductors provides the five-level voltage output with a reduction of the switching devices. The level making of the inverter is done through the L_{M1} and L_{M2} with the identical sum. The inductor $L_{M1} = L_{M2} = L$ is connected between leg v and w. The voltage equation can be expressed as,

$$L\frac{di_v}{dt} - M\frac{di_w}{dt} = V_{in} - V_{bn} \tag{7}$$

$$L\frac{di_w}{dt} - M\frac{di_v}{dt} = V_{wn} - V_{bn} \tag{8}$$

where, V_{in} = input voltage, V_{bn} = load voltage, and V_{wn} = w leg voltage

Applying current law of Kirchhoff's', the leg current can be written as,

$$i_u + i_v + i_w = 0 \tag{9}$$

Hence,

$$V_{bn} = \frac{V_{vn} + V_{wn} + (L - M)\frac{di_u}{dt}}{2} \tag{10}$$

The inverter leakage inductance L_{M1} and L_{M2} are designed approximately equal to the mutual inductance value; hence leakage inductance equals mutual inductance ($L = M$). Therefore, the voltage equation on V_{bn} can be written as,

$$V_{bn} = \frac{V_{vn} + V_{wn}}{2} \tag{11}$$

The voltage output of the inverter can be delivered as,

$$V_{ab} = V_{an} - V_{bn} = V_{an} - \left(\frac{V_{vn} + V_{wn}}{2}\right) \tag{12}$$

$$V_{ab} = \frac{V_{ab}}{2} - \frac{\left(+\frac{V_{ab}}{2} + \frac{V_{ab}}{2}\right)}{2} = 0 \tag{13}$$

The different modes of operation of active inverter switching for level making are explained as follows;

Mode-1: During Mode-1, the inverter switches S_1, S_4, and S_6 are turned ON, and S_2, S_3, and S_5 are turned OFF (see Figure 10). Hence, from Equation (12) the load voltage is derived as, $V_{ab} = \frac{V_{ab}}{2} - \frac{\left(-\frac{V_{ab}}{2} - \frac{V_{ab}}{2}\right)}{2} = V_{dc}.$

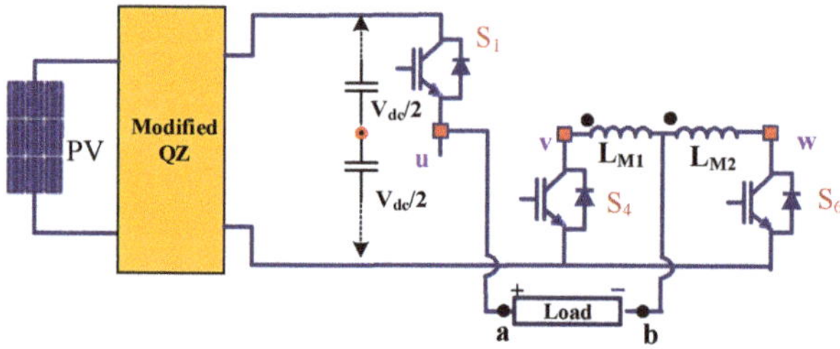

Figure 10. During mode 1 the inverter switches S_1, S_4, and S_6 are turned ON.

Mode-2: During Mode-2, the inverter switches S_1, S_4, and S_5 are turned ON, and S_2, S_3, and S_6 are turned OFF (see Figure 11). Hence, the inverter produces half of the V_{dc} (Vs = inverter input voltage). From Equation (12) the load voltage is derived as, $V_{ab} = \frac{V_{ab}}{2} - \frac{\left(-\frac{V_{ab}}{2} + \frac{V_{ab}}{2}\right)}{2} = \frac{V_{dc}}{2}.$

Mode-3: During the Mode-3, the inverter switches S_1, S_3, and S_6 are turned ON, and S_2, S_4, and S_5 are turned OFF (see Figure 12). Hence, the inverter produces half of the V_{dc}. From Equation (12) the load voltage is derived as, $V_{ab} = \frac{V_{ab}}{2} - \frac{\left(+\frac{V_{ab}}{2} - \frac{V_{ab}}{2}\right)}{2} = \frac{V_{dc}}{2}.$

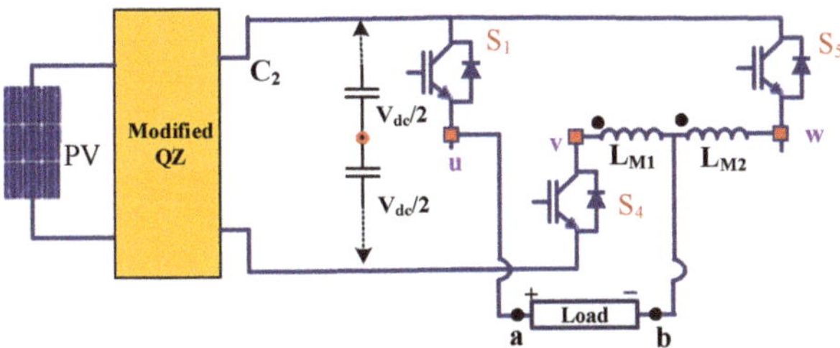

Figure 11. During Mode-2, the inverter switches S_1, S_4, and S_5 are turned ON.

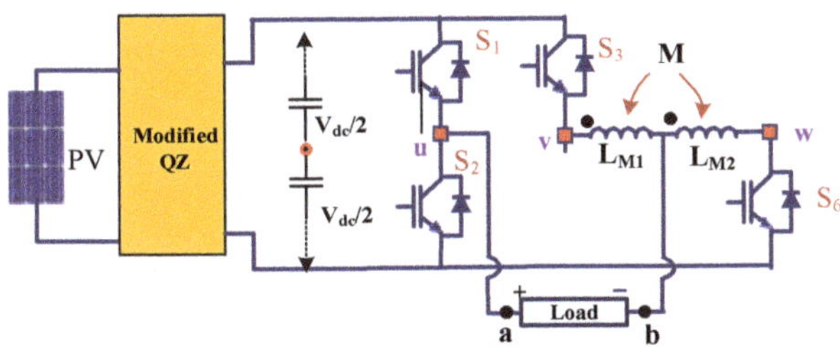

Figure 12. During mode 3 the inverter switches S_1, S_3, and S_6 are turned ON.

Mode-4: During Mode-4 the inverter switches S_1, S_3, and S_5 (all upper switch) is turned ON, and S_2, S_4, and S_6 (all lower switch) are turn OFF showing in Figure 13. Hence, the inverter produces zero output voltage. From Equation (12) the load voltage is derived as, $V_{ab} = \frac{V_{ab}}{2} - \frac{\left(+\frac{V_{ab}}{2}+\frac{V_{ab}}{2}\right)}{2} = 0$.

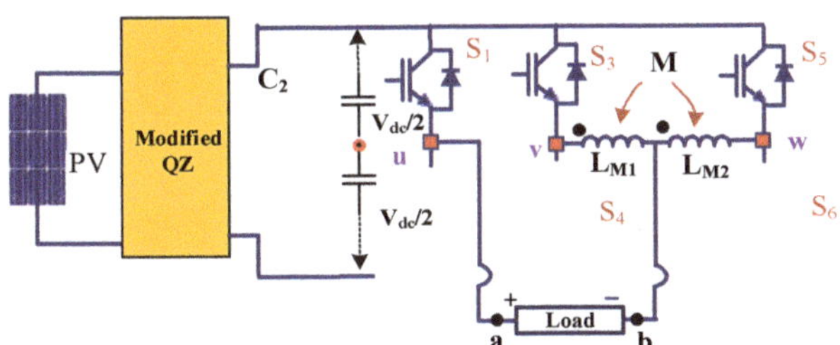

Figure 13. During Mode-4 the inverter switches S_1, S_3, and S_5 are turned ON.

Mode-5: During Mode-5 the inverter switches S_2, S_4, and S_6 (all lower switch) are turned ON, and S_1, S_3, and S_3 (all upper switch) are turned OFF as shown in Figure 14. Hence, the inverter produces zero output voltage. From Equation (12) the load voltage is derived as, $V_{ab} = -\frac{V_{ab}}{2} - \frac{\left(-\frac{V_{ab}}{2}-\frac{V_{ab}}{2}\right)}{2} = 0$.

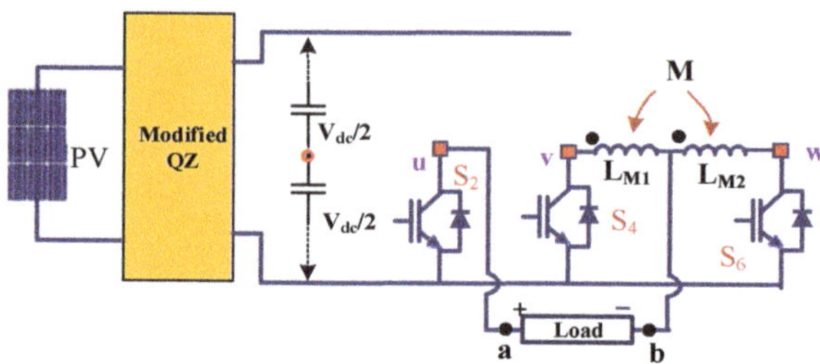

Figure 14. During mode 5 the inverter switches S_2, S_4, and S_6 are turned ON.

Mode-6: During Mode-6, the inverter switches S_2, S_4, and S_5 are turned ON, and S_1, S_3, and S_6 are turned OFF (see Figure 15). Hence, the inverter produces half of the V_{dc}. From Equation (12) the load voltage is derived as, $V_{ab} = -\frac{V_{ab}}{2} - \frac{\left(-\frac{V_{ab}}{2} + \frac{V_{ab}}{2}\right)}{2} = -\frac{V_{dc}}{2}$.

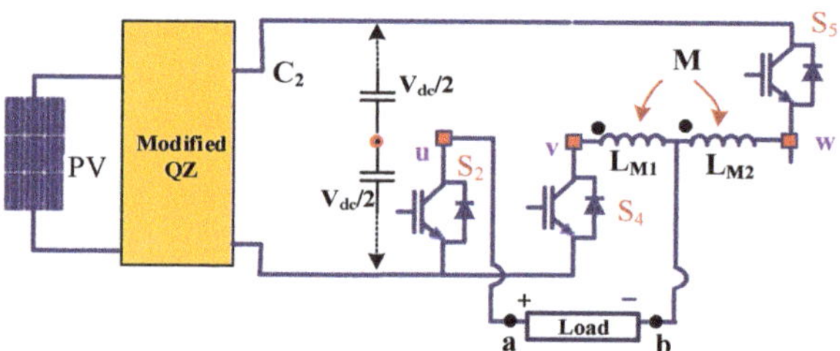

Figure 15. During mode-6 the inverter switches S_2, S_4, and S_5 are turned ON.

Mode-7: During Mode-7, the inverter switches S_2, S_3, and S_6 are turned ON, and S_1, S_4, and S_5 are turned OFF (see Figure 16). Hence, the inverter produces half of the V_{dc}. From the Equation.12 the load voltage is derived as, $V_{ab} = -\frac{V_{ab}}{2} - \frac{\left(+\frac{V_{ab}}{2} - \frac{V_{ab}}{2}\right)}{2} = -\frac{V_{dc}}{2}$.

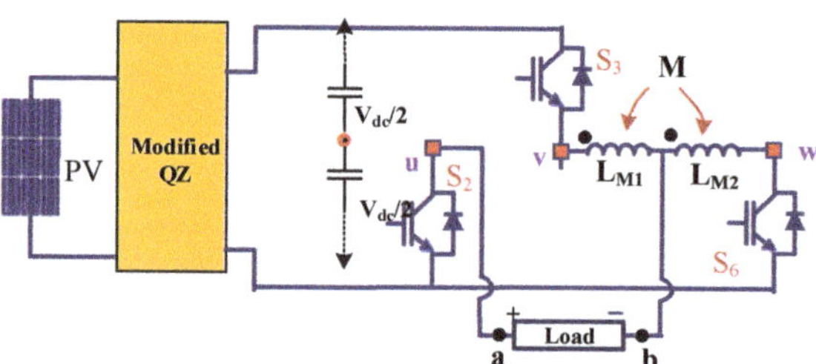

Figure 16. During Mode-7 the inverter switches S_2, S_3, and S_6 are turned ON.

Mode-8: During Mode-8, the inverter switches S_2, S_3, and S_5 are turned ON, and S_1, S_4, and S_6 are turned OFF (see Figure 17). Hence, the inverter produces full of the V_{dc} in negative. From Equation (12) the load voltage is derived as, $V_{ab} = -\frac{V_{ab}}{2} - \frac{\left(+\frac{V_{ab}}{2} + \frac{V_{ab}}{2}\right)}{2} = -V_{dc}$.

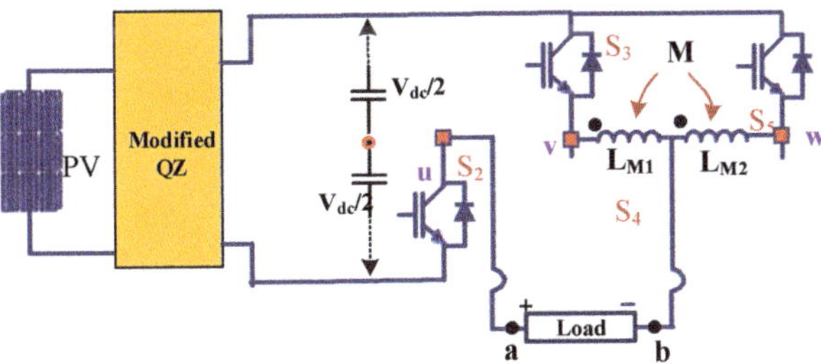

Figure 17. During Mode-8 the inverter switches S_2, S_3, and S_5 are turned ON.

Table 1 shows the consolidation of the eight modes of operations of the inverter. Of these eight modes, six modes are producing the voltage in different forms between $+V_{dc}$ to $-V_{dc}$. The Mode-4 and Mode-5 are producing the zero voltages, and hence, any one of the modes can be used for zero voltage. Figure 18 illustrates the all mode operation output voltage structure for proposed MLI.

Table 1. Switching table for the proposed topology.

Mode	Conducting Switches	Output
1	S_1, S_4, S_6	$V_{ab} = \frac{V_{ab}}{2} - \frac{\left(-\frac{V_{ab}}{2} - \frac{V_{ab}}{2}\right)}{2} = V_{dc}$
2	S_1, S_4, S_5	$V_{ab} = \frac{V_{ab}}{2} - \frac{\left(-\frac{V_{ab}}{2} + \frac{V_{ab}}{2}\right)}{2} = \frac{V_{dc}}{2}$
3	S_1, S_3, S_6	$V_{ab} = \frac{V_{ab}}{2} - \frac{\left(+\frac{V_{ab}}{2} - \frac{V_{ab}}{2}\right)}{2} = \frac{V_{dc}}{2}$
4	S_1, S_3, S_5	$V_{ab} = \frac{V_{ab}}{2} - \frac{\left(+\frac{V_{ab}}{2} + \frac{V_{ab}}{2}\right)}{2} = 0$
5	S_2, S_4, S_6	$V_{ab} = -\frac{V_{ab}}{2} - \frac{\left(-\frac{V_{ab}}{2} - \frac{V_{ab}}{2}\right)}{2} = 0$
6	S_2, S_4, S_5	$V_{ab} = -\frac{V_{ab}}{2} - \frac{\left(-\frac{V_{ab}}{2} + \frac{V_{ab}}{2}\right)}{2} = -\frac{V_{dc}}{2}$
7	S_2, S_3, S_6	$V_{ab} = -\frac{V_{ab}}{2} - \frac{\left(+\frac{V_{ab}}{2} - \frac{V_{ab}}{2}\right)}{2} = -\frac{V_{dc}}{2}$
8	S_2, S_3, S_5	$V_{ab} = -\frac{V_{ab}}{2} - \frac{\left(+\frac{V_{ab}}{2} + \frac{V_{ab}}{2}\right)}{2} = -V_{dc}$

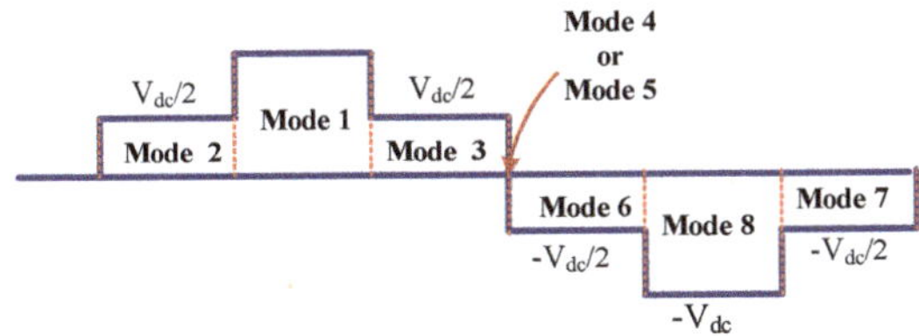

Figure 18. Output voltage structure for proposed MLI.

5. Modified Space Vector PWM

The Space Vector PWM is a continuous switching PWM technique, which explicitly selects the active and zero states placed within the carrier period [30]. While designing for the boost converter circuitry, the vector-based algorithm possesses the additional switching states which must be imposed to acquire a higher voltage gain in the traditional impedance source inverters. It may lead to impact the switches in the form of stress or failure of a switch [8]. While looking into the traditional strategy, the upper and lower switch combination must be short as the voltage gain may impose distortion on voltage. The proposed space vectors consist of normalized state operation, which generates the voltage gain by operating u, v, w legs upper or lower switches, as shown in Figure 10. The condition ST is incorporated with the regular zero vector operation. The projected control algorithm realizes the least number of switching operations to improve the efficiency of an inverter over one switching cycle, as indicated in Figure 18. The operation time for each period and every switching cycle of the dead time for short-zero as well as open-zero states, should be pre-calculated. The six modes (Mode-1 to Mode-6) of switching states are aligned with active vector and Mode-4, and Mode-5 are placed in zero vector.

In a three-phase balanced system, the voltage equation of Space Vector PWM is predefined as,

$$V_{ref}T_s = V_1 T_1 + V_2 T_2 + V_0 \frac{T_0}{2} \tag{14}$$

Here, V_{ref} is a reference vector (target vector), From Figure 19b in the sector-1, vector V_{ref} can be synthesized as,

$$V_{ref} = V_1 \frac{T_1}{T_s} + V_2 \frac{T_2}{T_s} + V_0 \frac{T_0}{2T_s} \tag{15}$$

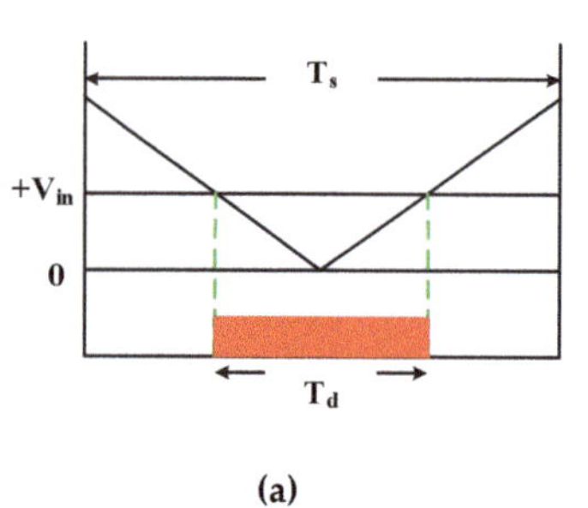

(a)

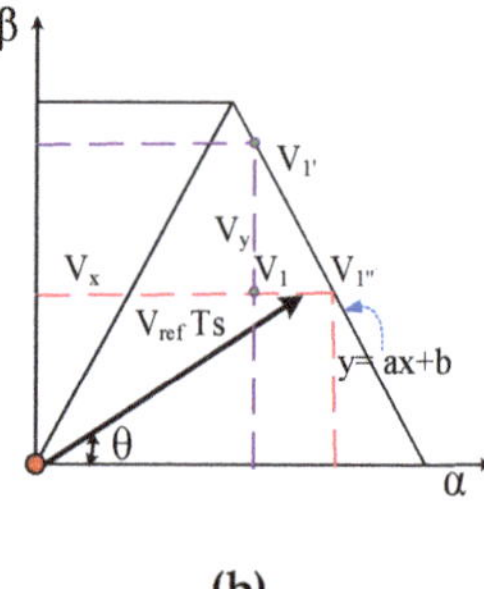

(b)

Figure 19. (a) Dwell time switching states synthesis, (b) proposed control strategy V_{ref} slope.

The V_1 and V_2 are the adjustment operating vectors at T_1 and T_2. The T_s is a switching period, and T_0 is zero vector time. The pulse period of the active vector is calculated from the below equation when operating in a given switching period T_s.

$$T_1 = \frac{2}{\sqrt{3}} + |V| \sin\left(\frac{2\pi}{3} - \theta\right) T_s \tag{16}$$

$$T_2 = \frac{2}{\sqrt{3}} |V| \sin\left(\theta - \frac{\pi}{3}\right) T_s \tag{17}$$

Hence, the zero-vector time:

$$T_0 = T_s - T_1 - T_2 = \left(1 - \frac{2}{\sqrt{3}} |V| \sin(\theta) T_s\right) \tag{18}$$

The proposed Space Vector PWM strategy is selecting the voltage vector switching sequence, according to V_{ref}. The V_{ref} is the reference vector of output V_{ab}. Hence, the output voltage of the inverter is selected by using switch S_1. According to Table 1, when S1 is turned ON, the inverter output voltage V_{ab} is either in positive voltage or zero. Hence the relation is framed between S_1 and V_{ab}, for the smooth implementation.

When S_1 is turned ON, the $V_{ab} \geq 0$ and S_1 is low, consequently $V_{ab} \leq 0$. Nevertheless, the other switching states should be appropriately combined with S_1 to make the desired voltage level of the inverter. In order to select the stable switching states, the dwell time (T_d) of the switch is calculated in every switching frequency T_s sampling period. The dwell time concerning inverter DC-link voltage V_{in} is defined as:

$$T_d = \frac{\left|V_{ref}\right| - kV_{in}}{V_{in}} \tag{19}$$

Here, the V_{ref} is related to the sampling period T_s. The ST duty ratio for inverter must be predefined for every switching cycle by adding the ST time within the T_S. Figure 19a shows the dwell time control of inverter, and it is related to T_S ($= 1/f_S$), and V_{in}.

To locate the selected switching vector for the V_{ref}, a new method of vector identification is proposed. The strategy is to locate the V_{ref} value and to find out the different switching states in the space vector diagram (SVD) hexagon (see Figure 20). Hence the control strategy is designed in two steps: (1) locate the real and imaginary equivalent of V_{ref}, and (2) find out the control vector within the sector.

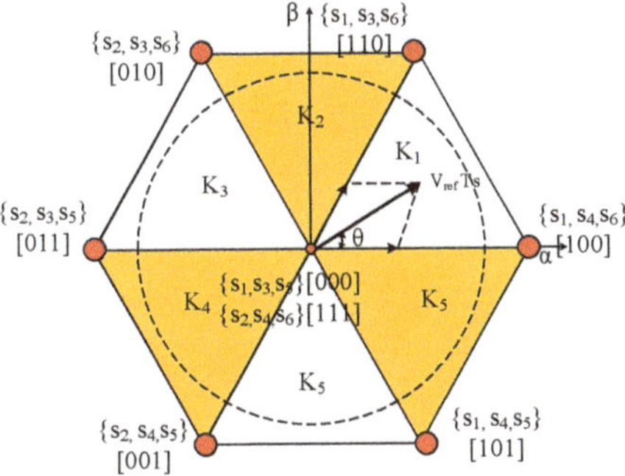

Figure 20. Space vector diagram and switching table.

Figure 19b represents the proposed control strategy in which, from the V_{ref} slope, the real and imaginary equivalent of V_x and V_y are determined and compared with the targeted switching vector V_1 and V_1'. Here, when the condition is $V_y < a\,V_x + b$, then V_1 is selected. Else V_1' is selected. Once this targeting vector is selected, the T_d is calculated and then active switching time, ST time and zero switching time is calculated according to the inverter output voltage requirement. Figure 20 represents the overall SVD for the proposed control strategy. Here, the entire active targeted vector is placed inside the SVD sector until the hexagonal boundary. The zero vectors [1, 1, 1] and [0, 0, 0] are placed at the origin. The control switching vector is directly related to the inverter modulation index (m_a), where V_{in} and V_{ref} are related to the inverter output. The maximum inverter control in the linear modulation range is allowed only until $\frac{2}{\sqrt{3}} V_{in}$ [23]. The switching pulse patent of the proposed PWM is prearranged in Figure 21. Once the active and zero states are done, the ST state patent is included in the switching sampling period. The ST state is calculated based on the V_{PV} value. Figure 22 represents the inverter switching pulse. The zero-state sharing ST state and ST time is calculated directly from the following equations:

$$T_{sh} = \frac{\left(\left|V_{ab}\right| - BV_{PV}\right)}{V_{PV}} \tag{20}$$

$$B = \frac{\left|V_{ab}\right|}{V_{PV}} \tag{21}$$

where B is a boosting factor.

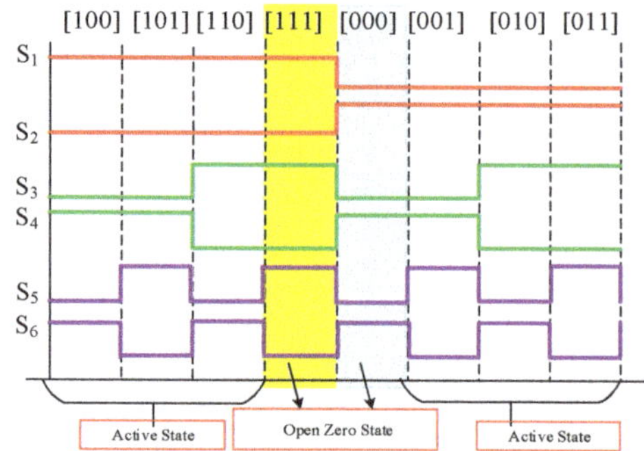

Figure 21. Space vector PWM algorithm and switching table.

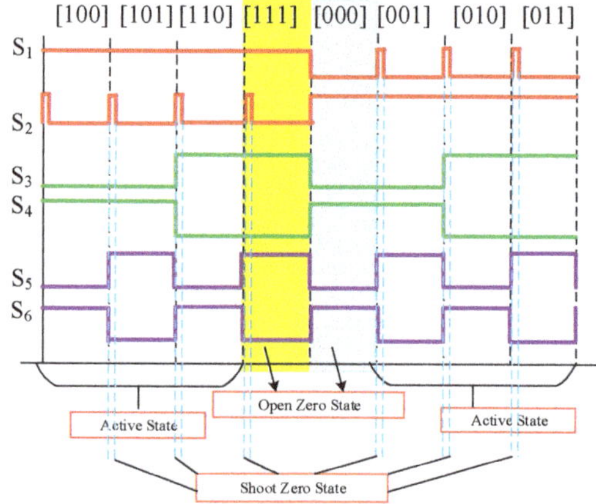

Figure 22. Proposed impedance source Space vector PWM and switching table.

6. Simulation Result

The PV powered Q-impedance network connected coupled inductor multilevel inverter and its control switching schemes strategy is designed in MATLAB/Simulink software simulation platform (Figure 23). The inverter is powered by 500 Watts peak power PV. The PV module is arranged to get 100 to 120 V to meet the 330 V DC-link voltage of the inverter. The insulation level of the PV array is 1000 W/m^2 for 10 s, 800 W/m^2 from 10 s to 30 s. The temperature of the PV array is 400°C for 10 s and 300 °C from 10 s onwards. The variation in PV array power input can be overcome by the Perturb and Observe the MPPT algorithm to obtain the constant DC voltage from the PV array. Table 2 shows the simulation parameter for the proposed inverter. The inverter performance is investigated with and without LC filters.

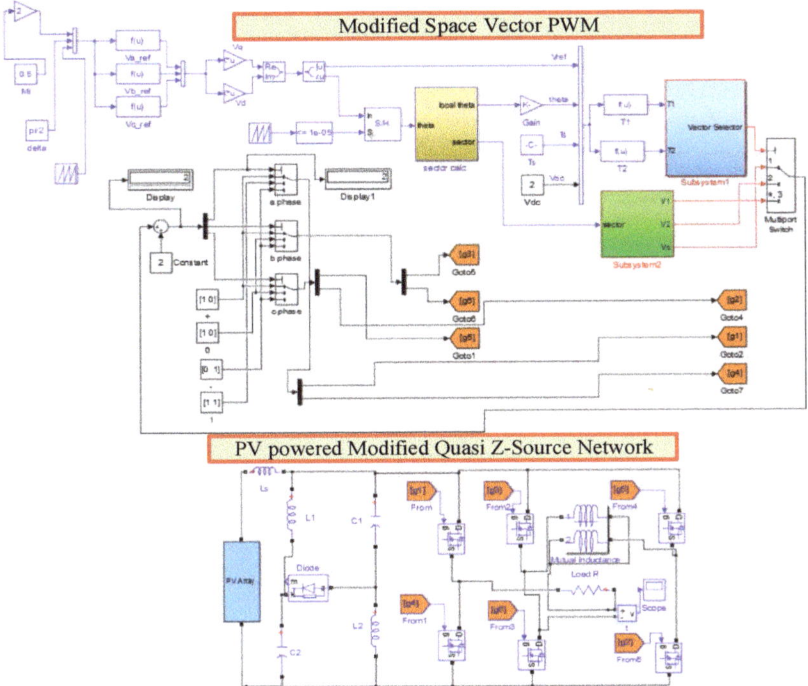

Figure 23. Simulation diagram of the proposed photovoltaic (PV) powered modified Q-impedance converter fed coupled inductor MLI.

Table 2. Simulation Parameters.

Input PV Voltage, V_{PV}	100 V–120 V
Impedance Network Inductor $L_s = L_2 = L_3$	2 mH
Impedance network inductor capacitors $C_1 = C_2$	200 µF
Switching frequency, f_s	10 KHz
Inverter coupled inductor $L_{M1} = L_{M2}$	5 mH
Mutual inductance	2.4 mH
Load resistance and inductance	10Ω, 5 mH
LC filter: inductance and capacitor	2.5 mH,50 µF

In order to validate the inverter performance simulation is carried out when the PV input power is kept at 500 W and the input PV voltage, V_{PV} is maintained at 120 V. The impedance network duty ratio (T_A and T_{ST}) is maintained at 20% to 25% to preserve the inverter input (DC-link) voltage 250 to 350 V. The inverter operation is investigated with their modulation index range $m_a = 0$ to 0.866. The simulation study is carried out for different impedance network duty ratio T_{ST} and inverter modulation index range m_a. Initially, the inverter is operated with a maximum modulation index of 0.886 with the ST switching time T_{ST} of 25%. Figure 24 shows the PWM pulse of inverter switches S_1 and S_6. The ST time between switch S_1 and S_2 is represented in Figure 25, in which the 25% switching time is used for ST event, and hence impedance network can boost input PV voltage nearly 290% and achieved 349 V in the output side of the impedance network (DC-link voltage of inverter). During the operation, the voltage of impedance network capacitors V_{C1} and V_{C2} shows uniform charging and discharging profile (see Figure 26) along with the uniform inductors current profile (see Figure 27). Hence, during the ST period, the impedance network can draw the constant current and provide a regulated boost DC-link voltage to the MLI. Figures 28 and 29 illustrates the captured the inverter input DC-link voltage and multi-level output voltage across the load (V_{ab}) respectively. From the results, it can be seen that the 120 V input PV voltage is boosted to 349 V. The inverter load voltage V_{ab} is observed as 247.3 V (RMS). The corresponding V_{ab} voltage THD spectrum is shown in Figure 30 (without filter). Here the inverter voltage THD is observed as 14.15%, which is higher due to the participation of passive elements in the impedance network. Hence the LC filter is connected across the load, and harmonics spectrum is captured (see Figures 31 and 32). The output voltage THD perceived is very less as 2.81%. The inverter load current waveform and its corresponding current THD spectrum are captured and shown in Figure 33a,b. As expected, the current THD is very less (1.7%).

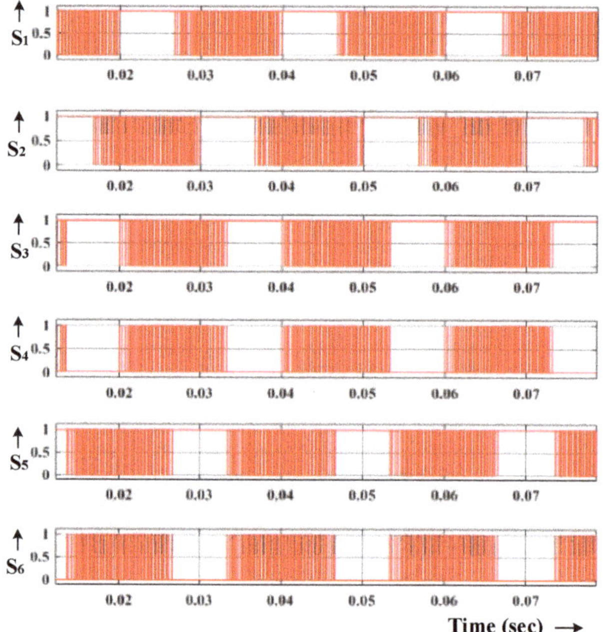

Figure 24. Modified Q- impedance converter fed coupled inductor MLI.

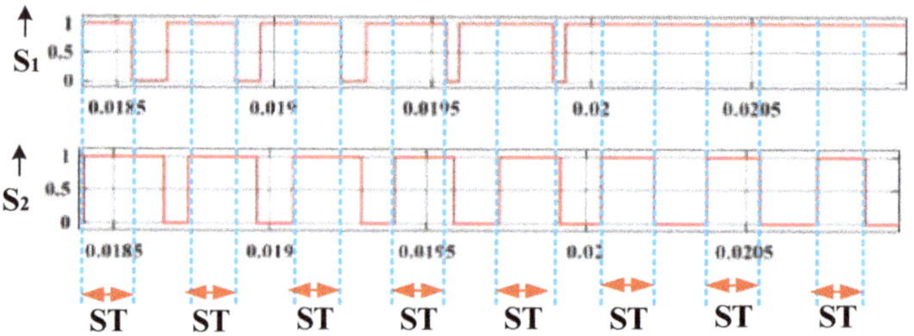

Figure 25. ST state representations of S_1 and S_2.

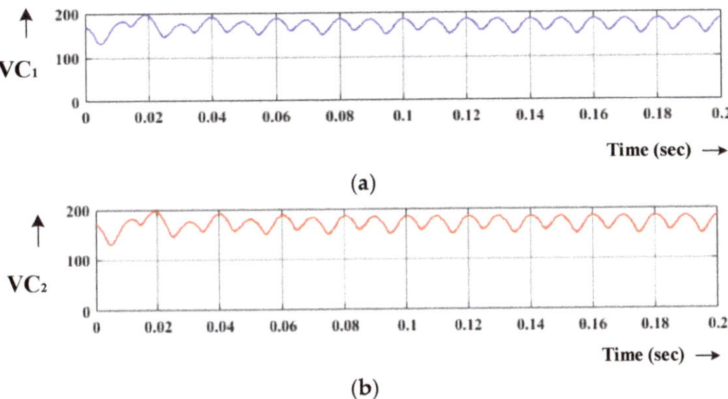

(a)

(b)

Figure 26. Voltage waveform of impedance network capacitors at T_{ST} = 25%; (**a**) V_{C1}, (**b**) V_{C2}.

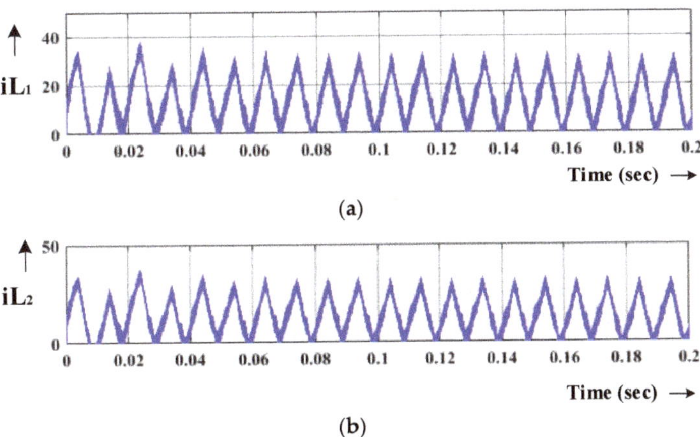

(a)

(b)

Figure 27. Current waveform of impedance network inductor at T_{ST} = 25%; (**a**) L_1, (**b**) L_2.

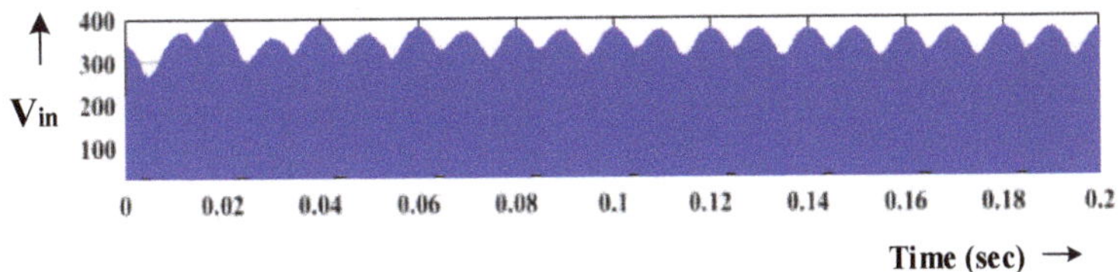

Figure 28. Inverter DC-link voltage at T_{ST} = 25%.

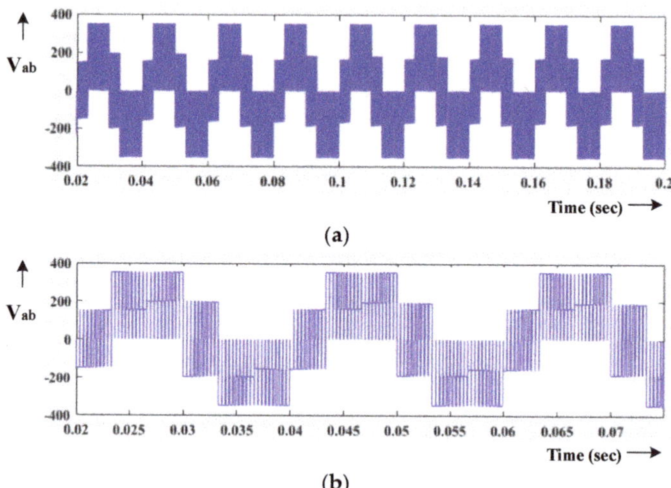

Figure 29. (**a**) Inverter output voltage at $T_{ST} = 25\%$ without filter, (**b**) zoomed view of the inverter output voltage at $T_{ST} = 25\%$ without a filter.

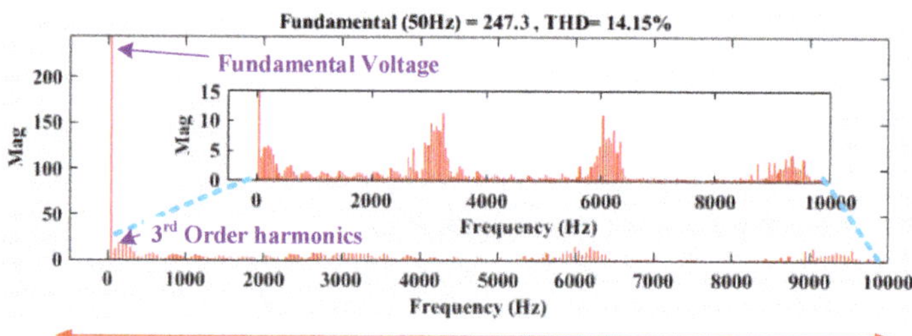

Figure 30. THD profile of inverter output voltage when $T_{ST} = 25\%$ without a filter.

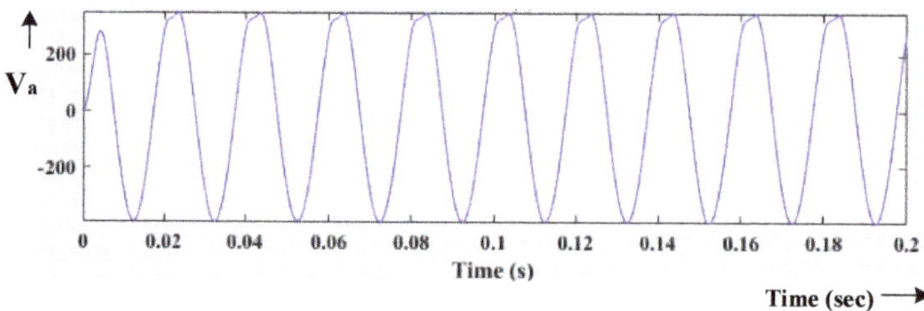

Figure 31. Inverter output current at $T_{ST} = 25\%$ with filter.

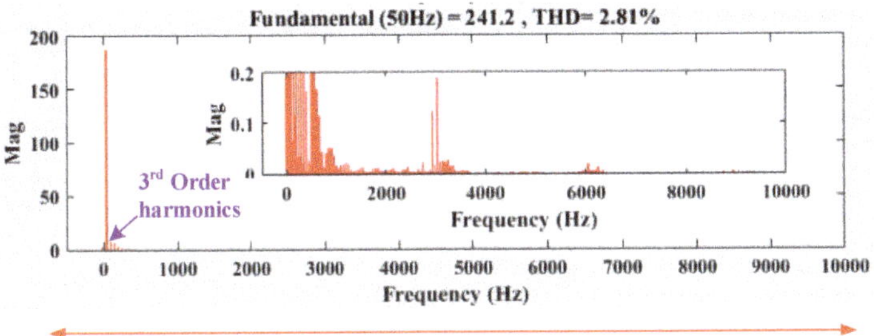

Figure 32. THD profile of inverter output voltage at $T_{ST} = 25\%$ with filter.

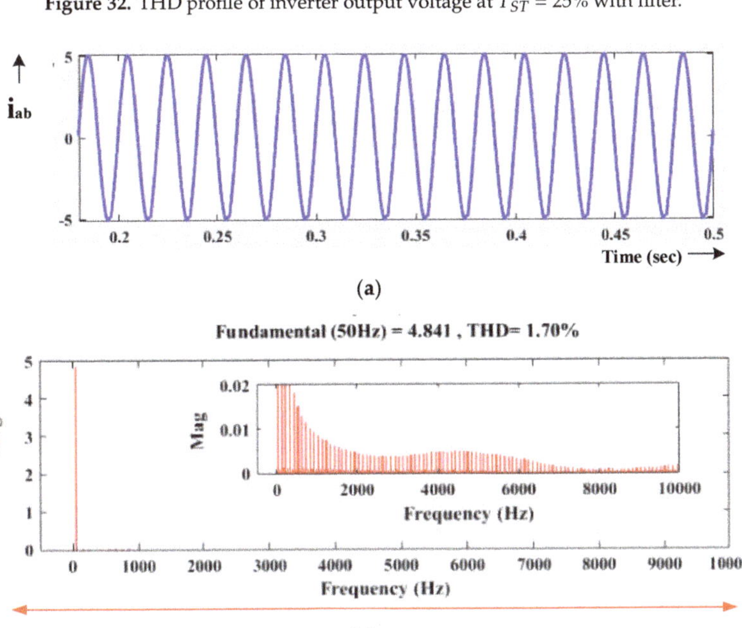

Figure 33. (**a**) Inverter output current at $T_{ST} = 25\%$, (**b**) THD profile of inverter output current at $T_{ST} = 25\%$.

Next, the simulation study is extended to $T_{ST} = 20\%$ with 0.866 m_a. In this operating condition, the impedance network has boosted the input PV voltage to nearly 230% and maintaining the inverter DC-link voltage at 280 V. The observed DC-link voltage and impedance network capacitor voltages are shown in Figures 34 and 35. While operating the inverter with a modulation index value of $m_a = 0.866$, the inverter has produced the output voltage of 197.23 V, as shown in Figure 36. The inverter output voltage THD is shown in Figure 37. Using LC filter in the inverter output terminal similar to 25% T_{ST} performance, the voltage THD for $T_{ST} = 20\%$ is maintained lesser value as 2.71%. Correspondingly the current THD is observed as 1.70%. In order to validate the higher ST switching time, the simulation is extended to $T_{ST} = 30\%$ with different modulation indices. During this operating condition, the impedance network passive elements are utilized fully and hence the voltage THD triumphs to a higher value. Figures 38 and 39 illustrates the inverter output voltage and its corresponding voltage THD (without LC filter) for $m_a = 0.6$. Though the inverter voltage is preserved as 292V nearly, the voltage THD is poor. Tables 3 and 4 illustrates inverter voltage and its THD performances for $m_a = 0.86$ and $m_a = 0.6$ through different duty ratio from 10% to 30% without and with an LC filter, respectively.

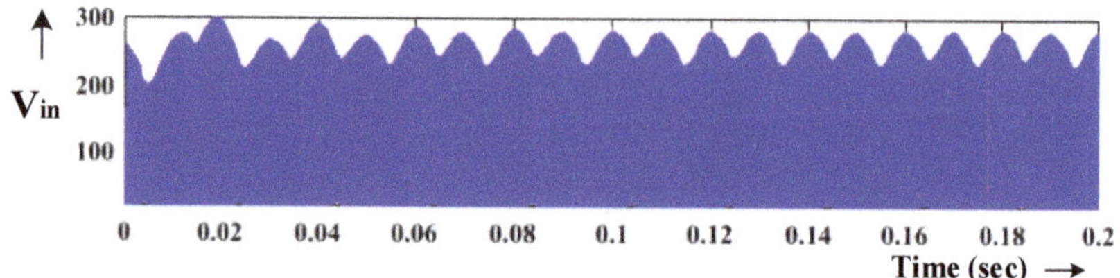

Figure 34. Inverter DC-link voltage at $T_{ST} = 20\%$.

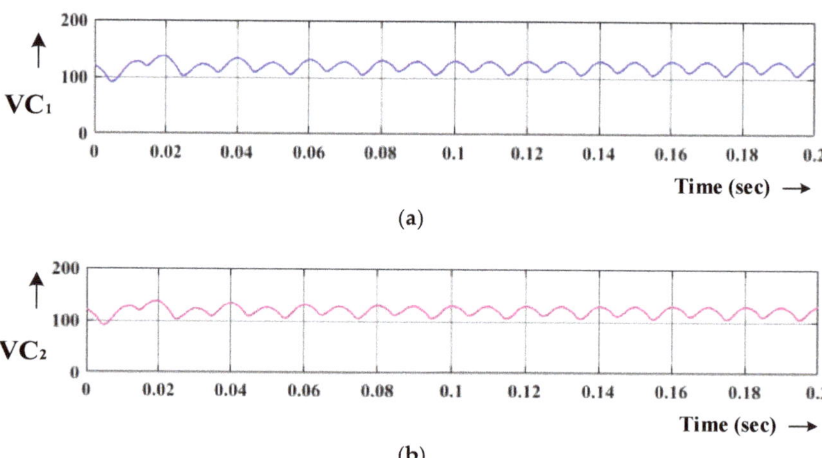

Figure 35. Voltage waveform of impedance network capacitors at $T_{ST} = 20\%$; (**a**) V_{C1}, (**b**) V_{C2}.

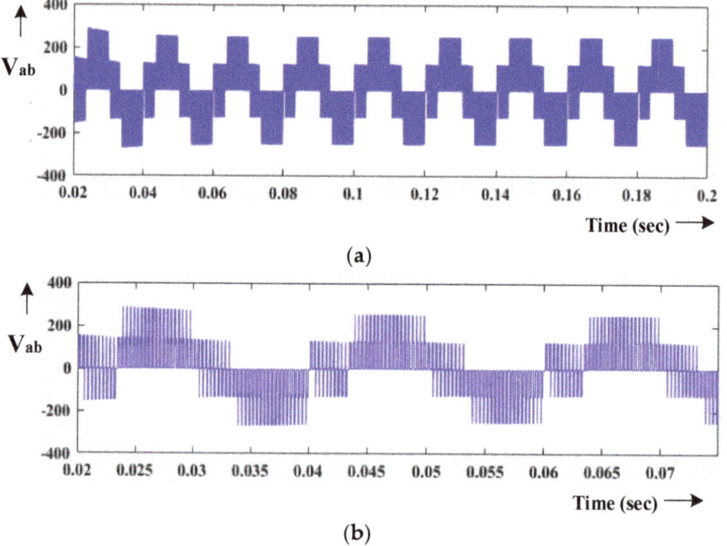

Figure 36. (**a**) Inverter output voltage at $T_{ST} = 25\%$ without filter, (**b**) zoomed view of inverter output voltage at $T_{ST} = 25\%$ without filter.

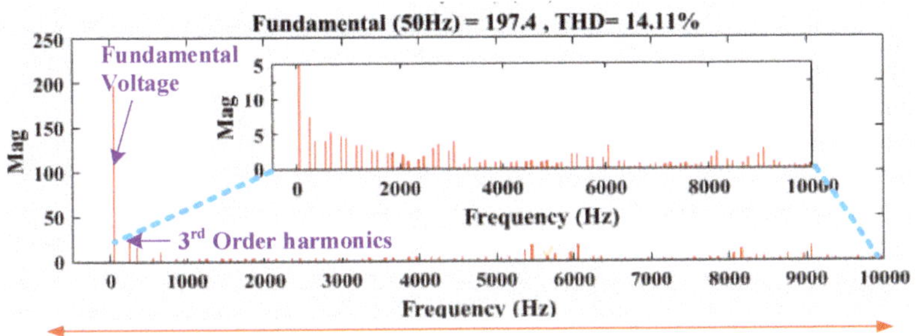

Figure 37. THD profile of inverter output voltage at $T_{ST} = 25\%$ without a filter.

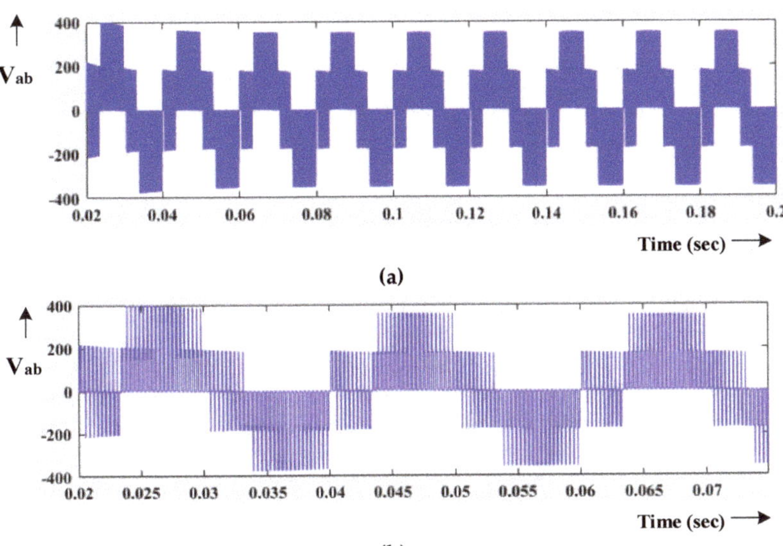

(a)

(b)

Figure 38. (**a**) Inverter output voltage at $T_{ST} = 30\%$ without filter, (**b**) zoomed view of the inverter output voltage at $T_{ST} = 30\%$ without a filter.

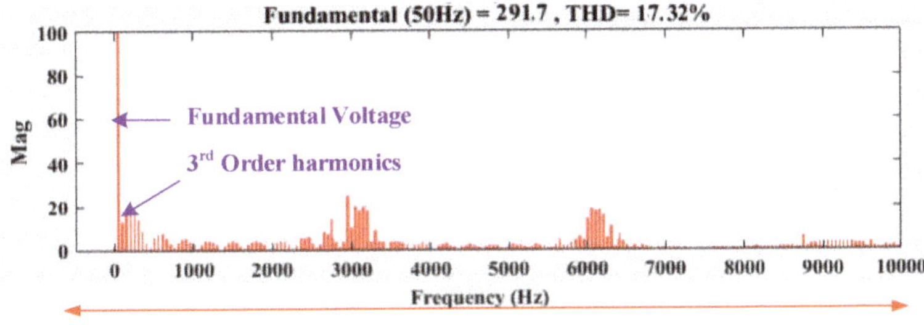

Figure 39. THD profile of inverter output voltage at $T_{ST} = 30\%$.

From the tabulated results, it could be seen that DC-link voltage has a linear variation with D_{ST}. However, the THD of the inverter voltage increases when increasing the D_{ST}. For any value less than or equivalent to D_{ST} 25% in any modulation index, the inverter provides a wide range of voltage variation with better voltage and current THD performance.

Table 3. Detailed simulation results for different duty ratio without an LC filter at $m_a = 0.86$ and 0.6.

D_{ST}	V_{in}	$V_o = V_{DC\text{-}link}$	V_{ab} for $m_a = 0.86$	% THD V_{ab} for $m_a = 0.86$	V_{ab} for $m_a = 0.60$	% THD V_{ab} for $m_a = 0.60$
0	120	120	86.4 V	13.6%	59.3 V	12.6%
10%	120	177	125.6 V	13.9%	88.5 V	13.8%
15%	120	210	149.2 V	14.06%	99.1 V	13.8%
20%	120	280	197.4 V	14.11%	135.3 V	13.54%
25%	120	349	247.3 V	14.15%	166.7 V	14.06%
30%	120	410	291.7 V	19.55%	201.6 V	17.32%

Table 4. Detailed simulation results for different duty ratio with an LC filter at $m_a = 0.86$ and 0.6.

D_{ST}	V_{in}	$V_o = V_{DC\text{-}link}$	V_{ab} for $m_a = 0.86$	% THD V_{ab} for $m_a = 0.86$	V_{ab} for $m_a = 0.60$	% THD V_{ab} for $m_a = 0.60$
0	120	120	85.1 V	2.69%	57.2 V	2.61%
10%	120	177	123.9 V	2.71%	86.9 V	2.68%
15%	120	210	147.1 V	2.72%	97.4 V	2.72%
20%	120	280	195.9 V	2.77%	133.9 V	2.74%
25%	120	349	241.2 V	2.81%	164.9 V	2.79%
30%	120	410	291.7 V	3.83%	200.1 V	3.68%

7. Experimental Result

The proposed PV powered Q-impedance fed coupled inductor multilevel inverter experimental setup was built using six MOSFETs IRF640. The switching signals were associated with the MLI through gate driver TLP250. The switching frequency, f_s of the inverter is fixed 10 kHz for the 50 Hz inverter output. The 500 W PV module is arranged to get 100 to 120 V to meet 330 V DC-link voltage to the MLI. The RL load (resistance = 10 Ω and inductance = 5 mH) and LC filter (inductance and capacitor values of 2.5 mH, and 50 μF respectively) are used in the inverter output terminal.

Figure 40 shows the experimental setup photograph. The impedance network and other parameters used for the inverter are the same as the simulation model given in Table 2. The control switching scheme strategy is designed in PIC16F778A microcontroller, and the collaborative results are shown in keyset two channels digital signal oscilloscope (DSO).

The experimentation is carried out for the 500W constant PV power for 100 V and 5 amps current. The impedance network duty ratio is maintained at 25% with the aim of inverter constant DC-link voltage 300V. The inverter operation is investigated with its modulation index range $m_a = 0$ to 0.866. Initially, the inverter is operated with maximum $m_a = 0.886$, and the results are captured. Figure 41 shows the PWM pulse of inverter switches S_1 and S_2. The ST time between switch S_1 and S_2 is represented, in which the 25% switching time is used for ST event, and hence impedance network can generate 300% boosting to maintain the DC-link voltage. Figure 42a shows the current waveform of impedance network inductor L_1 and L_2. Figure 42b displays the voltage waveform of impedance network capacitors C_1 and C_2. The voltage profile across the impedance network capacitors V_{C1} and V_{C2} are indicating that variation in the V_{C1} and V_{C2} are identical. It can be observed that voltages V_{C1} and V_{C2} are equally charging the voltage since both are connected in series. Hence the proposed impedance network can provide regulated DC-link voltage to the inverter. Figure 43a,b shows the inverter output voltage (without filters) and input voltage waveform at $m_a = 0.866$. The results show the measured five-level inverter voltage with symmetrical set output voltage 0V, $\pm$ 125 V, $\pm$ 250 V. The THD value of load voltage is captured using power analyzer and is found to be about 17.1% (shown in Figure 44). The same experimentation is investigated further with filter, and the results are captured. Figure 45a,b shows the filtered output load and load current respectively, where the current and voltage are maintaining their THD lesser as 3.1% and 1.4% respectively (Figure 46a,b).

(**a**) (**b**)

Figure 40. Experimental setup; (**a**) overall laboratory-scale 500 W PV powered modified Q-impedance fed coupled inductor MLI, (**b**) modified Q-impedance fed coupled inductor MLI.

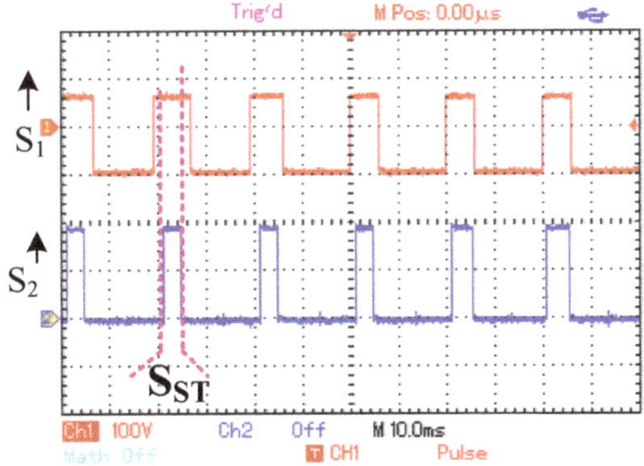

Figure 41. PWM pulse generation (S_1 and S_2).

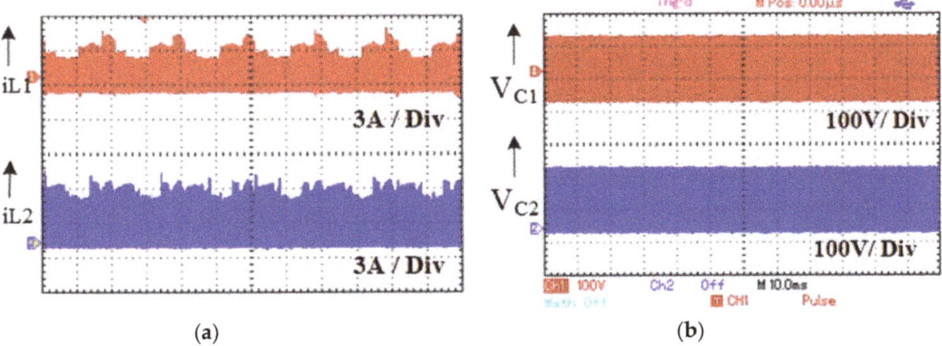

(**a**) (**b**)

Figure 42. Experimental result; (**a**) boost inductor current (i_{L1} and i_{L2}), (**b**) voltage waveform of impedance network capacitors C_1 and C_2.

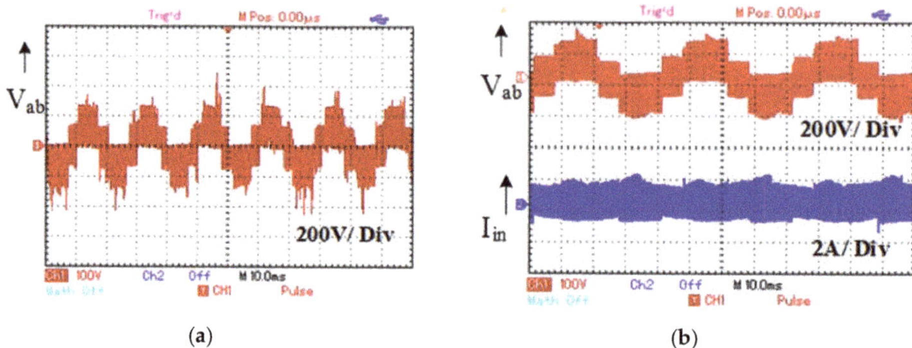

(**a**) (**b**)

Figure 43. Experimental result; (**a**) inverter five-level output voltage, (**b**) quasi Z-source inverter input current waveform.

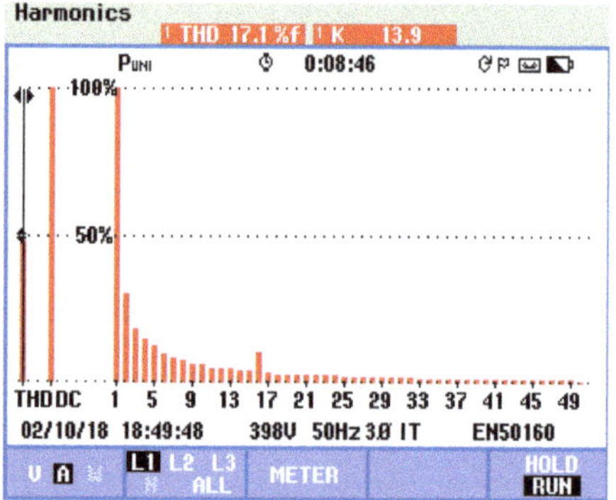

Figure 44. Experimental voltage THD spectrum without a filter.

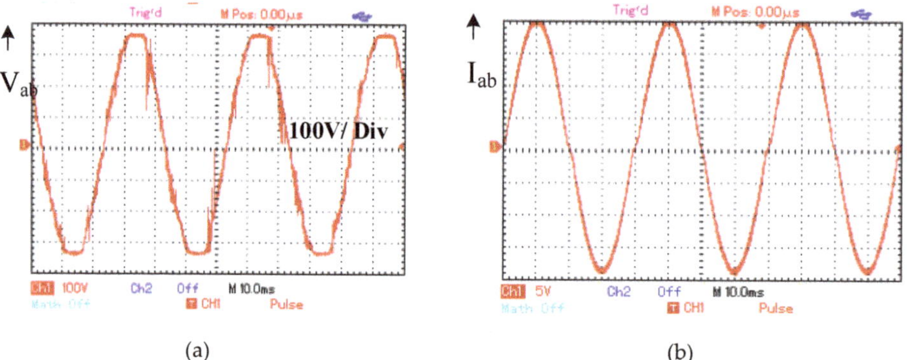

(**a**) (**b**)

Figure 45. Experimental result; (**a**) inverter output voltage with filter, (**b**) inverter output current waveform with filter.

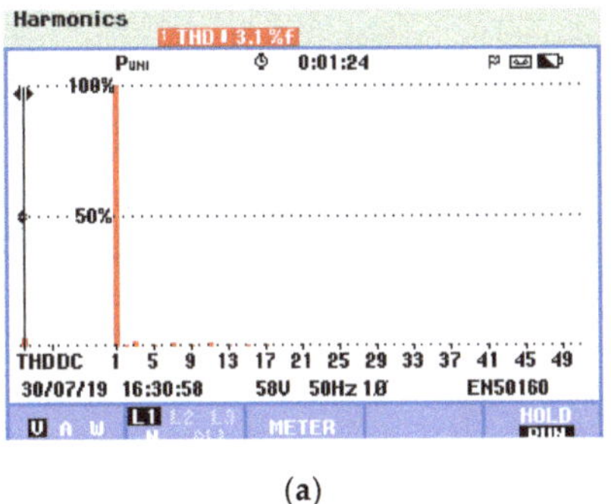

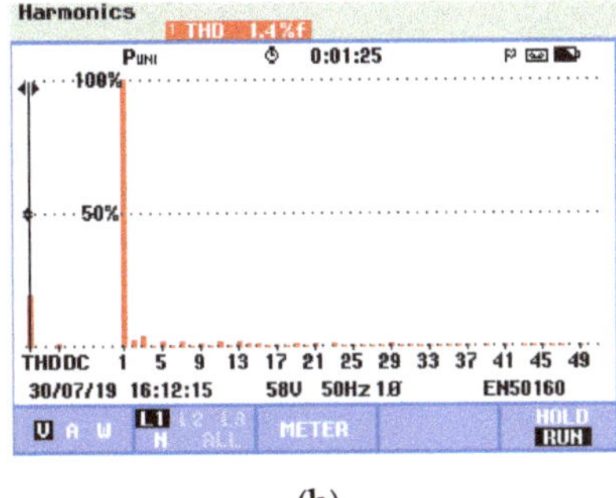

(a)   **(b)**

Figure 46. Experimental result; (**a**) voltage THD spectrum with filter, (**b**) current THD spectrum with filter.

To understand the inverter input current control, the duty cycle is varied to 25% and 30%, and results are observed. As expected, the inverter is maintaining the input current regulation at 25% D_{ST}. However, when D_{ST} is applied to 30%, the impedance network starts losing its input current regulation. Figure 47 illustrates the output voltage of MLI and impedance input current at 25% and 30% D_{ST}. From the results, it could be seen that the input PV voltage is boosted via the impedance network to achieve the load voltage of 230 V peak to peak. The DC-link voltage is regulated with a minor ripple of 3%, and hence inverter can maintain its half symmetry and THD of the output voltage is perceived as less. The interesting point to notice at this stage is that the inverter can draw constant current since the impedance network input inductor L_S limits the current. The proposed inverter reliability study is conducted for different inverter operating conditions. Table 5 illustrations the switching loss, inverter efficiency, and THD of the proposed inverter for D_{ST} 10% to 30% and m_a = 0.86. From the results, it can seem that the efficiency is higher in all duty cycle. During the 10% D_{ST}, the inverter efficiency is about 95.68%. However, there is a small dip inefficiency at 20% and 30% duty cycle.

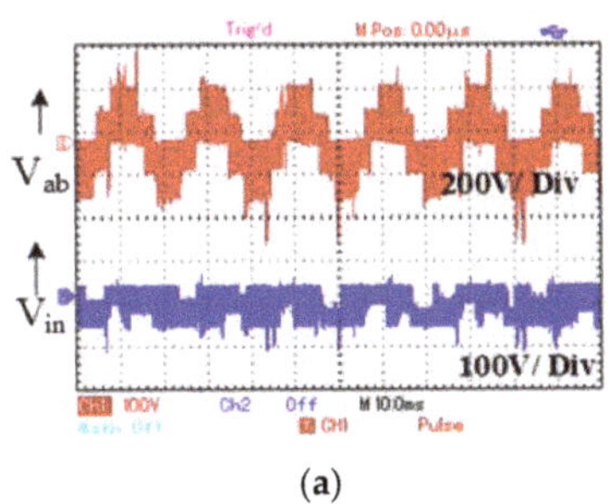

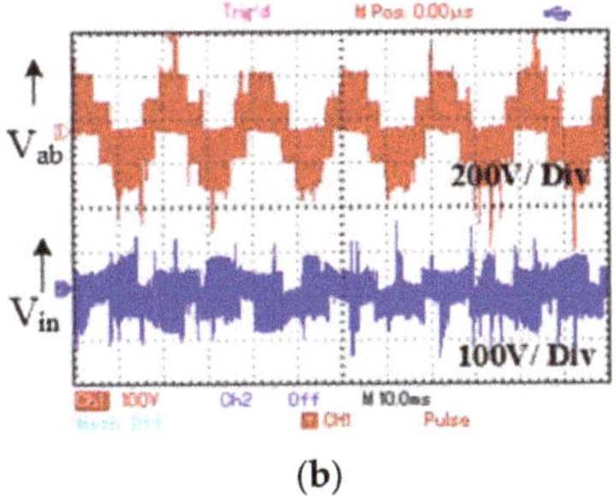

(a)   **(b)**

Figure 47. Experimental result; (**a**) quasi Z-source coupled MLI output voltage and input current at D_{ST} = 25%, (**b**) quasi Z-source coupled MLI output voltage and input current at D_{ST} = 30%.

Table 5. Switching loss, inverter efficiency, and THD of the proposed inverter for m_a = 0.86 and f_s = 10 kHz concerning D_{ST} from 10% to 30%.

D_{ST}	Switching Loss in Watts	Conduction Loss in Watts	Inverter Efficiency in %	THD in %(without Filter)	THD in %(with Filter)
10%	6.4	15.2	95.68%	13.9%	2.72%
20%	9.4	16.6	94.8%	14.11%	2.77%
30%	15.2	17.3	93.5%	19.55%	3.83%

The proposed inverter topology, in comparison with other reported topologies for the gain accomplishments, are deliberated in Table 6. In terms of passive element usage and maximum achievable voltage gain, the proposed topology is better than topology presented in [45–47]. By comparing the proposed topology with inverter proposed in [36], though the proposed inverter used one extra inductor, the voltage gain is high (3 times). Figure 48 illustrates the passive components rating, cost, boosting, and THD comparisons with other similar topologies [31,39,48]. The proposed inverter topology attains fewer passive elements usage, higher voltage gain conversion and better voltage THD than [31,39,48]. Thus, the proposed inverter topology presents its efficiency and suitability for PV standalone and grid-connected systems.

Table 6. Switching loss, inverter efficiency, and THD of the proposed inverter for m_a = 0.86 and f_s = 10 kHz concerning D_{ST} from 10% to 30%.

Topology Proposed in	Number of Passive Elements Used		Number of Switches Used	A Coupled Inductor or Transformer Type	Maximum Achievable Voltage Gain in %
	L	**C**			
[39]	2	2	6	Coupled inductor	≤ 2 times
[6]	2	Intergraded winding	8	NA	≤ 2 times
[45]	4	2	2 only for converter	Coupled inductor	≥ 2 times
[46]	4	4	8	NA	
[47]	4	4	8	NA	≤ 2 times
[48]	2	2	10	NA	≤ 2 times
Proposed QZ Inverter	3	2	6	Coupled inductor	≥ 2 times

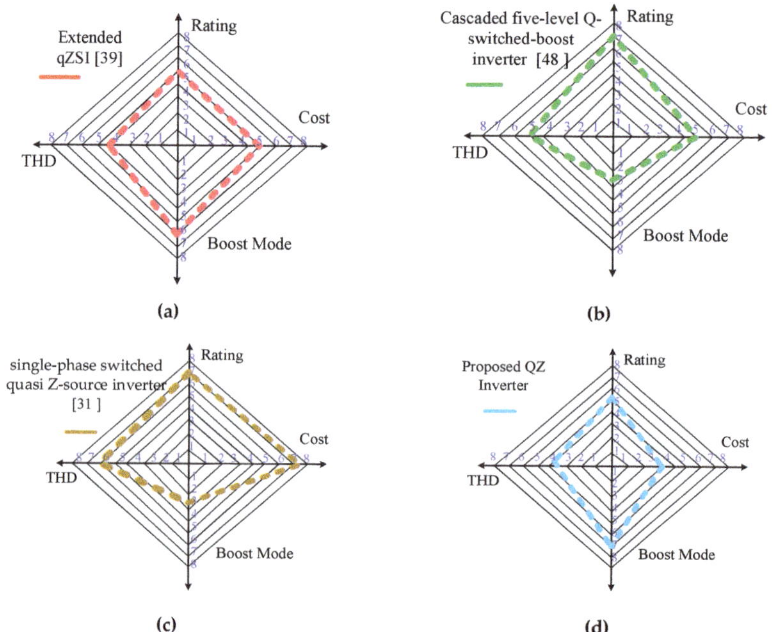

Figure 48. Passive components rating, cost, boosting, and THD comparisons; (**a**) Reference [39], (**b**), Reference [41], (**c**) Reference [39], (**d**) proposed QZ Inverter.

8. Conclusions

The proposed Q-source MLI coupled inverter ZSI is a combination of modified Q-source impedance network with six switches coupled inductor connected single-phase five-level MLI. The quasi-Z source coupled inductors MLI tied photovoltaic system with modified space vector PWM produces a maximum voltage gain of 140%. The suggested topology generates the five-level output voltage with the higher voltage gain (maximum voltage gains of 310%) with exceptionally low voltage and current THD. Besides, the proposed MLI reduces the switching stress on the inverter for all the duty cycles in the switching algorithm, when increasing the duty cycle, the boost factor also increases. The proposed quasi Q-ZSI has the merits of operation such as reliability, reduced passive components, voltage boosting capability and reduction in switching stress. The modified space vector PWM is proposed to integrate the boosting and regular inverter switching state within one sampling period.

In comparison with other MLI, it utilizes only half of the power switch, lower modulation index to acquire high voltage gain. The performance of the proposed boost MLI topology and control algorithm is theoretically investigated and validated through MATLAB/Simulink software and experimental upshots. The proposed topology is an attractive solution for stand-alone and grid-connected PV applications.

Author Contributions: M.P. and B.C. have developed the proposed research concept, and they both are involved in studying the execution and implementation with statistical software by collecting information from the real environment and developed the simulation model for the same. S.K.V., S.P., L.M.-P., Y.A. shared their expertise and validation examinations to confirm the concept theoretically with the obtained numerical results for its validation of the proposal. All authors are to frame the final version of the manuscript as a full. Moreover, all authors involved in validating and to make the article error-free technical outcome for the set investigation work.

Funding: No source of funding for this research activity.

Conflicts of Interest: The authors declare no conflict of interest.

References

1. Nayak, P.K.; Mahesh, S.; Snaith, H.J.; Cohen, D. Photovoltaic solar cell technologies: Analyzing state of the art. *Nat. Rev. Mater.* **2019**, *4*, 269–285. [CrossRef]

2. Heydari Gharahcheshmeh, M.; Tavakoli, M.M.; Gleason, E.F.; Robinson, M.T.; Kong, J.; Gleason, K.K. Tuning, optimization, and perovskite solar cell device integration of ultrathin poly (3,4-ethylene dioxythiophene) films via a single-step all-dry process. *Sci. Adv.* **2019**, *5*, 0414. [CrossRef] [PubMed]

3. Cui, Y.; Yao, H.; Zhang, J.; Zhang, T.; Wang, Y.; Hong, L.; Xian, K.; Xu, B.; Zhang, S.; Peng, J.; et al. Over 16% efficiency organic photovoltaic cells enabled by a chlorinated acceptor with increased open-circuit voltages. *Nat. Commun.* **2019**, *10*, 2515. [CrossRef] [PubMed]

4. Kavya Santhoshi, B.; Mohana Sundaram, K.; Padmanaban, S.; Holm-Nielsen, J.B. Critical Review of PV Grid-Tied Inverters. *Energies* **2019**, *12*, 1921. [CrossRef]

5. Ellabban; Abu-Rub, H. Z-source Inverter: Topology Improvements Review. *IEEE Ind. Electron. Mag.* **2016**, *10*, 6–24. [CrossRef]

6. Siwakoti, Y.P.; Peng, F.Z.; Blaabjerg, F.; Loh, P.C. Impedance-Source Networks for Electric Power Conversion Part I: A Topological Review. *IEEE Trans. Power Electron.* **2015**, *30*, 699–716. [CrossRef]

7. Peng, F.Z. Z-source inverter. *IEEE Trans. Ind. Appl.* **2003**, *39*, 504–510. [CrossRef]

8. Peng, F.Z.; Yuan, X.; Fang, X.; Qian, Z. Z-source inverter for adjustable speed drives. *IEEE Trans. Power Electron.* **2013**, *1*, 33–35. [CrossRef]

9. Peng, F.Z.; Joseph, A.; Wang, J.; Shen, M.; Chen, Z.L.; Pan, E.; Rivera, O.; Huang, Y. Z-source inverter for motor drives. *IEEE Trans. Power Electron.* **2005**, *20*, 857–863. [CrossRef]

10. Zhou, Z.J.; Zhang, X.; Xu, P.; Shen, W.X. Single-phase uninterruptible power supply based on Z-source inverter. *IEEE Trans. Ind.Electron.* **2008**, *55*, 2997–3004. [CrossRef]

11. Liu, J.; Jiang, S.; Cao, D.; Peng, F.Z. A digital current control of quasi-Z-source inverter with battery. *IEEE Trans. Ind. Informat.* **2013**, *9*, 928–936. [CrossRef]

12. Peng, F.Z.; Shen, M.; Holland, K. Application of Z-source inverter for traction drive of fuel cell-battery hybrid electric vehicles. *IEEE Trans. Power Electron.* **2007**, *22*, 1054–1061. [CrossRef]

13. Zhang, Y.; Shi, J.; Fu, C.; Zhang, W.; Wang, P.; Li, J. An Enhanced Hybrid Switching-Frequency Modulation Strategy for Fuel Cell Vehicle Three-Level DC-DC Converters with Quasi-Z Source. *Energies* **2018**, *11*, 1026. [CrossRef]
14. Loh, P.C.; Gao, F.; Blaabjerg, F.; Feng, S.Y.C.; Soon, N.J. Pulsewidth-modulated Z-source neutral-point-clamped inverter. *IEEE Trans. Ind. Appl.* **2007**, *43*, 1295–1308. [CrossRef]
15. Ge, B.; Li, Q.; Qian, W.; Peng, F.Z. A family of Z-source matrix converters. *IEEE Trans. Ind. Electron.* **2012**, *59*, 35–46. [CrossRef]
16. Vinnikov, D.; Roasto, I. Quasi-Z-source-based isolated DC/DC converters for distributed power generation. *IEEE Trans. Ind. Electron.* **2011**, *58*, 192–201. [CrossRef]
17. Siwakoti, Y.P.; Loh, P.C.; Blaabjerg, F.; Town, G. Y-source impedance network. *IEEE Trans. Power Electron.* **2014**, *29*, 3250–3254. [CrossRef]
18. Ge, B.; Abu-Rub, H.; Peng, F.Z.; Li, Q.; de Almeida, A.T.; Ferreira, F.J.T.E.; Sun, D.; Liu, Y. An energy stored quasi-Z-source inverter for application to photovoltaic power system. *IEEE Trans. Ind. Electron.* **2013**, *60*, 4468–4481. [CrossRef]
19. Yang, S.; Peng, F.Z.; Lei, Q.; Inoshita, R.; Qian, Z. Current-Fed Quasi-Z-Source Inverter with Voltage Buck. *IEEE Trans. Ind. Appl.* **2011**, *47*, 882–892. [CrossRef]
20. Chub, A.; Vinnikov, D.; Liivik, E.; Jalakas, T. Multiphase Quasi-Z-Source DC–DC Converters for Residential Distributed Generation Systems. *IEEE Trans. Ind. Electron.* **2018**, *65*, 8361–8371. [CrossRef]
21. Lei, Q.; Peng, F.Z. Space Vector Pulse width Amplitude Modulation for a Buck–Boost Voltage/Current Source Inverter. *IEEE Trans. Power Electron.* **2014**, *29*, 266–274. [CrossRef]
22. Bharatiraja, C.; Jeevananthan, S.; Munda, J.L.; Latha, R. Improved SVPWM vector selection approaches in OVM region to reduce common-mode voltage for three-level neutral point clamped inverter. *Int. J. Electron. Power Energy Syst.* **2016**, *79*, 285–297. [CrossRef]
23. Manjrekar, M.D.; Steimer, P.K.; Lipo, T.A. Hybrid multilevel power conversion system: A competitive solution for high-power applications. *IEEE Trans. Ind. Appl.* **2000**, *36*, 834–841. [CrossRef]
24. Barbosa, P.; Steimer, P.; Steinke, J.; Meysenc, L.; Winkelnkemper, M.; Celanovic, N. Active neutral-point-clamped multilevel converters. In Proceedings of the Power Electronics Specialists Conference, Recife, Brazil, 16 June 2005; pp. 2296–2301.
25. Malinowski, M.; Gopakumar, K.; Rodriguez, J.; Perez, M.A. A survey on cascaded multilevel inverters. *IEEE Trans. Ind. Electron.* **2010**, *57*, 2197–2206. [CrossRef]
26. Ruiz-Caballero, D.A.; Ramos-Astudillo, R.M.; Mussa, S.A.; Heldwein, M.L. Symmetrical hybrid multilevel DC-AC converters with reduced number of insulated dc supplies. *IEEE Trans. Ind. Electron.* **2010**, *57*, 2307–2314. [CrossRef]
27. Abdelhakim, A.; Blaabjerg, F.; Mattavelli, P. Modulation Schemes of the Three-Phase Impedance Source Inverters—Part I: Classification and Review. *IEEE Trans. Ind. Electron.* **2018**, *65*, 6309–6320. [CrossRef]
28. Salmon, J.; Ewanchuk, J.; Knight, A.M. PWM inverters using split-wound coupled inductors. *IEEE Trans. Ind. Appl.* **2009**, *45*, 2001–2009. [CrossRef]
29. Ueda, F.; Matsui, K.; Asao, M.; Tsuboi, K. Parallel-connections of pulse width modulated inverters using current sharing reactors. *IEEE Trans. Power Electron.* **1995**, *10*, 673–679. [CrossRef]
30. Bharatiraja, C.; Padmanaban, S.; Siano, P.; Leonowicz, Z.; Iqbal, A. A hexagonal hysteresis space vector current controller for single Z-source network multilevel inverter with capacitor balancing. In Proceedings of the IEEE International Conference on Environment and Electrical Engineering and 2017 IEEE Industrial and Commercial Power Systems Europe, Milan, Italy, 6–9 June 2017.
31. Nguyen, M.K.; Lim, Y.C.; Park, S.J. Improved trans-Z-source inverter with continuous input current and boost inversion capability. *IEEE Trans. Power Electron.* **2010**, *28*, 4500–4510. [CrossRef]
32. Bharatiraja, C.; Sanjeevikumar, P.; Mahesh; Swathimala, A.S.; Raghu, S. Analysis, design and investigation on a new single-phase switched quasi Z-source inverter for photovoltaic application. *Int. J. Power Electron. Drive Syst.* **2017**, *8*, 853–860.
33. Banaei, M.R.; Dehghanzadeh, A.R.; Salary, E.; Khounjahan, H.; Alizadeh, R. Z-source-based multilevel inverter with reduction of switches. *IET Power Electron.* **2011**, *5*, 385–392. [CrossRef]
34. Battiston, A.; Miliani, E.-H.; Pierfederici, S.; Meibody-Tabar, F. A novel quasi-z-source inverter topology with special coupled inductors for input current ripples cancellation. *IEEE Trans. Power Electron.* **2015**, *31*, 2409–2416. [CrossRef]

35. Li, D.; Loh, P.C.; Zhu, M.; Gao, F.; Blaabjerg, F. Cascaded multicell trans-Z-source inverters. *IEEE Trans. Power Electron.* **2013**, *28*, 826–835.

36. Qin, L.; Dong, C.; Fang, Z.P. Novel loss and harmonic minimized vector modulation for a current-fed quasi-z-source inverter in HEV motor drive application. *IEEE Trans. Power Electron.* **2013**, *29*, 1344–1357. [CrossRef]

37. Bharatiraja, C.; Jeevananthan, S.; Latha, R. Vector Selection Approach-based Hexagonal Hysteresis Space Vector Current Controller for a Three-phase Diode Clamped MLI with Capacitor Voltage Balancing. *IET Power Electron.* **2016**, *9*, 1350–1361. [CrossRef]

38. Bharatiraja, C.; Munda, J.L. Simplified SVPWM for Z Source T-NPC-MLI Including Neutral Point Balancing. In Proceedings of the 4th IEEE Symposium on Computer Applications & Industrial Electronics (ISCAIE2016), Penang, Malaysia, 30–31 May 2016.

39. Banaei, M.R.; Dehghanzadeh, A.; Oskouei, A.B. Extended switching algorithms based space vector control for five-level quasi-Z-source inverter with coupled inductors. *IET Power Electron.* **2014**, *7*, 1509–1518. [CrossRef]

40. Kumar Gupta, A.; Khambadkone, A.M. A space vector modulation scheme to reduce common mode voltage for cascaded multilevel inverters. *IEEE Trans. Ind. Electron.* **2007**, *22*, 1672–1681.

41. Li, Q.; Wolfs, P.A. review of the single phase photovoltaic module integrated converter topologies with three different dc link configuration. *IEEE Trans. Power Electron.* **2008**, *23*, 1320–1333.

42. Liang, Y.; Liu, B.G.; Abu-Rub, H. Investigation on pulse-width amplitude modulation-based single-phase quasi-Z-source photovoltaic inverter. *IET Power Electron.* **2017**, *10*, 1810–1818. [CrossRef]

43. Tariq, M.; Iqbal, M.T.; Meraj, M.; Iqbal, A.; Maswood, A.I.; Bharatiraja, C. Design of a proportional resonant controller for packed U cell 5 level inverter for grid-connected applications. In Proceedings of the 2016 IEEE International Conference on Power Electronics, Drives and Energy Systems (PEDES), Trivandrum, India, 14–17 December 2016.

44. Kitson, J.; McNeill, N. A High Efficiency Voltage-Fed Quasi Z-Source Inverter with Discontinuous Input Current Using Super-Junction MOSFETs. In Proceedings of the 8th IET International Conference on Power Electronics, Machines and Drives (PEMD), Glasgow, UK, 19–21 April 2016.

45. Liu, H.; Li, F.A. Novel High Step-up Converter with a Quasi Active Switched-Inductor Structure for Renewable Energy Systems. *IEEE Trans. Power Electron.* **2016**, *31*, 5030–5039. [CrossRef]

46. Shults, T.E.; Husev, O.; Blaabjerg, F.; Roncero-Clemente, C.; Romero-Cadaval, E.; Vinnikov, D. Novel Space Vector Pulsewidth Modulation Strategies for Single-Phase Three-Level NPC Impedance-Source Inverters. *IEEE Trans. Power Electron.* **2019**, *34*, 4820–4830. [CrossRef]

47. Ho, A.V.; Chun, T.W. A Buck-Boost Multilevel Inverter for PV Systems in Smart Cities. Industrial Networks and Intelligent Systems. *Springer Int. Pub.* **2018**, 295–306. [CrossRef]

48. Tran, T.T.; Nguyen, M.K. Cascaded five-level quasi-switched-boost inverter for single-phase grid-connected system. *IET Power Electron.* **2017**, *10*, 1896–1903. [CrossRef]

49. Aly, M.; Ahmed, E.M.; Shoyama, M. Thermal Stresses Relief Carrier-Based PWM Strategy for Single-Phase Multilevel Inverters. *IEEE Trans. Power Electron.* **2017**, *32*, 9376–9388. [CrossRef]

Article

H$_\infty$ Mixed Sensitivity Control for a Three-Port Converter

Jiang You [1], Hongsheng Liu [1], Bin Fu [2,*] and Xingyan Xiong [2]

[1] Electrical Engineering Institute, College of Automation, Harbin Engineering University, Harbin 150001, China; youjiang@hrbeu.edu.cn (J.Y.); liuhongsheng@hrbeu.edu.cn (H.L.)
[2] Department of Robot Engineering, School of Light Industry, Harbin University of Commerce, Harbin 150001, China; xiongxingyan2019@163.com
* Correspondence: fubinhcu@hotmail.com; Tel.: +86-137-9665-8929

Received: 8 May 2019; Accepted: 4 June 2019; Published: 12 June 2019

Abstract: The three-port converter (TPC) obtains major attention due to its power density and ability to dispose different electric powers flexibly. Since the control models of the TPC are derived from particular steady state work point through small signal modeling method, the model parameters usually be deviated from their normal values with the change of operation and load conditions. Furthermore, there are couplings and interactions in power delivery between different ports, which have a significant influence in the dynamic control performance of the system. In this paper, the H$_\infty$ mixed sensitivity method is employed to design robust controllers for a TPC control system. Simulation results are given to demonstrate the effectiveness of the proposed scheme, and experimental studies are conducted on a prototype circuit to further validate the developed method. Compared to a traditional PI controller, it shows that a mixed sensitivity based robust controller manifest balanced performance in model parameters changes attenuation and dynamic control performance.

Keywords: three-port converter; parameters change; mixed sensitivity; H$_\infty$ control

1. Introduction

Sustainable energy generation, such as photovoltaics, wind and fuel cells, etc. have been widely investigated in the last decade. As shown in Figure 1, various energy sources can be interfaced to a DC microgrid or a common DC bus for direct utilization or further conversion through their separated power converters. These converters are linked together at the DC bus and controlled independently. In some systems, a communication bus might be included to transmit information and instruction for power management between different power conversion subsystems [1]. However, this structure has drawbacks in complexity, cost and power density due to utilization of a number of different power converters (the amount of power converters depends on the size of the whole power conversion system) and communication devices between individual subsystems. With the development of power electronics, the demand for light and compact power converters for renewable generation and industrial applications has been increasing steadily. The configuration of a DC microgrid using a multi-input converter in renewable energy generation system is shown in Figure 2, compared to the conventional structure presented in Figure 1. The power conversions for different energy sources are integrated into a single power converter which is denoted as a multi-port converter. This multi-input topology for combining diverse power sources can be a non-isolated direct connection [2–4] or an isolated magnetic coupling [5–7].

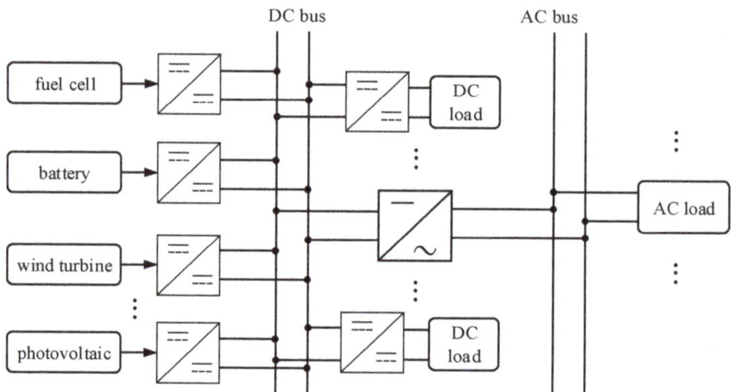

Figure 1. Conventional structure of a DC microgrid.

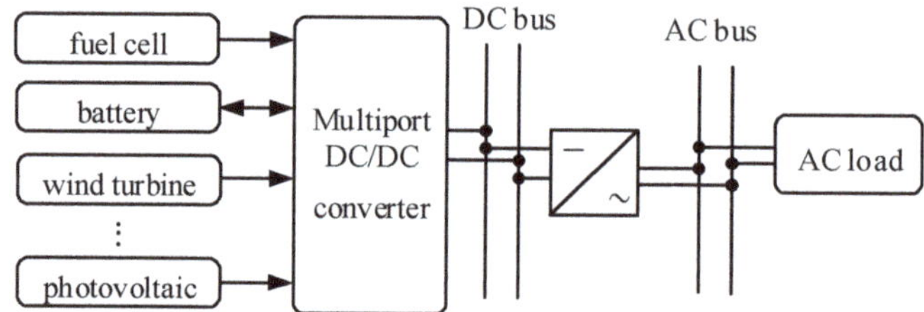

Figure 2. Multiport DC/DC converter based structure of a DC microgrid.

An isolated three-port converter is one of the recently developed multi-port power converters which have become attractive due to their compact structure, power density and flexibility in power conversion. Since the three windings of the high frequency transformer (HFT) share a common magnetic core, interaction and coupling of power delivery among the three different ports are unavoidable, and it is necessary to reduce the interactions between different ports through the reasonable control method. The decoupling control method is usually used in three-port converter control, while two single-input single-output (SISO) subsystems can be obtained by introducing appropriate decoupling compensations [8–11], and frequency control theory can be utilized to design controllers for each subsystem, respectively. However, the isolated three-port converter is a multiple-input multiple-output (MIMO) system, several phase-shifting angles and equivalent duty cycles can be used as control variables, and several voltages and currents of different ports can be used as output variables. Therefore, from the view of a MIMO system, a linear quadratic regulator (LQR) based method for three-port converter controller design is proposed in reference [12]. Theoretically, it seems that the LQR method has the capability to achieve a balanced control performance for different ports, however, it has relatively high sensitivity to the accuracy of system parameters (while the small signal models used for control system design are derived at a specific steady state operation point, and the parameters of the models will be varied with the change of operation point), moreover, the parameters design of the time domain based LQR method is relatively complex, and compared to traditional frequency domain design method, LQR method lacks physical meaning.

The topology of a typical full-bridge isolated three-port converter is shown in Figure 3a. For control system design, the linear small signal model can be derived by calculating the partial differential of current in each port, then two independent SISO subsystems can be obtained by feedforward decoupling compensation, and then classical frequency domain control theory can be adopted for control system design [13–15]. However, for those cases without considering the inductor, L_{d1}, the double switching frequency component in i_{d1} has a negative impact on the power source (e.g., fuel cell and photovoltaic panel). Therefore, an LC circuit is utilized to suppress the high frequency current ripple. However, this can deteriorate the current control performance and can even cause a stability issue due to the resonant peak introduced by the LC circuit [16]. Although the negative impact of resonant issue can be relieved by decreasing the current control bandwidth, the desirable control performance of the system cannot be guaranteed.

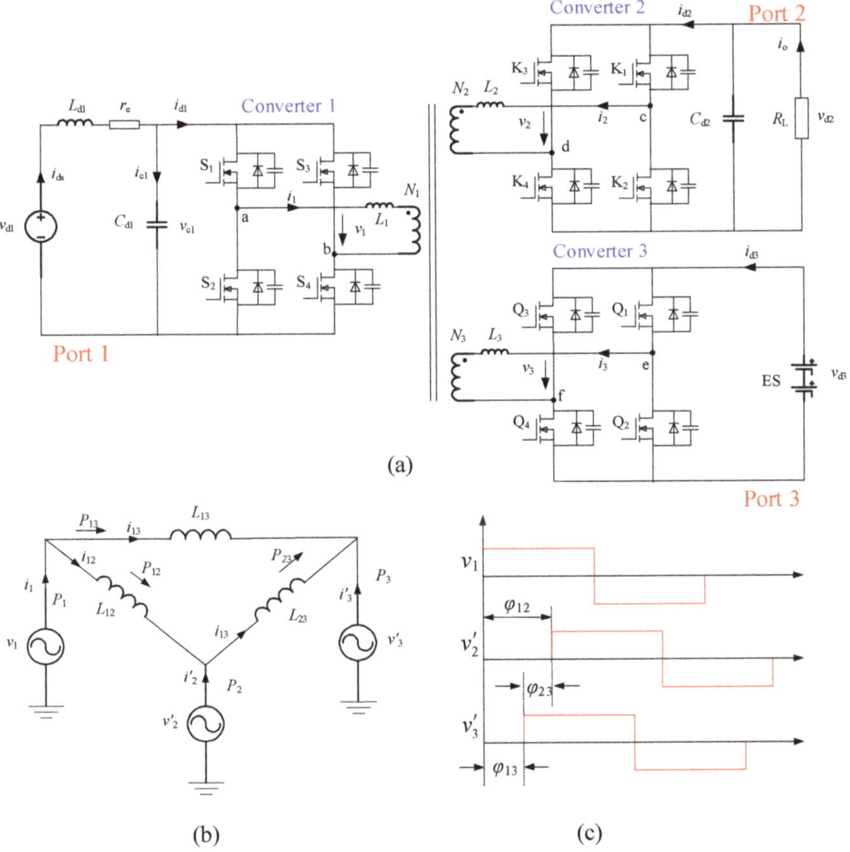

Figure 3. Topology, equivalent circuit and modulation scheme of isolated three-port converter (a) topology of the isolated three-port converter, (b) equivalent Δ-connection circuit, (c) modulation scheme.

The H$_\infty$ mixed sensitivity control is employed in this paper to address the aforementioned issues. This controller design method has prominent characteristic to balance control performance (e.g., disturbance rejection) and stability with consideration of model parameter deviations. In this method, the performance requirements and parameter uncertainties can be taken into account by designing appropriate weight functions with particular amplitude-frequency characteristic. Compared to the TPC control system designed by a traditional SISO based frequency method, the advantages of the proposed method mainly lie in two aspects, (1) the corrected TPC control system using H$_\infty$ mixed sensitivity method has stronger abilities in load disturbance rejection with operation point change and model parameters variation. (2) Furthermore, the resonant peak of the corrected current control subsystem of TPC can be attenuated effectively, which is very helpful for improving the stability and dynamic performance of the current control system.

The rest of the paper is organized as follows. In Section 2, the topology, modulation scheme, power delivery relationship, and control-oriented small signal models are presented. H∞ mixed sensitivity method and design results along with analysis are given in Section 3, a brief illustration of the conventional decoupling control method is also provided in this section for comparison purpose. The simulation and experiment results are presented in Section 4. Finally, the conclusion is drawn in Section 5.

2. Modeling of Isolated TPC

2.1. Power Delivery

The topology of an isolated TPC is shown in Figure 3a. As shown in this figure, the Port 1 is connected to a DC power source, such as a photovoltaic panel or a fuel cell. The load is supplied by Port 2, the batteries connected to Port 3 are used as energy storage (ES). L_{d1} and C_{d1} form an LC circuit to reduce the double switching frequency component in i_{d1}. C_{d2} is used to smooth v_{d2} at Port2. The number of turns in the three windings of the high frequency transformer are N_1, N_2 and N_3, respectively. L_1, L_2 and L_3 are the equivalent series inductances (including winding leakage and additional inductances) of the three transformer windings respectively. v_{d1}, v_{d2} and v_{d3} are the voltages corresponding to the three ports. S_1-S_4, K_1-K_4 and Q_1-Q_4 are power switches of the full bridge converters in the three different ports. Taking the converter of Port 1 (converter 1) as an example, switches S_1 and S_3 are complementary to the switches S_2 and S_4 respectively and their duty cycles are all 50%, the phase shifting between legs a and b is 180°. The switching patterns of the switching devices in converter 2 (the converter in Port 2) and converter 3 (the converter in Port 3) and the phase shifting between their two legs are identical to those of converter 1.

By transferring the parameters of converter 2 and converter 3 to converter 1, the simplified equivalent Δ-circuit of the TPC can be depicted as Figure 3b. In this figure, v_2', v_3', i_2' and i_3' are the transferred voltages and currents of Port 2 and Port 3 respectively. The expressions of L_{12}, L_{13} and L_{23} are given in (1).

$$\begin{cases} L_{12} = L_1 + L_2' + \dfrac{L_1 L_2'}{L_3'} \\[2mm] L_{23} = L_2' + L_3' + \dfrac{L_2' L_3'}{L_1} \\[2mm] L_{13} = L_3 + L_1 + \dfrac{L_1 L_3'}{L_2'} \end{cases} \tag{1}$$

in (1)

$$\begin{cases} L_2' = \dfrac{N_1^2 L_2}{N_2^2} \\[2mm] L_3' = \dfrac{N_1^2 L_3}{N_3^2} \end{cases} \tag{2}$$

Taking v_1 as reference, the relationships of v_1, v_2' and v_3' are shown in Figure 3c. The phases shifting between v_1 and v_2', v_1 and v_3' are ϕ_{12} and ϕ_{13} respectively, and the phase shifting between v_2' and v_3' is ϕ_{23}. By regulating ϕ_{12}, ϕ_{13} and ϕ_{23}, the direction and values of power transfer between different ports can be controlled.

According to Figure 3b, the powers of Port 1, Port 2 and Port 3 can be expressed as (3).

$$\begin{cases} P_1 = P_{12} + P_{13} \\ P_2 = P_{21} + P_{23} \\ P_3 = P_{31} + P_{32} \end{cases} \tag{3}$$

In (3), P_1, P_2 and P_3 are the powers of Port 1, Port 2 and Port 3, respectively. P_{12} is the power transferred from Port 1 to Port 2, while P_{21} is the power transferred from Port 2 to Port 1, and $P_{12} = -P_{21}$

is always held. The relationships between P_{13} and P_{31}, P_{23} and P_{32} are $P_{13} = -P_{31}$ and $P_{23} = -P_{32}$. The power equation for P_{12} and P_{21} can be written as (4). Since the fundamental power is close to the total power in each switching period, the fundamental power is used to formulate the power delivery model of TPC in this paper.

According to Figure 3c and Fourier series expansion, the fundamental phasor of v_1, v_2' and v_3' can be formulated as (4).

$$\begin{cases} \dot{v}_{1f} = \dfrac{4V_1}{\pi\sqrt{2}}\angle 0 = V_{1f}\angle 0 \\[2mm] \dot{v}_{2f} = \dfrac{4N_1 V_2}{\pi N_2 \sqrt{2}}\angle -\varphi_{12} = V_{2f}'\angle -\varphi_{12} \\[2mm] \dot{v}_{3f} = \dfrac{4N_1 V_3}{\pi N_3 \sqrt{2}}\angle -\varphi_{13} = V_{3f}'\angle -\varphi_{13} \end{cases} \tag{4}$$

V_1, V_2 and V_2 are the amplitudes of v_1, v_2' and v_3' respectively.

Neglecting the influence of Port 3, Port1 and Port2 behave like a dual active bridge (DAB), the equivalent simplified circuit is shown in Figure 4.

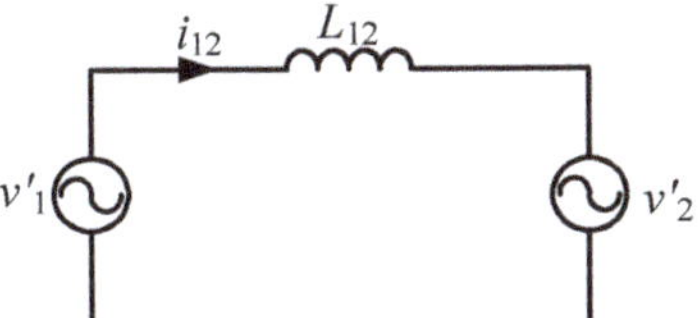

Figure 4. Equivalent circuit model of a dual active bridge.

The current, i_{12} can be expressed as (5).

$$i_{12} = \frac{V_{2f}' \sin\varphi_{12}}{\omega L_{12}} - \frac{j(V_{1f} - V_{2f}' \cos\varphi_{12})}{\omega L_{12}} \tag{5}$$

Combining (5) and considering the power factor angle, α, the fundamental power can be written as (6).

$$P_{12} = -P_{21} = V_{1f}i_{12}\cos\alpha = \frac{N_1 V_{1f} V_{2f} \sin\varphi_{12}}{N_2 \omega L_{12}} \tag{6}$$

Similarly, the power transfer equations of Port 2 and Port 3 are given in (7)

$$\begin{cases} P_{13} = -P_{31} = \dfrac{N_1 V_{1f} V_{3f} \sin\varphi_{13}}{N_3 \omega L_{13}} \\[3mm] P_{23} = -P_{32} = \dfrac{N_1^2 V_{2f} V_{3f} \sin(\varphi_{13} - \varphi_{12})}{N_2 N_3 \omega L_{23}} \end{cases} \tag{7}$$

From (3), the power of Port 3 can be derived as $P_3 = -(P_1 + P_2)$, it indicates that the power of ES port is determined by the power of Port 1 and Port 2. Therefore Port 3 can be set as a free port. Combining (3) with (6) and (7), the powers of Port 1 and Port 2 can be represented as (8).

$$\begin{cases} P_1 = \dfrac{N_1 V_{1f} V_{2f} \sin\varphi_{12}}{N_2 \omega L_{12}} + \dfrac{N_1 V_{1f} V_{3f} \sin\varphi_{13}}{N_3 \omega L_{13}} \\[3mm] P_2 = -\dfrac{N_1 V_{1f} V_{2f} \sin\varphi_{12}}{N_2 \omega L_{12}} + \dfrac{N_1^2 V_{2f} V_{3f} \sin(\varphi_{13} - \varphi_{12})}{N_2 N_3 \omega L_{23}} \end{cases} \tag{8}$$

2.2. Small Signal Model of TPC

According to (8), the average values of i_{d1} and i_{d2} are written as (9)

$$\begin{cases} \bar{i}_{d1} = \dfrac{N_1 V_{2f} \sin \varphi_{12}}{N_2 \omega L_{12}} + \dfrac{N_1 V_{3f} \sin \varphi_{13}}{N_3 \omega L_{13}} \\ \bar{i}_{d2} = -\dfrac{N_1 V_{1f} \sin \varphi_{12}}{N_2 \omega L_{12}} + \dfrac{N_1^2 V_{2f} \sin(\varphi_{13} - \varphi_{12})}{N_2 N_3 \omega L_{23}} \end{cases} \tag{9}$$

The small signal disturbance of $\bar{i}_{d1}$ and $\bar{i}_{d2}$ can be obtained by calculating the partial derivatives of $\bar{i}_{d1}$ and $\bar{i}_{d2}$ at a steady state work point A (ϕ_{120}, ϕ_{130}) in (9), the result is shown in (10).

$$\begin{bmatrix} \hat{i}_{d2} \\ \hat{i}_{d1} \end{bmatrix} = \begin{bmatrix} G_{11} & G_{12} \\ G_{21} & G_{22} \end{bmatrix} \begin{bmatrix} \hat{\phi}_{12} \\ \hat{\phi}_{13} \end{bmatrix} = G_A \begin{bmatrix} \hat{\phi}_{12} \\ \hat{\phi}_{13} \end{bmatrix} \tag{10}$$

where

$$\begin{cases} G_{11} = -\dfrac{N_1 V_{1f} \cos \varphi_{120}}{N_2 \omega L_{12}} - \dfrac{N_1^2 V_{2f} \cos(\varphi_{130} - \varphi_{120})}{N_2 N_3 \omega L_{23}} \\ G_{12} = \dfrac{N_1^2 V_{2f} \cos(\varphi_{130} - \varphi_{120})}{N_2 N_3 \omega L_{12}} \\ G_{21} = \dfrac{N_1 V_{2f} \cos \varphi_{120}}{N_2 \omega L_{12}} \\ G_{22} = \dfrac{N_1 V_{3f} \cos \varphi_{130}}{N_3 \omega L_{13}} \end{cases} \tag{11}$$

In (10), it can be seen that there are cross couplings between $\hat{i}_{d2}$ and $\hat{\phi}_{13}$ and also between $\hat{i}_{d1}$ and $\hat{\phi}_{12}$ which are caused by G_{12} and G_{21} respectively. Therefore, decoupling is needed to improve dynamic control performance of the three-port converter. The small signal model diagram with decoupling compensations is shown in Figure 5. In this figure, the feedforward decoupling terms H_{12} and H_{21} are given in (12).

$$\begin{cases} H_{12} = -\dfrac{G_{12}}{G_{11}} \\ H_{21} = -\dfrac{G_{21}}{G_{22}} \end{cases} \tag{12}$$

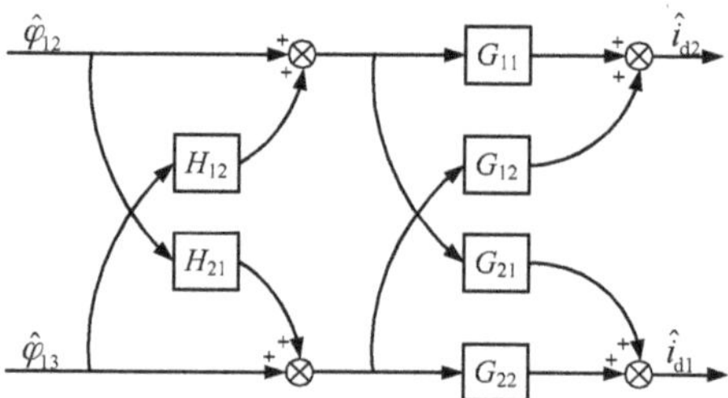

Figure 5. Decoupled model.

Therefore, the small signal model for currents $\hat{i}_{d1}$ and $\hat{i}_{d2}$ with decoupling compensations can be simplified as in (13)

$$\begin{cases} \hat{i}_{d2} = G_{11} \hat{\phi}_{12} \\ \hat{i}_{d1} = G_{22} \hat{\phi}_{13} \end{cases} \tag{13}$$

By utilizing Kirchhoff's circuit laws and (13), the small signal differential equation group that represents the dynamic behavior of Port1 and Port2 can be obtained as (14).

$$\begin{cases} \dfrac{d\hat{v}_{d2}}{dt} = -\dfrac{1}{R_L C_{d2}}\hat{v}_{d2} - \dfrac{G_{11}}{C_{d2}}\hat{\varphi}_{12} \\ \dfrac{d\hat{i}_{ds}}{dt} = \dfrac{1}{L_{d1}}\hat{v}_{d1} - \dfrac{1}{L_{d1}}\hat{v}_{c1} - \dfrac{r_e}{L_{d1}}\hat{i}_{ds} \\ \dfrac{d\hat{v}_{c1}}{dt} = \dfrac{1}{C_{d1}}\hat{i}_{ds} - \dfrac{G_{22}}{C_{d1}}\hat{\varphi}_{13} \end{cases} \tag{14}$$

The corresponding state space model can be written as (15)

$$\begin{cases} \dot{x} = Ax + Bu \\ y = Cx \end{cases} \tag{15}$$

where x, u and y are state vector, input vector and output vector respectively shown in (16).

$$\begin{cases} x^T = [\hat{v}_{d2}\ \hat{i}_{ds}\ \hat{v}_{c1}]^T \\ u^T = [\hat{\varphi}_{12}\ \hat{\varphi}_{13}]^T \\ y^T = [x_1\ x_2]^T \end{cases} \tag{16}$$

And the coefficient matrices A, B and C are given in (17).

$$A = \begin{bmatrix} -\dfrac{1}{C_{d2}R_L} & 0 & 0 \\ 0 & -\dfrac{r_e}{L_{d1}} & -\dfrac{1}{L_{d1}} \\ 0 & \dfrac{1}{C_{d1}} & 0 \end{bmatrix},\ B = \begin{bmatrix} -\dfrac{G_{11}}{C_{d2}} & 0 \\ 0 & 0 \\ 0 & -\dfrac{G_{22}}{C_{d1}} \end{bmatrix},\ C = \begin{bmatrix} 1 & 0 & 0 \\ 0 & 1 & 0 \end{bmatrix},\ D = 0 \tag{17}$$

The objectives of the control system are to obtain the desired current, i_{ds} at Port 1 (for example, the output current of a photovoltaic panel with maximum power point tracking), and stabilize the output voltage, v_{d2} of Port 2. The energy storage port, Port 3 works as a free port and the batteries are charged or discharged automatically depending on the power exchange between Port 1 and Port 2. The control block diagram of the three-port converter is shown in Figure 6. In this figure, the phase shifting, ϕ_{12} is used to control the Port 2 voltage, v_{d2}, while the phase shifting, ϕ_{13} is used to control the Port 1 current, i_{ds}. In dynamic situations (e.g., with load or reference signal change), the transient operating point may deviate greatly from the designed steady-state operating point, which might discount the control performance significantly, therefore, both ϕ_{12} and ϕ_{13} should be limited between 0 and $\pi/2$.

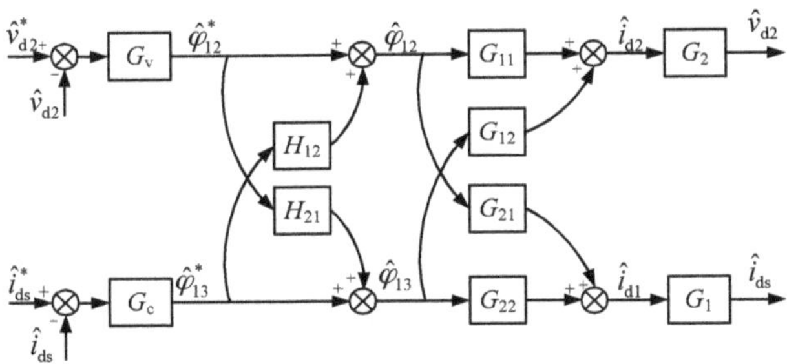

Figure 6. Control diagram block of the TPC system.

According to this figure and (14), the opened loop transfer functions of the decoupled voltage and current subsystems are shown in (18) and (19) respectively.

$$G_{ov} = \frac{G_{11}R_L}{C_{d2}R_L s + 1} \tag{18}$$

$$G_{oc} = \frac{G_{22}}{L_{d1}C_{d1}s^2 + r_e C_{d1}s + 1} \tag{19}$$

3. Controller Design of TPC

The general design procedure using H_∞ mixed sensitivity method is shown in Figure 7. The control plant modeling of TPC is derived in Section 2. Weight functions selection, which is an important step, will be discussed in this section. This selection has a significant influence on control performance and some instances is used to illustrate the design method. As shown in Figure 7, the step named "Generating an augmented LTI (linear time-invariant) plant" is used to create a state space model of augmented control plant with weight functions which is utilized for an H_∞ controller design (this task can be completed using the "augw" function in Matlab Robust Control Toolbox). The final step shown in Figure 5 is to solve the H_∞ controller using state space method [17]. Since the last two steps are relatively complex and mathematic, the details of the steps will not be discussed further in this section.

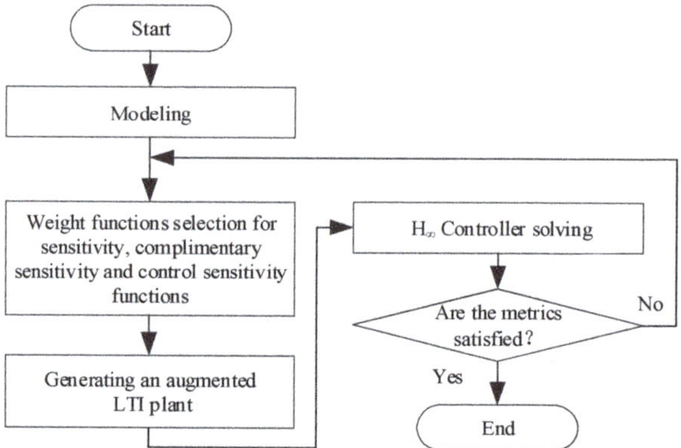

Figure 7. Design procedure of H_∞ mixed sensitivity based controller.

3.1. Fundamental of H_∞ Mixed Sensitivity Design

The standard model for H_∞ mixed sensitivity design is shown in Figure 8 [18]. W_1, W_2 and W_3 are weight function matrices, z_1, z_2 and z_3 are performance evaluation signal vectors, d is disturbance vector. r, e, u and y are reference vector, error vector, control signal vector and output vector respectively. K is the controller matrix and G is the control plant matrix.

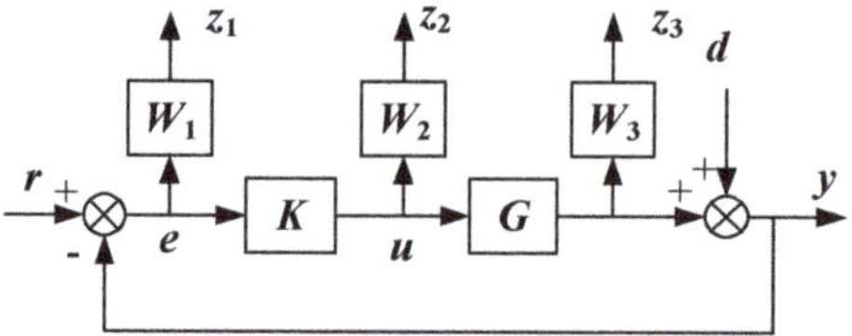

Figure 8. H_∞ mixed sensitivity standard design model.

The augmented control plant (20) can be obtained according to Figure 8.

$$\begin{bmatrix} z_1 \\ z_2 \\ z_3 \\ \hline e \end{bmatrix} = \begin{bmatrix} W_1 e \\ W_2 u \\ W_3 y \\ \hline e \end{bmatrix} = \begin{bmatrix} W_1 & -W_1 G \\ 0 & W_2 \\ 0 & W_3 G \\ \hline I & -G \end{bmatrix} \begin{bmatrix} r \\ u \end{bmatrix} \tag{20}$$

Weight functions selection is the most important step in H$_\infty$ mixed sensitivity design process, which will determine the control performance, such as disturbance attenuation, robustness and dynamic response ability.

A generalized closed loop transfer function matrix in (21) can be obtained by substituting $u = Ke$ into (20)

$$P = \begin{bmatrix} W_1 S \\ W_2 R \\ W_3 T \end{bmatrix} \tag{21}$$

where $S = (I + GK)^{-1}$ is called sensitivity function matrix, it is the transfer function matrix from d to e. $T = GK(I + GK)^{-1}$ represents transfer function matrix from u to y. Since $S + T = I$, T is called complimentary sensitivity function matrix, $R = K(I + GK)^{-1}$ is the transfer function matrix from e to u. In (21) $W_1 S$ is used to represents control performance requirements for disturbance rejection, and this performance metric can be designed by selecting appropriate W_1. The strength or the effectiveness of the control signal, u is restricted by W_2. $W_3 T$ represents requirements for robust stability, W_3 reflects the design constraints for multiplicative model uncertainty, it depends on the parameter deviations of control plants. In the frame of H$_\infty$ mixed sensitivity control, the performance requirements for disturbance rejection and robust stability are interactive, however, performance balance (or tradeoff) can be obtained by proper weight functions design. Generally, the following factors should be considered in weight functions selection (scalars are used in following text for ease statement).

(1) Considerations of W_1 Selection. W_1 represents the performance metric of the control system for disturbance rejection. For the sensitivity function matrix, S denotes the relationship between tracking error e and external disturbance d, while W_1^* influences the tracking performance. It is desired that W_1 has a high gain in low frequency to reduce steady state error. And a steep declining slope of W_1 in high frequency is required for interference attenuation. Therefore, W_1 is usually selected as a high gain first order transfer function. Furthermore, the crossover frequency, f_{w1} of W_1 should be lower than the desired crossover frequency of the corrected control subsystem.

(2) Considerations of W_2 Selection. The strength or effectiveness of control signal u in Figure 8 can be limited by W_2, which is beneficial for keeping u in its allowable range, therefore controller saturation and overshooting can be effectively avoided. The amplitude of u will be reduced if the gain of W_2 is increased. The gain of W_2 can be rationally high according to the required control performance. The bandwidth of control system can be influenced by W_2, the control bandwidth will be reduced if the gain of W_2 is increased and vice versa. Therefore, W_2 should be appropriately designed by taking into account the effectiveness of the control signal and control bandwidth requirement. In order to avoid high order of the resulted controller, W_2 is often selected as a constant in practices.

(3) Considerations of W_3 Selection. W_3 is selected as a metric for multiplicative perturbation. Generally, the nominal transfer function can used to represent the characteristics of control plant accurately in low frequency, while the accuracy will be degraded in high frequency range, deviations in gain and phase will be resulted accordingly. This type of deviation can be expressed as multiplicative uncertainty, which is usually used to describe parameter uncertainty and high

frequency unmodeled dynamic of the system. Multiplicative uncertainty, $\Delta(s)$ can be obtained by solving (22).

$$\Delta(s) = \frac{G_\Delta(s) - G_0(s)}{G_0(s)} = \frac{G_\Delta(s)}{G_0(s)} - 1 \tag{22}$$

In (22), G_0 and G_Δ are nominal transfer function and practice transfer function, respectively.

In a mixed sensitivity design method, the gain W_3 should be designed to guarantee that $\Delta(s)$ in (22) is properly covered by W_3 (as shown in Figures 9 and 10). W_3 always has a high pass characteristic to make sure that the corrected control system has a favorable performance to attenuate high frequency disturbance. And the crossover frequency W_3, f_{w3} should be higher than the desired crossover frequency of the corrected control subsystem. Taking into account the crossover frequency of W_1, the crossover frequency of the corrected control subsystem will be located between f_{w1} and f_{w3}. More systematic and detailed discussions about weight functions selection can be found in references [19,20].

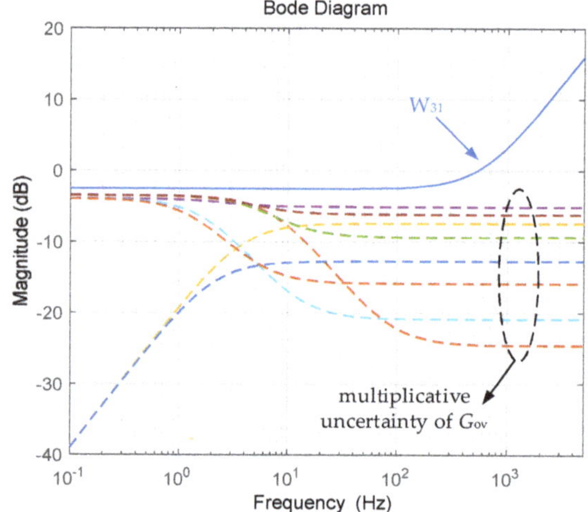

Figure 9. Bode plots of W_{31} and multiplicative uncertainty of G_{ov}.

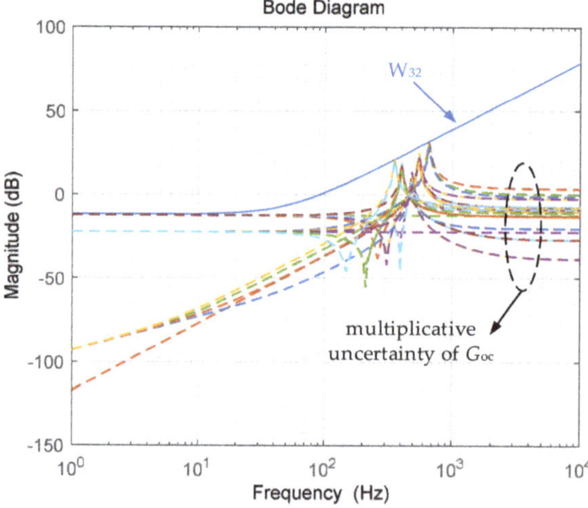

Figure 10. Bode plots of W_{32} and multiplicative uncertainty of G_{oc}.

3.2. H_∞ Controller Design

For (18) and (19), the values of G_{11} and G_{22} will be changed with the variation of steady state operation point, the load change can be represented by different values of R_L, and there are deviations between the nominal values of C_{d1}, C_{d2}, L_{d1}, r_e and their corresponding actual values. By using (22), the model deviation caused by the mentioned factors can be described through multiplicative uncertainty.

According to the parameters listed in Table 1, assuming C_{d2} has $\pm 25\%$ deviation from its nominal value. The equation in (23) can be selected as W_{31}.

$$W_{31} = 2.0250 \times 10^{-4}s + 0.75 \tag{23}$$

Table 1. Simulation Parameters.

Parameter/Unit.	Value
v_{d1}/V	24
v_{d2}/V	50
v_{d3}/V	36
Turns ratio N_1:N_2:N_3	1:1:1
L_1/μH	55
L_2/μH	55
L_3/μH	55
L_{d1}/μH	100
R_L/Ω	90/30
C_{d1}/μF	1200
C_{d2}/μF	1000
Switching frequency/kHz	20

The possible multiplicative uncertainties of G_{ov} (denoted by dashed lines) and the Bode plot (solid line) of W_{31} are shown in Figure 9. It can be seen that the crossover frequency of W_{31} is about 520 Hz. Similarly, $\pm 25\%$ parameter deviation for C_{d1} and L_{d1} are assumed respectively, W_{32} shown in (24) is selected to cover the multiplicative uncertainty of G_{oc}.

$$W_{32} = 2.34 \times 10^{-6}s^2 + 0.00156s + 0.26 \tag{24}$$

The Bode plots of W_{32} (solid line) and multiplicative uncertainties (dashed lines) with different parameter values are shown in Figure 10. The crossover frequency of W_{32} in this case is about 90 Hz.

The weight functions, W_{11} and W_{12} shown in (25) are designed for the sensitivity functions of current and voltage subsystem respectively, the crossover frequencies of W_{11} and W_{12} are 127 Hz and 26 Hz, respectively.

$$W_1 = \begin{bmatrix} W_{11} & 0 \\ 0 & W_{12} \end{bmatrix} = \begin{bmatrix} \dfrac{800}{s + 0.001} & 0 \\ 0 & \dfrac{800}{5s + 0.001} \end{bmatrix} \tag{25}$$

The weight functions, W_{21} and W_{22} given in (26) are used to restrict the control signals.

$$W_2 = \begin{bmatrix} W_{21} & 0 \\ 0 & W_{22} \end{bmatrix} = \begin{bmatrix} 0.1 & 0 \\ 0 & 0.9 \end{bmatrix} \tag{26}$$

According to the parameters listed in Table 1, the resulted robust controllers for current and voltage control subsystems are obtained in (27) and (28) respectively (the controller can be solved using "hinfsyn" function in Matlab Robust Control Toolbox).

$$G_c = \frac{4234s + 7.057 \times 10^4}{s^2 + 6043s + 6.043} \tag{27}$$

$$G_v = \frac{5777s^2 + 5.777 \times 10^5 s + 4.814 \times 10^{10}}{s^3 + 1.367 \times 10^5 s^2 + 1.376 \times 10^8 s + 2.751 \times 10^4} \tag{28}$$

The Bode plot of the corrected voltage control subsystem is shown in Figure 11, the crossover frequency is about 243 Hz and the phase margin is about 76°. The Bode plot of the corrected current control subsystem is presented in Figure 12, the crossover frequency of the current control loop is about 59 Hz, the gain margin is about 51 dB and the phase margin is about 70°. It can be seen from Figure 12 that there is a notch in the Bode plot of the resulted H_∞ controller by which the resonant peak of the uncorrected system can be cancelled accordingly, therefore, a smooth Bode plot of the corrected system is obtained.

As shown in (19), G_{oc} will manifest a weak damping characteristic if r_e is very small and there is a significant resonant peak in the Bode plot of G_{oc}. If G_v and G_c are designed as a proportional-integral (PI) controller, and the crossover frequency of the corrected current loop by PI controller is expected to be lower than the resonant frequency (about 460 Hz) of G_{oc}, in this condition, in order to avoid 0 dB axis intersecting with the corrected current control around the resonant frequency of G_{oc}, the crossover frequency should be greatly reduced. As seen in Figure 13, $G_c = 0.0144 + 36/s$ is utilized, the crossover frequency of the corrected current control system is about $f_c = 7$ Hz, the gain margin is about 9dB and the phase margin is about 89°. For comparison, if a higher crossover frequency of the corrected current control subsystem is wanted with a PI controller, for example, with about $f_c = 60$ Hz (like that in Figure 12 using H_∞ control method) with $G_c = 0.17 + 425/s$, as presented by the dashed line shown in Figure 13, the resulted current control subsystem will be unstable under this condition, because the Bode plot of the corrected current control system crosses 0 dB axis twice around its resonant peak. In contrast, since the resulted current controller, G_c designed by H_∞ mixed sensitivity method in Figure 12 has a natural notch at the resonant frequency of G_{oc}, which can cancel the negative impact of G_{oc} resonance effectively, the improved current control performance can be obtained accordingly.

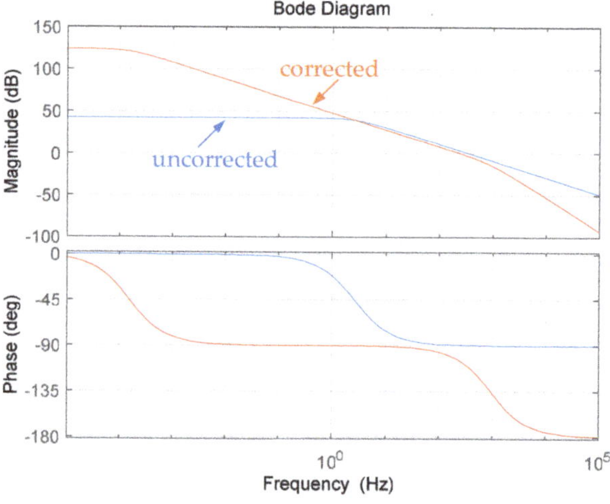

Figure 11. Bode of voltage control loop with H_∞ robust controller.

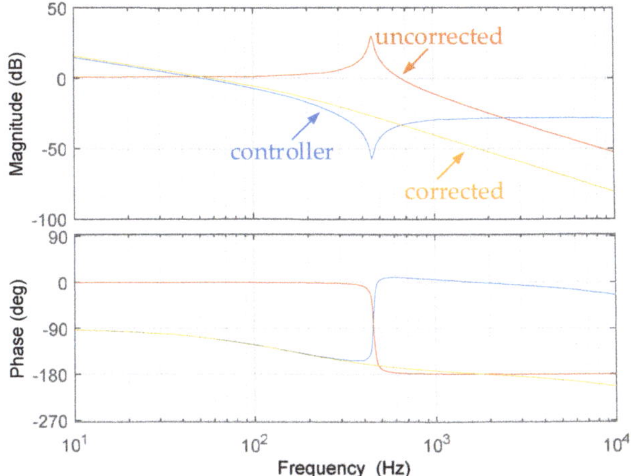

Figure 12. Bode of current control loop with H$_\infty$ robust controller.

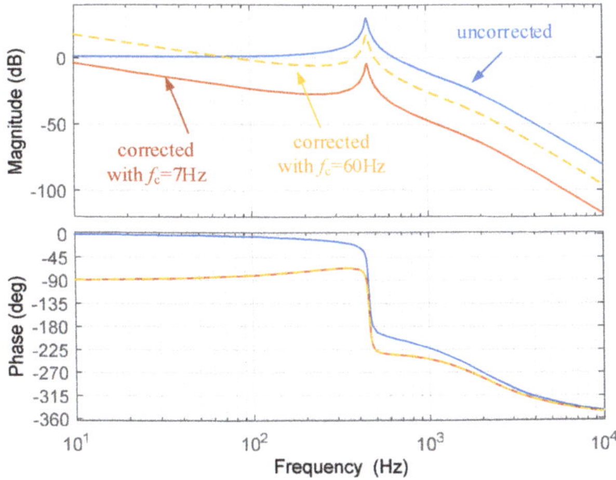

Figure 13. Bode of current control loop with a PI controller.

4. Simulation and Experimental

4.1. Simulation Results

A simulation model of the proposed design method is developed in MATLAB/Simulink environment to verify the effectiveness of the proposed method. The parameters used in the simulation model are presented in Table 1. The simulation results for changing the current reference, i_{ds}^* with different control methods are presented in Figure 14. The simulation results using a PI controller are shown in Figure 14a. In this figure, the current reference value is reduced from 2.1 A to 1 A (at 0.25 s) and then suddenly increased again to 2.1 A (at 0.4 s). It can be seen the actual current, i_{ds} can track its reference signal. However, since the current control bandwidth is relatively low in this condition, it takes about 30 ms to reach the reference current value. And the output voltage, v_{d2} is almost kept constant in this process. With the same reference change condition, the simulation results using H$_\infty$ controller are shown in Figure 14b, as it can be concluded from this figure that the dynamic response speed of i_{ds} is much faster than that in Figure 14a, the transient state process time is decreased to about 7 ms. And it can be seen in Figure 14 that the battery current i_{d3} is increased from about −0.22 A to about 0.5 A and then to -0.22 A with the corresponding change of i_{ds}^*, the dynamic response speed with H_∞ controller in Figure 14b is much faster than that with a PI controller in Figure 14a.

Figure 15 shows the simulation results of a sudden load change test. In Figure 15a, the load resistor value is reduced from 60 Ω to 30 Ω (the load is increased from about 42 W to about 83 W) at 0.25 s and then suddenly increased to 60 Ω at 0.4 s again. In the transient state process, due to the interaction of voltage and current control subsystems, there are fluctuations in i_{ds} and v_{d2} simultaneously at the load change moment. With the same load change condition, the simulation results with H_∞ robust controller are given in Figure 15b. Though the fluctuations in v_{d2} and i_{d3} are similar to that in Figure 15a, the amplitudes of fluctuations and the transient recovery time of i_{ds} in Figure 15b are significantly reduced compared to that in Figure 15a.

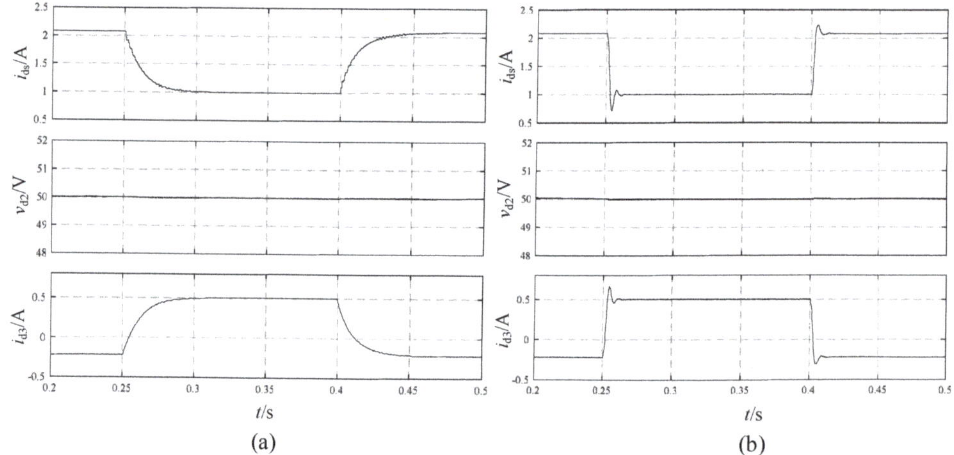

Figure 14. Simulation results with current reference change using a (**a**) PI controller, (**b**) H_∞ controller.

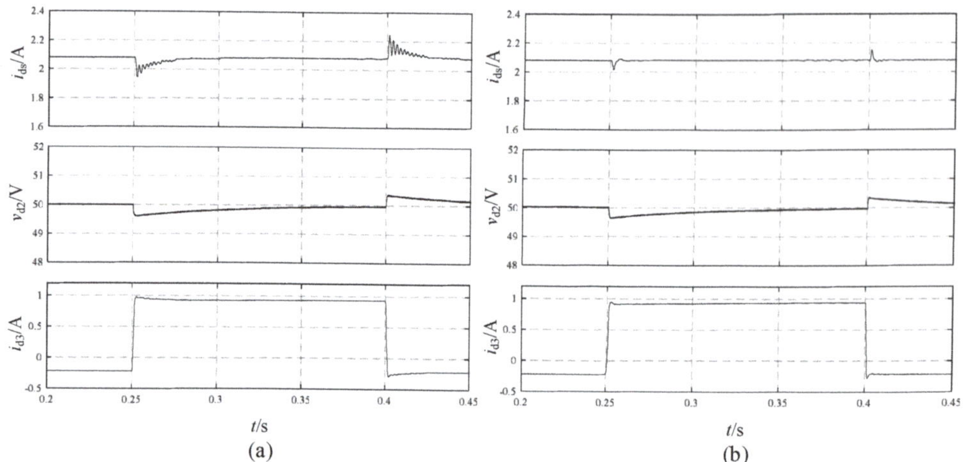

Figure 15. Simulation results with load change using a (**a**) PI controller, (**b**) H_∞ controller.

4.2. Experiment Results

An experimental hardware shown in Figure 16 is developed to validate the theoretical design and simulation results. The parameter deviations of C_{d1} (about 1130 μF), L_{d1} (about 107 μH) and C_{d2}

(about 1046 μF) are limited to 10% of their nominal value (this condition can be easily guaranteed in practice). The other parameters used in the experimental tests are approximately identical to the simulation parameters listed in Table 1. The experiment results are presented in Figures 17–22.

Figure 17 shows the experiment results of v_{d2} and i_{ds} using the PI controller. The current reference value is changed from 1 A to 2.1 A and reduced to 1 A again suddenly in Figure 17a,b respectively, and the corresponding transient state process times are about 32 ms and 31 ms. Under the same reference value change condition, the experiment results using H_∞ controller are shown in Figure 18. Since the current control bandwidth using H_∞ controller (59 Hz) is higher than that using a PI controller (7 Hz), the transient state process times shown in Figure 18a,b are 19 ms and 23 ms respectively, which are much shorter than that in Figure 17.

With the same reference value change condition being used in Figure 17, the corresponding experiment results of battery current, i_{d3} using PI and H_∞ controllers are shown in Figures 19 and 20, respectively. It can be seen that the changes of battery current are much faster in Figure 20 than that in Figure 19. Compared to Figure 19a, less time is taken in Figure 20a for the battery current to reach its steady state value due to a higher control bandwidth of the H_∞ controller. A similar result can also be obtained by comparing Figure 19b with Figure 20b.

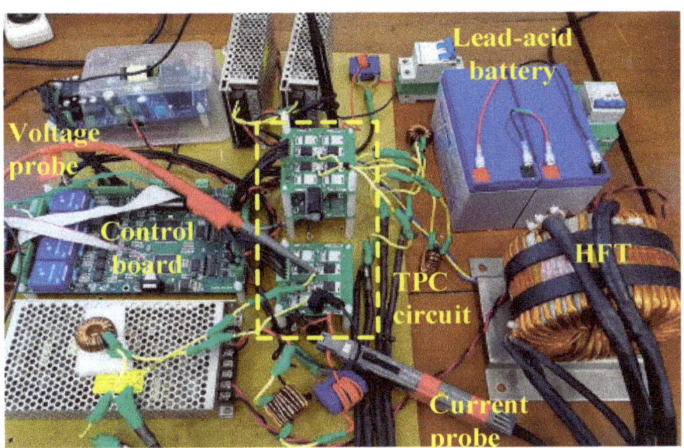

Figure 16. Hardware experiment circuit of three-port converter.

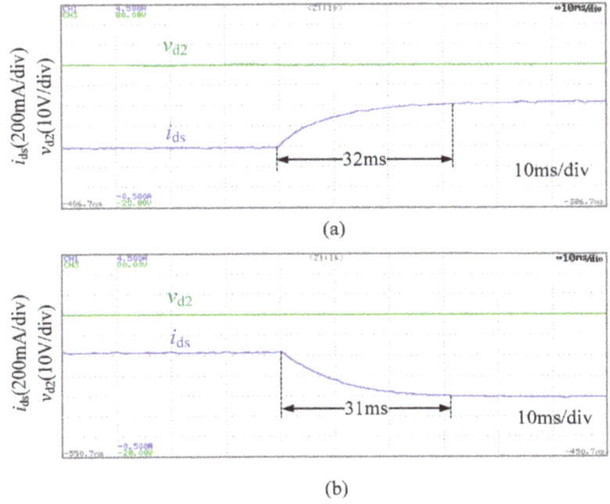

Figure 17. v_{d2} and i_{ds} using a PI controller with i_{ds}^* is (**a**) increased, (**b**) decreased.

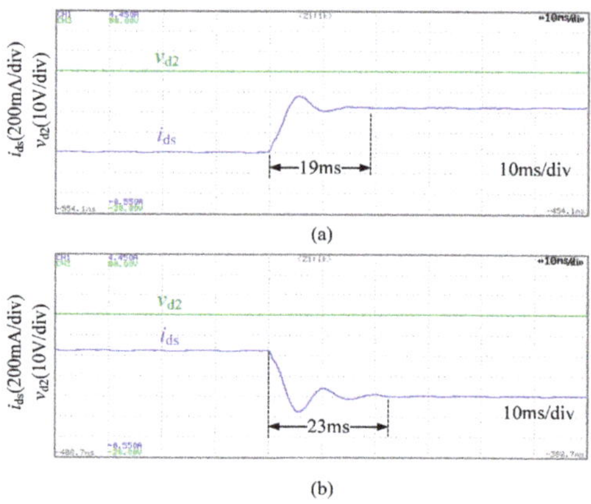

(a)

(b)

Figure 18. v_{d2} and i_{ds} using H$_\infty$ controller with i_{ds}^* is (**a**) increased, (**b**) decreased.

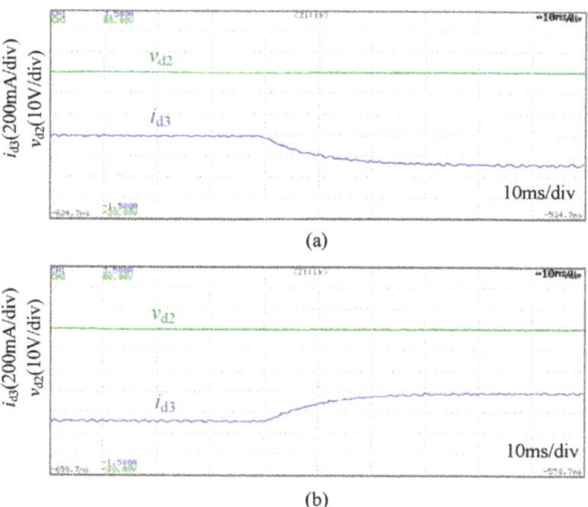

(a)

(b)

Figure 19. v_{d2} and i_{d3} using a PI controller with i_{ds}^* is (**a**) increased, (**b**) decreased.

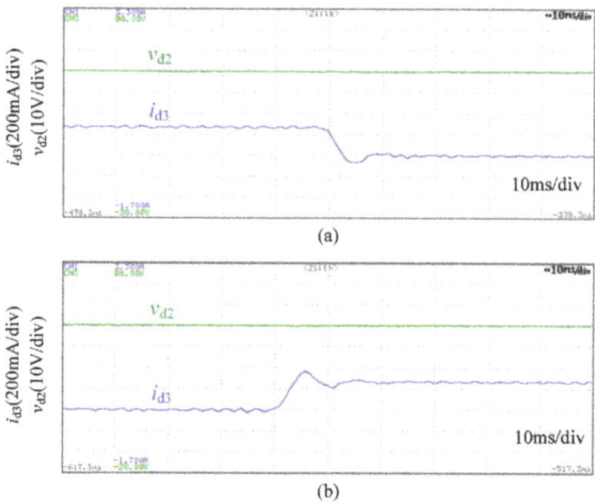

(a)

(b)

Figure 20. v_{d2} and i_{d3} using H$_\infty$ controller with i_{ds}^* is (**a**) increased, (**b**) decreased.

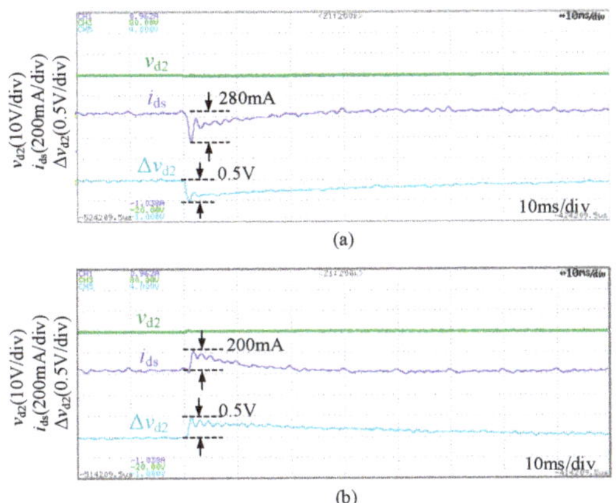

(a)

(b)

Figure 21. v_{d2}, i_{ds} and Δv_{d2} using a PI controller with R_L is (**a**) decreased, (**b**) increased.

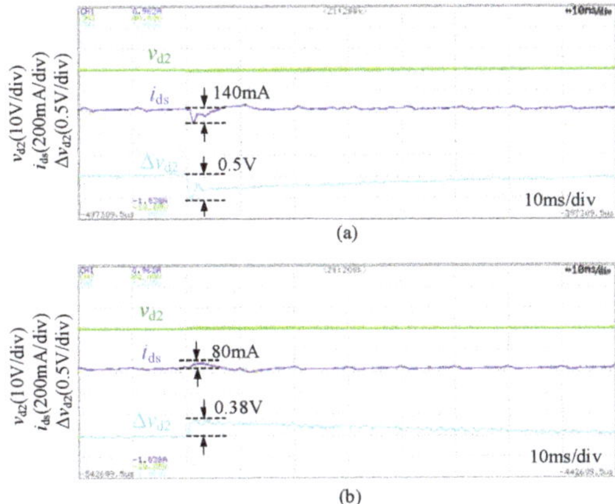

(a)

(b)

Figure 22. v_{d2}, i_{ds} and Δv_{d2} using H$_\infty$ controller with R_L is (**a**) decreased, (**b**) increased.

The load change experiment results using PI and H$_\infty$ controllers are shown in Figures 21 and 22 respectively. In Figure 21a, the load resistor is changed from 90 Ω to 30 Ω (the load is increased from about 28 W to about 83 W) instantaneously, which causes about a 0.5 V voltage drop in v_{d2} (the peak value of Δv_{d2}) and about a 280 mA current drop in i_{ds}. In Figure 22a, the corresponding voltage and current drops are 140 mA and 0.5 V respectively with the same load change. When the load resistor is again increased from 30 Ω to 90 Ω (the load is decreased from about 83 W to 28 W), a 200 mA current increment is produced in i_{ds} using a PI controller in Figure 21b, and an 80 mA current increment is caused in i_{ds} using an H$_\infty$ controller in Figure 22b. It can be concluded that the interaction between voltage control subsystem and current control subsystem is better attenuated by adopting an H$_\infty$ controller. Furthermore, the voltage fluctuation is reduced from 0.5 V in Figure 21b to about 0.38 V in Figure 22b with the H$_\infty$ controller when the load resistor is increased.

In contrast to the traditional PI controller, the experiment results indicate that the negative impact caused by the couplings between current control subsystem and voltage control subsystem can be effectively suppressed by using the H$_\infty$ controller. And the dynamic response performance of the control system can also be improved with the proposed method.

5. Conclusions

An H_∞ mixed sensitivity method is introduced in this paper for three-port converter control. The H_∞ mixed sensitivity method has an inherent characteristic to balance performance and stability of a control system through the use of an appropriate weight functions selection. The H_∞ current controller manifests a superior characteristic of effectively damping the resonant peak of the current control subsystem that can reduce limitations in the control bandwidth design, and this is beneficial for enhancing the stability of the current control subsystem. The simulation and experiment results show that the resulted optimal H_∞ controller has advantages in dynamic response performance and load disturbance rejection compared to a traditional proportional-integral (PI) controller.

Author Contributions: Conceptualization and experiment, J.Y., H.L. and B.F.; All the authors contributed equally to the other parts of work.

Funding: This work is sponsored by the fundamental research funds for the central universities of China (No. HEUCFG201822), and the National Natural Science Foundation of China (No.51479042).

Conflicts of Interest: The authors declare no conflict of interest.

References

1. Tao, H.; Kotsopoulos, A.; Duarte, J.L.; Hendrix, M.A.M. Family of multiport bidirectional DC-DC converters. *IEE Proc. Electr. Power Appl.* **2006**, *153*, 451–458. [CrossRef]
2. Vázquez, N.; Sánchez, C.M.; Hernández, C.; Vázquez, E.; Lesso, R. A three port converter for renewable energy applications. In Proceedings of the IEEE International Symposium on Industrial Electronics, Gdansk, Poland, 27–30 June 2011; pp. 1736–1740.
3. Wu, H.; Sun, K.; Ding, S.; Xing, Y. Topology derivation of non-isolated three-port DC-DC converters from DIC and DOC. *IEEE Trans. Power Electron.* **2013**, *28*, 3297–3307. [CrossRef]
4. Zhu, H.; Zhang, D.; Zhang, B.; Zhou, Z. A non-isolated three-port DC-DC converter and three-domain control method for PV-battery power systems. *IEEE Trans. Ind. Electron.* **2015**, *62*, 4937–4947. [CrossRef]
5. Phattanasak, M.; Ghoachani, R.G.; Martin, J.P.; Mobarakeh, B.N.; Pierfederici, S.; Davat, B. Control of a hybrid energy source comprising a fuel cell and two storage devices using isolated three-port bidirectional DC-DC converters. *IEEE Trans. Power Electron.* **2015**, *51*, 491–497. [CrossRef]
6. Liu, R.; Xu, L.; Kang, Y.; Hui, Y.; Li, Y. Decoupled TAB Converter with Energy Storage System for HVDC Power System of More Electric Aircraft. *J. Eng.* **2018**, *2018*, 593–602. [CrossRef]
7. Jakka, V.N.S.R.; Shukla, A.; Demetriades, G.D. Dual-Transformer-Based Asymmetrical Triple-Port Active Bridge (DT-ATAB) Isolated DC–DC Converter. *IEEE Trans. Ind. Electron.* **2017**, *64*, 4549–4560. [CrossRef]
8. Zhao, C.; Kolar, J.W. A novel three-phase three-port UPS employing a single high-frequency isolation transformer. In Proceedings of the 2004 IEEE 35th Annual Power Electronics Specialists Conference, Aachen, Germany, 20–25 June 2004.
9. Zhao, C.; Round, S.D.; Kolar, J.W. An Isolated Three-Port Bidirectional DC-DC Converter with Decoupled Power Flow Management. *IEEE Trans. Power Electron.* **2008**, *23*, 2443–2453. [CrossRef]
10. Tao, H.; Kotsopoulos, A.; Duarte, J.L.; Hendrix, M.A.M. Transformer-Coupled Multiport ZVS Bidirectional DC–DC Converter with Wide Input Range. *IEEE Trans. Power Electron.* **2008**, *23*, 771–781. [CrossRef]
11. Wang, L.; Wang, Z.; Li, H. Asymmetrical Duty Cycle Control and Decoupled Power Flow Design of a Three-port Bidirectional DC-DC Converter for Fuel Cell Vehicle Application. *IEEE Trans. Power Electron.* **2012**, *27*, 891–904. [CrossRef]
12. Falcones, S.; Ayyanar, R. LQR control of a quad-active-bridge converter for renewable integration. In Proceedings of the 2016 IEEE Ecuador Technical Chapters Meeting (ETCM), Guayaquil, Ecuador, 12–14 October 2016.
13. Tao, H.; Kotsopoulos, A.; Duarte, J.L.; Hendrix, M.A.M. A soft-switched three-port bidirectional converter for fuel cell and supercapacitor applications. In Proceedings of the IEEE 36th Power Electronics Specialists Conference, Recife, Brazil, 12–16 June 2005; pp. 2487–2493.
14. Falcones, S.; Ayyanar, R.; Mao, X. A DC–DC multiport-converter-based solid-state transformer integrating distributed generation and storage. *IEEE Trans. Power Electron.* **2013**, *28*, 2192–2203. [CrossRef]

15. Zhang, J.; Wu, H.; Cao, F.; Zhu, L.; Xing, Y. Modeling and Decoupling Control of Three-port Half-bridge Converters. *Proc. CSEE* **2015**, *35*, 671–678.
16. You, J.; Liao, M.; Chen, H.; Ghasemi, N.; Vilathgamuwa, M. Disturbance Rejection Control Method for Isolated Three-Port Converter with Virtual Damping. *Energies* **2018**, *11*, 3204. [CrossRef]
17. Doyle, J.C.; Glover, K.; Khargonekar, P.P.; Francis, B.A. State-space solutions to standard H_2 and H_∞ control problems. *IEEE Trans. Autom. Control* **1989**, *34*, 831–847. [CrossRef]
18. Zhou, K.; Doyle, J.C.; Glover, K. *Robust and Optimal Control*; Prentice Hall: Upper Saddle River, NJ, USA, 1996.
19. Wu, X.; Xie, X. Weight Function Matrix Selection in H_∞ Robust Control. *J. Tsinghua Univ.* **1997**, *37*, 27–30.
20. Hu, J.; Unbehauen, H.; Bohn, C. A practical approach to selecting weighting functions for H_∞ control and its application to a pilot plant. In Proceedings of the UKACC International Conference on Control '96, Exeter, UK, 2–5 September 1996; pp. 998–1003.

MDPI

St. Alban-Anlage 66

4052 Basel

Switzerland

Tel. +41 61 683 77 34

Fax +41 61 302 89 18

www.mdpi.com

Energies Editorial Office

E-mail: energies@mdpi.com

www.mdpi.com/journal/energies